CHEMICAL ENGINEERING METHODS AND TECHNOLOGY

MEMBRANE PROCESSES FOR SUSTAINABLE GROWTH

CHEMICAL ENGINEERING METHODS AND TECHNOLOGY

Additional books in this series can be found on Nova's website under the Series tab.

Additional e-books in this series can be found on Nova's website under the e-books tab.

CHEMICAL ENGINEERING METHODS AND TECHNOLOGY

MEMBRANE PROCESSES FOR SUSTAINABLE GROWTH

ANGELO BASILE
AND
ALFREDO CASSANO
EDITORS

nova publishers
New York

For permission to use material from this book please contact us:
Telephone 631-231-7269; Fax 631-231-8175
Web Site: http://www.novapublishers.com

NOTICE TO THE READER

Additional color graphics may be available in the e-book version of this book.

Library of Congress Cataloging-in-Publication Data

ISBN: 978-1-62618-446-6

Published by Nova Science Publishers, Inc. † New York

Contents

Preface		**vii**
Chapter 1	Polymeric Membranes for Sustainable Gas Separation *Angelo Basile and Annarosa Gugliuzza*	**1**
Chapter 2	Membrane Contactor Technology: Fundamentals, Interfaces and Prospects *Annarosa Gugliuzza and Angelo Basile*	**31**
Chapter 3	Pervaporation as Sustainable Process *Alberto Figoli, Pieter Vandezande, Francesco Galiano and Sina Zereshki*	**65**
Chapter 4	Membranes for Sustainable Food Packaging *Alberto Figoli and Silvia Simone*	**101**
Chapter 5	Membrane and Membrane Reactors for Sustainable Growth *Angelo Basile, Sudip Chakraborty and Adolfo Iulianelli*	**121**
Chapter 6	Pressure-Driven Membrane Applications: Microfiltration and Ultrafiltration *Alfredo Cassano and Carmela Conidi*	**153**
Chapter 7	Application of Membrane Technology for the Development of Sustainable Biotechnology Processes *Faisal I. Hai and Oskar Modin*	**193**
Chapter 8	Pressure-Driven Membrane Applications: Nanofiltration *Bart Van der Bruggen and Yang Zhang*	**225**
Chapter 9	High Temperature Performance Polymer Electrolyte Membranes *Suwabun Chirachanchai, Autchara Pangon and Chatchai Jarumaneeroj*	**247**
Chapter 10	Advances in Reverse Osmosis Technology for the Processing of Fruit Juices *Navin K. Rastogi and K. S. M. S. Raghavarao*	**289**
Index		**313**

PREFACE

The purpose of this book is to provide some relevant examples of how membrane operations can contribute to an industrial sustainable growth. In many industrial fields, membrane operations present many advantages over conventional systems in terms, for example, of energy consumption, reduction of environmental impact and product quality. Furthermore, successful examples of conventional operations integrated with membrane processes have been also reported in different industrial productions within the logic of the process intensification strategy.

Following the same logic, for what concerns the content of this book, each chapter deals with a specific area of membrane technology and includes information on the basic principles, commercial applications and an overview of current research and development.

The book starts with Chapter 1 (Basile and Gugliuzza) in which the use of polymeric membranes in gas separation, including basic concepts, mechanisms, manufacturing approaches, materials along with module assembly and applications is discussed. It is followed (Chapter 2 - Gugliuzza and Basile) by an overview of principles and applications of key membrane contactors operations such as membrane emulsification, membrane distillation, osmotic membrane distillation and membrane crystallization is given.

Basic aspects of the pervaporation process as well as the applications, classified as solvent dehydration, organic/water separation and organic/organic separation, are discussed in detail in Chapter 3 (Figoli, Vandezande, Galiano and Zereshki).

Chapter 4 (Figoli and Simone) shows how membranes can be successfully used in food packaging, where concepts of gas permeability are important in order to meet the new demands of food safety and storage.

Chapter 5 (Basile, Chakraborty and Iulianelli) describes how the inorganic membrane technology can be applied fruitfully both to generate/separate hydrogen and concentrate CO_2 to be further stored. This chapter also pays attention to the membrane bioreactor technology, pointing out the new findings in this area and describing its potential application in true life.

Basic principles of pressure-driven membrane operations and their applications in different industrial fields including drinking water, water recycling, wastewater treatments, food and beverage are described in Chapters 6, 8 and 10. In particular, Chapter 6 (Cassano and Conidi) provides an overview of microfiltration and ultrafiltration processes together with well-established applications in food and beverages industries. The fundamentals of nanofiltration used as a sustainable method for application in water production and recycling of wastewater for secondary use are discussed in Chapter 8 (Van der Bruggen and Zhang), whereas Chapter 10 (Rastogi and Raghavarao) addresses the significant progress of reverse osmosis in fruit juice concentration and its impact on product quality.

Chapter 7 (Hai and Modin) provides an overview of membrane applications in selected established and emerging biotechnology processes highlighting some approaches to overcoming the technology bottlenecks that impede the scale-up of such systems.

Chapter 9 (Chirachanchai, Pangon and Jarumaneeroj) focuses on the development of high temperature resistance polymer electrolyte membranes which function in water retention ability, enhancing reaction kinetics, and reducing electrode CO poisoning.

The Editors would like to take also this opportunity to thank the all chapter Authors for their expert contribution to this volume.

Angelo Basile

Alfredo Cassano

In: Membrane Processes for Sustainable Growth
Editors: Angelo Basile and Alfredo Cassano
ISBN: 978-1-62618-446-6

Chapter 1

POLYMERIC MEMBRANES FOR SUSTAINABLE GAS SEPARATION

Angelo Basile **[*]** ***and Annarosa Gugliuzza***
Research Institute on Membrane Technology-
National Research Council (ITM-CNR), Rende (CS), (Italy)

ABSTRACT

This chapter offers an overview to the use of polymeric membranes in gas separation. Basic concepts, mechanisms, manufacturing approaches, materials along with module assembly and applications, with a particular emphasis to CO_2 separation, are discussed. A detailed description of the equations that govern the mass transport through dense and porous membranes, along with glassy and rubbery polymeric membranes, below and up the critical temperature and the glass transition temperature, is also given.

The chapter deals also with the various membrane fabrication and assembly procedures, yielding indication about the most used module configuration in gas separation.

Keywords: gas separation; membrane; materials; composites; transport mechanisms; permeability; selectivity; diffusivity; solubility

LIST OF SYMBOLS AND ACRONYMS

A membrane area
b ratio of rate coefficients of adsorption and desorption
C concentration
C'_H maximum adsorption capacity
C_D concentration of gas in the membrane matrix
D diffusion ocefficient
ΔH_C enthalpy of condensation

[*] Corresponding author: a.basile@itm.cnr.it.

ΔH_l enthalpy of mixture
ΔH_S enthalpy of solution
E_d activation energy for diffusion
E_p apparent activation energy for permeation
f fugacity of the gas considered
G the geometrical factor
I refers to the feed
II refers to the permeate
J flux across the membrane
k proportionally coefficient
M molecular weight
n slope of the limit in the Robeson diagram
P permeability
p pressure
R the universal gas constant
S solubility
SF separation factor
S_H Henry's law constant
T absolute temperature
t time
Tc critical temperature
Tg glass transition temperature
x molar concentration in the the feed side
y molar concentration in the the permeate side
z axial coordinate

List of Acronyms

RO reverse osmosis
NF ultrafiltration
UF ultrafiltration
MF microfiltration

Subscripts

p particle
o reference state
i species (or component)
j species (or component)

Greek Symbols

α selectivity
Δ difference
δ membrane thickness
ϕ pore diameter

1. INTRODUCTION

Following the IUPAC definition, a membrane is a structure with lateral dimensions much greater than its thickness through which mass transfer can occur under a variety of driving forces, such as gradient of concentration, pressure, temperature, electric potential, etc. (Koros et al., 1996). A schematic representation of a two-phase system separated by membrane is given in Figure 1, in which Phase I is usually considered as the feed, while Phase II as the permeate. Practically, in separating one or more components from a feed mixture and in generating a permeate stream, the membrane acts as a "filter" able to allow the preferential passage of certain species. Membranes can be classified in various ways.

For example, as schematically summarised in Figure 2, membranes can be classified according to their nature, geometry and separation regime (Khulbe et al., 2007).

The classification based on their nature distinguishes membranes into *biological* and *synthetic* ones.

The latter are still subdivided into *organic* and *inorganic* membranes (Figure 3):

- *Biological membranes* are easy to manufacture, but they present many drawbacks such as limited operating temperature (below 100 °C) and pH range; and problems related to clean-up and susceptibility to microbial attack due to their natural origin (Xia et al., 2003).
- *Synthetic membranes* can be subdivided into organic (polymeric) and inorganic (ceramic, metallic, carbon, etc.) ones according to their operative temperature limit: *polymeric membranes* commonly operate in the range 100 - 200 °C, whereas *inorganic* ones above 200 °C (Catalytica, 1998). Generally, inorganic membranes are stable between 200 - 800 ºC and in some cases they can operate at elevated temperatures (ceramic membranes over 1000 ºC) (Van Veen et al., 1996).

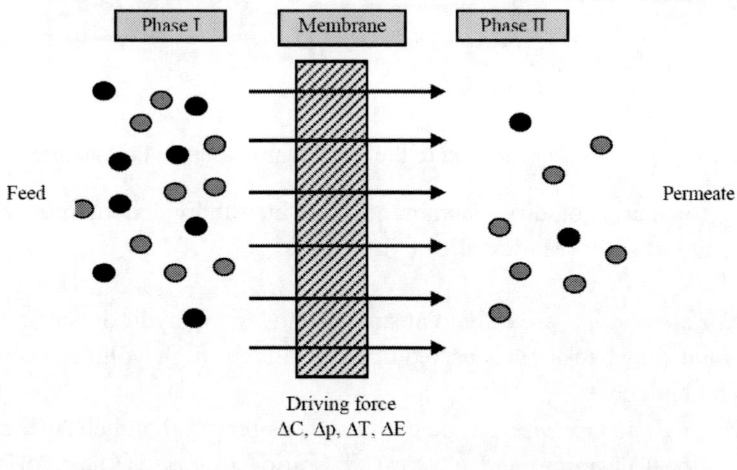

Figure 1. Schematic representation of a two-phase system separated by a membrane.

From the viewpoint of the morphology and/or membrane structure categorization, inorganic membranes can also be subdivided into porous and dense membranes. In particular, *porous membranes* differ, according to their pore diameter (Koros et al., 1996), in:

- microporous (ϕ_{pore} < 2 nm),
- mesoporous (2 nm < ϕ_{pore} < 50 nm) and
- macroporous (ϕ_{pore} > 50 nm).

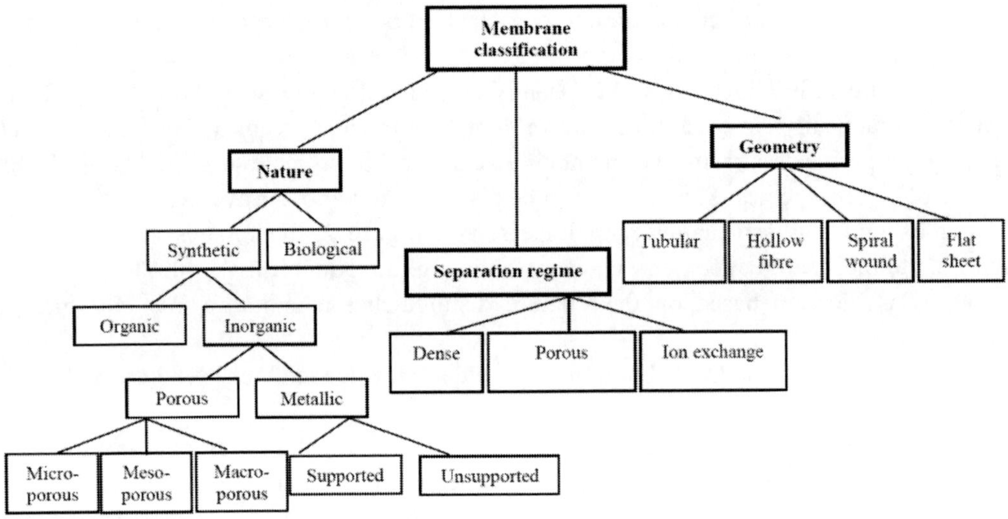

Figure 2. Scheme of a general classification of the membranes.

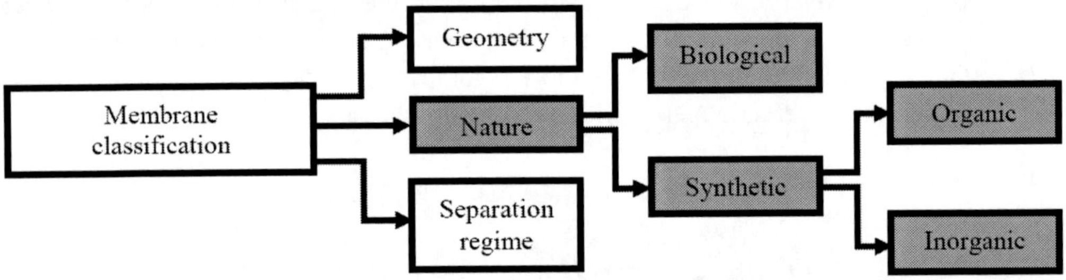

Figure 3. Scheme of a general classification of the membranes based on their nature.

According to their geometry, membranes can be subdivided in *tubular*, *hollow fiber*, *spiral wound* and *flat sheet* (Mallevialle et al., 1988):

- *Tubular membranes* are easy to clean and show good hydrodynamic control, but with important drawbacks such as requiring relatively high volume per membrane area unit and high costs.
- *Hollow fiber membranes* can be considered as practical and cheaper alternatives than conventional chemical and physical separation processes. They offer high packing densities and can withstand relatively high pressure owing to their structural integrity. They also allow flexibility in system design and operation.

- *Spiral wound membranes* offer advantages such as compactness, good membrane surface/volume and low capital/operating cost ratios. Nevertheless, they are not suitable for viscous fluid and are difficult to clean.
- *Flat sheet membranes* offer moderate membrane surface/volume ratios. However, they are susceptible to plugging due to flow stagnation points, difficult to clean and expensive.

A further membrane classification is based on the separation mechanism. There are three main separation mechanisms, which are determined by specific properties of the components (Mulder, 1996):

- separation based on molecules/membrane surface interactions (e.g. multi-layer diffusion) and/or difference between the average pore diameter and the average free path of fluid molecules (e.g. Knudsen mechanism);
- separation based on the difference of diffusivity and solubility of substances in the membrane: the so-called *solution-diffusion* mechanism;
- separation owing to the difference in charge of the species to be separated: *electrochemical effect.*

Based on these separation mechanisms, membranes can be further classified in *porous*, *dense* and *ion-exchange*.

In Table 1, the different diffusion mechanisms for porous and dense membranes are resumed; whilst ion-exchange membranes are not included. In particular, in the case of porous membranes:

- *Poiseuille (viscous flow) mechanism* occurs when the average pore diameter is bigger than the average free path of fluid molecules. In this case, the collisions among the different molecules are more frequent than those among the molecules and the porous wall: as a consequence no separation takes place (Saracco and Specchia, 1994).
- *Knudsen mechanism* occurs when the average pore diameter is similar to the average free path of fluid molecules. In this case, the collisions of the molecules with the porous wall are very frequent and the flux of the component permeating through the membrane is calculated by means of the following equation (Saracco and Specchia, 1994).

$$J_i = \frac{G}{\sqrt{2 \cdot M_i \cdot R \cdot T}} \cdot \frac{\Delta p_i}{\delta} \qquad (1)$$

where J_i is the flux of the i-species across the membrane, G the geometrical factor, which takes into account the membrane porosity and the pore tortuosity, M_i the molecular weight of the i-species, R the universal gas constant, T the absolute temperature and Δp_i the pressure difference of i-species, and δ *the* membrane thickness.

- *Surface diffusion* is achieved when one of the permeating molecules is adsorbed on the pore wall due to the active sites presented in the membrane (Knozinger and

Ratnasamy, 1978). This type of mechanism can reduce the effective pore dimensions, consequently not favouring the transfer of different molecular species (Kapoor et al., 1989). However, this diffusion can take place also in the presence of a Knudsen transport. This separation mechanism becomes less significant by increasing the temperature owing to the progressive decrease of the bond strength between molecules and surface.

- *Capillary condensation* occurs when one of the components condenses within the pores due to capillary forces, which are sufficiently strong only at low temperature and in the presence of small pores. If the pore dimension is small and homogeneous and the pores are uniformly distributed on the membrane, this mechanism can offer high selectivity (Falconer et al., 1995; Sperry et al., 1991). Generally, the capillary condensation favours the transfer of relatively large molecules (Lee and Hwang, 1986).
- *Multi-layer diffusion* is developed when the molecule/surface interactions are strong. This mechanism is like an intermediate flow regime between surface diffusion and capillary condensation (Ulhorn et al., 1992).
- *Molecular sieve* takes place when the pore diameters are very small, allowing the permeation of only smaller molecules.

Vice versa, in the case of dense membranes, the transport mechanism takes place according to the *solution-diffusion* mechanism, in which the molecules on the gas/membrane interface are firstly adsorbed on one membrane surface, afterwards they diffuse through the membrane network and finally desorbe on the opposite gas/membrane interface.

Figure 4 illustrates some of the most important mechanisms of gas permeation through porous and dense membranes.

Among all types of membranes, the inorganic dense membranes have attracted the interest of many researchers due to their capacity to separate a product from gaseous mixtures completely (Adhikari and Fernand, 2006).

Table 1. Diffusion mechanisms in porous and dense membranes

Membrane	**ϕ_{pore} (nm)**	**Diffusion mechanism**	
Macroporous	> 50	Poiseuille (viscous flow)	
Mesoporous	2 - 50	Knudsen	
Microporous	< 2	Activated process	Surface diffusion Capillary condensation Multi-layer diffusion Molecular sieve
Dense	-	Fick	

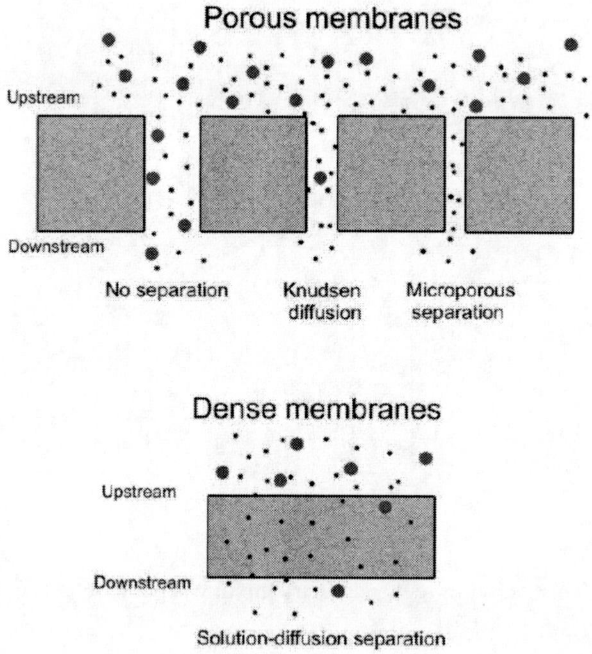

Figure 4. A schematic view of the most important mechanisms of gas permeation through porous and dense membranes.

In particular, dense palladium-based membranes are used because of their complete hydrogen perm-selectivity. In the last few years, the increasing interest towards this type of membranes has also been due to hydrogen application as energy carrier.

The key parameters, strictly connected to membrane performances, are:

- *Permeability (P_i)*, which is related to the amount of flux of a specific gas *i* transported through the membrane.
- *Ideal gas selectivity (α_{ij})*: the membrane preference to pass one gas species *i* and not another one *j*.

For many membranes, there is a trade-off between these two parameters: selectivity and permeability. In fact, a highly permeable membrane tends to have low selectivity and vice versa. For example, in the case of polymeric membranes, Robeson (1991; 2008) suggested that this trade-off can be represented by a graph, where the selectivity is represented against the permeability of the more permeable gas for a binary mixture, as shown schematically in Figure 5. In this figure, the upper-bound represents the limit for achieving a high selectivity combined with a high permeability. The upper-bound relationship is expressed by the following equation:

$$P_i = \qquad (2)$$

where P_i is the permeability of the more permeable gas, *k* a proportionally coefficient, and *n* ~~is~~ the slope of the limit.

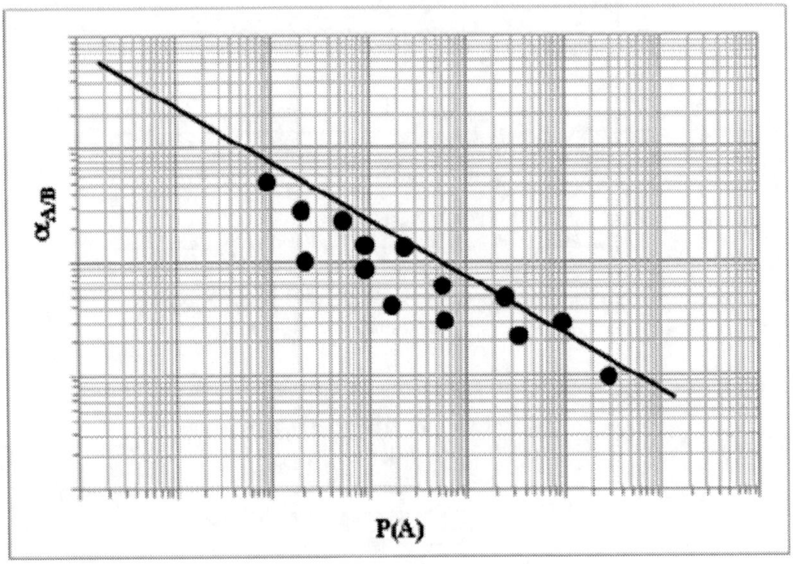

Figure 5. Upper bound correlation for general binary mixture separation.

1.2. Principles of Gas Separation through Polymeric Membranes

The permeation of the various species through the membrane is determined by their diffusivity and their concentration gradient along the membrane. The driving force for the selective transport of a component across a membrane is typically associated with a gradient of concentration, pressure, temperature, electric potential, etc. Many different processes could be cited according to pressure and membrane pore size, such as, for example: reverse osmosis (RO), nanofiltration (NF), ultrafiltration (UF) and microfiltration (MF).

In the following, only the partial pressure difference, which can be expressed conveniently in terms of mole fractions of the retentate and permeate phases, is considered. von Wroblewski (1879) proposed the following equation for the gas permeation rate (or flux) J_i (mol/m^2 s) of a generic component i through a membrane:

$$J_i = P_i\,(\Delta p_i/\delta) \qquad (3)$$

where Δp_i is the pressure drop across the membrane, δ the membrane thickness and P_i the proportionally coefficient called *permeability* or *permeability coefficient*.

Another important equation is Fick's law, which relates the mass flow through the membrane area (A) perpendicular to the flow direction to the concentration gradient of the species i:

$$J_i = -D_i\, dC_i/dz \qquad (4)$$

where: D_i is the diffusion coefficient (m^2/s), dC_i/dz the concentration gradient (mol/m^3 m) for component i over the length z.

The integration of Fick's law along the membrane thickness gives: $dx = \delta$, and $dC_i = \Delta C$, the concentration difference (related to the partial pressures for gases) over the membrane; D_i varies according to the dominating transport mechanism.

Membrane gas permeability, a measure of the quantity of a component that permeates through the membrane, is defined in a phenomenological way as the flow rate across the membrane surface normalized by the thickness and driving force of the process:

$$P_i = \frac{J_i \delta}{A(p_I^i - p_{II}^i)} \quad (5)$$

where P_i is the gas permeability ($mol \cdot m/m^2 \cdot s^- Pa$), J_i the molar flow rate of the species i (mol/s), δ the membrane thickness (m), A the exchange surface (m^2) and and the partial pressure of species i (P_a) at retentate (I) and permeate (II) sides, respectively.

It must be stressed that, being a phenomenological definition, the gas permeability P_i is determined both by the membrane characteristics and the operative conditions of the particular experiment.

Often, the membrane thickness is not known because of the difficulty to measure it or because it is not constant throughout the membrane. In these cases another parameter is used instead of the permeability: the *permeance*, defined as the flux through the membrane normalised by the driving force. Using the formula, the permeance P_i/δ, for a given gas *i*, is given by:

$$Permeance = P_i/\delta = J_i/\Delta p_i \quad (6)$$

P_i/δ is also referred to as *permeability flux* and expressed as ($mol/m^2\ s\ Pa$) or ($m^3(STP)/m^2\ h\ bar$), Δp_i the partial pressure difference (in *Pa* or *bar*) of the species i across the membrane.

This equation also shows that the flux through the membrane is directly proportional to the pressure difference across the membrane and inversely proportional to the membrane thickness.

In particular, the gas permeability is related to two different parameters. In fact, it can be expressed as the product of diffusion, D_i (m^2/s), and solubility, S_i ($mol/m^3 \cdot Pa$), of the gas i through the membrane:

$$P_i = D_i S_i \quad (7)$$

In general, this formula is also alid in the case of species different from gases and it is written as:

$$P = D\, S \quad (8)$$

The product $D\,S$ is always the membrane permeability. The importance of D and S varies according to both the type of component and the type of membrane material considered. In the case of gases, both the coefficients depend on the temperature, while a pressure dependency is only observed for certain gases and materials.

In particular, the diffusion coefficient takes into account of the mobility of the individual molecules in the membrane; whereas the solubility coefficient is related to the number of molecules dissolved in the membrane. Solubility mostly increases with increasing molecular weight, while the diffusivity is decreased.

As a consequence, also the permeability depends, in the most general case, on the temperature, the nature of gas and the chemical and physical structure of the polymer. In particular, it increases with increasing condensability of the gas, because of the higher sorption in the polymer.

Finally, it is worth noting that for a multi-component system, Eq. 3 is valid for each one of them. No other equations are requested if additional transport phenomena, such as concentration polarization, longitudinal mixing and pressure loss across the membrane, are negligible. Longitudinal dispersion always tends to produce a more uniform concentration along the channel and, therefore, in many cases, it reduces the degree of separation (Hwang and Kammermayer, 1975; Vansant and Dewolf, 1990; Baker, 2004; Javaid, 2005).

1.3. Key Parameters: Permeability and Selectivity

The performance of a gas separation process is mainly determined by two parameters: the purity of the product gas and the recovery (*i.e.* the fraction of the feed gas recovered as product). The crucial requirements for gas separation membranes are high permeability and selectivity for the component considered.

The challenge to research is the development of materials with excellent selectivity and the production of ultra-thin layers without defects. The presence of a few defects is very difficult to avoid during membrane preparation and even a single small membrane defect can dramatically decrease the selectivity of gas separation membranes.

There are several factors that affect the performance of a membrane, among which: permeability, selectivity, pressure ratio, stage-cut and membrane thickness. In the next section a discussion on the selectivity is considered in details.

1.3.1. Ideal Selectivity to Pure Gases

As said, selectivity is defined as the ratio of the individual gas permeabilities and therefore it measures the ability of a membrane to separate two gases. And, as a rule, the number of membrane stages required for a given purity and/or the quantity of product that is lost during the process is related to selectivity, whereas the productivity of a membrane is associated with the permeability.

Based on single gas permeability of two components, the ideal selectivity of the component *i* with respect to *j* is defined by:

$$\alpha_{ij} = \frac{P_i}{P_j} \qquad (9)$$

The permeability being equal to the product of the diffusivity and solubility coefficients of the gas species (Eq. 8), the selectivity can be also written as:

$$\alpha_{ij} = \frac{S_i}{S_j}\frac{D_i}{D_j} \quad (10)$$

In real systems, the diffusion coefficient D_i (or D_j) and the solubility coefficient S_i (or S_j) can be both a function of concentration, so the theoretical analysis could be more complicated.

In particular, the ratio of the diffusion coefficients of the two gases, D_i/D_j, is often also called *diffusivity selectivity* and is related to the different size of the molecules. On the other hand, the ratio of the solubility coefficients of the two gases, S_i/S_j, reflects the relative sorption of the gases and can be also viewed as the *solubility selectivity*. For example, this term determines the overall selectivity in rubber polymers.

To summarise, the following relations are always valid:

$$\alpha_{ij} = P_i/P_j = (S_i D_i)/(D_j S_j) = (S_i /S_j) (D_i /(D_j) = \alpha_{Sij}\, \alpha_{Dij} \quad (11)$$

where:

α_{Sij} = solubility controlled factor (favours condensables)

α_{Dij} = mobility controlled factor (favours small molecules).

As can be seen in Table 2 (below), apart from the cellulose acetate membrane, in the other cases $\alpha_{Dij} > \alpha_{Sij}$ is observed. This *anomaly* is explained considering the plastic effect of CO_2 on the cellulose acetate membrane.

The diffusion coefficient decreases with increasing molecular size in polymer materials.

In fact, large molecules interact with more segments of the polymer chain with respect to small molecules; whereas the mobility selectivity always favours the transport of small molecules with respect to the large ones.

1.3.2. Selectivity in Mixture of Gases - Separation Factor

The ideal selectivity is related to the permeation of pure gases through a membrane.

This concept is convenient because, in the absence of strong interactions between the permeating gases, the permeability coefficients of the pure gases can be used.

Vice versa, in the case of mixtures, strong interactions between the permeating components are generally present and so, another parameter is generally considered more important for the design of a membrane plant: the separation factor, SF_{ij}.

For a binary mixture it is defined as:

$$SF_{ij} = (y_i /y_j)/(x_i /x_j) \quad (12)$$

where y and x are the molar concentrations in the permeate and feed sides, respectively, and the subscripts i and j usually refer to the two components in the mixture considered.

Table 2. Diffusivity, solubility and selectivities of CO_2 and CH_4 in various polymers

	CO_2		CH_4		Selectivity		
Polymer	$Dx10^8$ (cm^2/s)	S ($cm^3/(cm^2$ atm))	$Dx10^8$ (cm^2/s)	S ($cm^3/(cm^2$ atm))	P_{CO2}/P_{CH4}	D_{CO2}/D_{CH4}	S_{CO2}/S_{CH4}
Polyphenilene oxide	22.0	2.10	3.2	0.95	15.1	6.7	2.2
Polycarbonate	3.2	1.47	0.5	0.41	24.4	6.8	3.6
Cellulose Acetate	5.9	2.05	1.4	0.28	30.8	4.2	7.3
Polysulphone	2.3	1.44	0.3	0.45	28.3	8.9	3.2
Kapton	0.1	1.53	0.007	0.37	63.6	15.4	4.1

When the membrane selectivity is too low to reach a desired purity, more than one membrane stage is necessary. This will increase both capital and operating cost owing to the additional membrane area required and the recompression between the stages. Investment costs for compressor, vacuum pump and membrane area have to be balanced with the operating costs (Vansant et al., 1990; Baker, 2004; Javaid, 2005). It should be stressed that it is generally not possible to predict the mass transport behaviour of a mixture starting from single component measurements. Future research on gas and vapour mixture separation is of great importance, because the separation efficiency of the membrane for practical applications is a crucial parameter. During experiments, both x_i and x_j are fixed by the experimental conditions, whereas y_i and y_j must be determined by gas chromatography or mass spectrometry. By definition, the separation factor is always greater than 1, and depends on the experimental conditions (for example: pressure difference or the absolute pressure of the supplied gas).

1.4. Transport Mechanisms through Dense Polymeric Membranes

Depending on the operating temperature (T) relative to the glass transition temperature (Tg) of the polymer, polymeric membranes are in rubbery or glassy form. In the following the main differences in terms of transport equations are shown.

1.4.1. Rubbery Polymers (T > Tg)

A rubbery membrane is an amorphous polymeric material that operates above its Tg under the condition of use: the membrane is usually considered to be in thermodynamic equilibrium. In these membranes, for cases in which the sorbed concentrations are low, the sorption of low molecular weight is typically described by Henry's law:

$$C_D = S_H f \qquad (13)$$

where C_D is the concentration of gas in the membrane matrix, S_H Henry's law constant and f the fugacity of the gas considered.

The flux of a component i through the membrane is expressed as:

$$J_i = D_i\,(C^i_I - C^i_{II})/\delta = D_i\,S_i\,\Delta p/\,\delta = P_i\,\Delta p/\delta \qquad (14)$$

where C^i_I and C^i_{II} are the concentration of component i at feed side (I) and permeate side (II), respectively; D_i the diffusivity coefficient that measures the mobility of the molecules within the membrane; S_i the solubility coefficient that measures the solubility of gas molecules within the membrane; P_i the permeability coefficient that measures the ability of the membrane to selectively transport one gas species and not another one.

For ideal gases: the permeability is related to the gas permeation rate (Q_i) through the membrane, the surface area (A) of the membrane, the thickness (δ) of the membrane and the driving force for separation, the pressure difference across the membrane (Δp_i):

$$\not{p}Permeance = P_i\,/\delta = Q_i/(A\cdot\Delta p_i) \qquad (15)$$

The ideal selectivity of gas i over another gas, j, as usually is defined with Eq. 9 .
For rubbery polymers and low concentrations of penetrant:

- D_i is typically constant, and
- P_i is independent of the feed pressure.

In the case of the presence of high-activity of gases or vapours, deviation from Henry's law sorption is observed. Of course, also in dense rubbery polymers the transport of molecules is typically described by a solution-diffusion mechanism and in the following the dependence of P from T and pressure (fugacity f) are resumed.

➢ Temperature: diffusivity, solubility, and so permeability, follow an Arrehnius-law:

$$D = D_o \, exp(- E_d/RT) \tag{16}$$

$$S = S_o \, exp(- \Delta H_S/RT) \tag{17}$$

where E_d is the activation energy for diffusion, and ΔH_S the enthalpy of solution.
From which:

$$P_o = D_o \, S_o \tag{18}$$

and

$$P = D \, S = P_o \, exp \, ((- E_d - \Delta H_S)/RT) \tag{19}$$

It also results:

$$E_p = (- E_d - \Delta H_S) \tag{20}$$

and

$$P = P_o \, exp \, (- E_p/RT) \tag{21}$$

where E_p is the apparent activation energy for permeation.

➢ Pressure

D and S are constant with respect to C. This corresponds to saying that P is independent of C and thus also of pressure.

In other words, P is independent of partial pressures and also independent of Δp_i (pressure difference between the two parts I and II of the membrane). Moreover, concerning the enthalpy of solution, also this relation is valid:

$$\Delta H_S = \Delta H_C + \Delta H_l \tag{22}$$

ΔH_S could be positive or negative depending on the enthalpy of condensation (ΔH_C is always positive because the process is exothemic) and the enthalpy of mixture (ΔH_l is always negative because the process is endothermic). To be also considered that *Ed* is always positive. Generally, it is $\Delta H_S > 0$ (when $\Delta H_S < 0$, the value is almost low, see Table 3) and so $E_p > 0$ (i.e. *P* increases exponentially with *T*). That's also to say that *Ln(P)* versus *1/T* gives a linear relationship (it is called *Arrhenius* plot). In Table 3, some data of ΔH_S and E_d for some gases are shown.

In Table 4, a comparison between solubility and transport in rubbery membranes at temperatures below and above the critical temperature (*Tc*) is reported.

1.4.2. Glassy Membranes (T < Tg)

A glassy membrane is an amorphous polymeric material that operates below its *Tg* under the conditions of use: the membrane is far from thermodynamic equilibrium. The polymer chains are packed imperfectly, leading to excess free volume in the form of microscopic voids in the polymeric matrix. Within these voids, *Langmuir adsorption* of gases occurs, which increases the solubility. Differently from rubbery membranes, glassy membranes are able to discriminate effectively between extremely small differences in the molecular dimensions of common gases (*e.g.* 0.2 to 0.5 Angstrom).

Table 3. Enthalpy of solution and activation energy for diffusion for some gases

Gas	ΔH_S (kcal/mol)	E_d (kcal/mol)
He	2.11	5.35
H2	2.47	7.5
Ne	1.05	8.5
O2	-1.10	14.5
Ar	-1.88	16.5
Kr	-1.2	19.4
CH4	0.46	19.3

Table 4. Solubility and transport in rubbery membranes (T>Tg)

T > Tc	T < Tc
Henry's law is obeyed: $S = S_o$ = constant	Henry's law is not obeyed: $S = S(C)$
D is independent of concentration $D = D_o$ = constant	*D* is a function of concentration $D = D(C)$
Activation energy for diffusion is constant: E_d = constant	Activation energy for diffusion is variable of *C* and time (*t*): $E_d = E_d(C, t)$
Permeability constant: $P = P_o$ = constant	Permeability strongly depends on pressure: $P = P(p)$

In glassy membranes, the transport of molecules is typically described by the so-called dual-mode model. In this model, part of the gas molecules are absorbed in the polymer matrix (and, like in the case of rubbery membranes, it follows Henry's law), whereas another part is adsorbed into microscopic voids and its concentration, C_H, is described by the following equation:

$$C_H = C'_H \, b \cdot f / (1 + b \cdot f) \tag{23}$$

where C'_H is the maximum adsorption capacity (*i.e.* it measures the "degree of saturation of holes"), b the ratio of rate coefficients of adsorption and desorption (it measures the degree of "affinity of the holes"), f the fugacity of the gas considered. The total sorption for glassy polymers is then described by the sum of the two components of gas molecules adsorbed in the polymeric matrix

$$C = C_D + C_H \tag{24}$$

and the overall solubility $S(f)$ (at T constant) is given by:

$$S(f) = C/f = (C_D + C_H) = S_H + (C'_H \, b)/(1+bf) \tag{25}$$

where S_H is the Henry's law constant.

The success of the dual-mode sorption model in describing penetrant sorption in glassy polymers is due to the physical significance that can be related to model parameters. In this model, D is independent of the concentration C ($D = Do$ = constant) and P decreases as f (or the pressure p) increases:

$$P = S(f) \, D_o \, (1 + F \, K(1 + b \, f)) \tag{26}$$

where:

$$F = D_H/D_o = \text{ratio between the diffusion coefficients} \tag{27}$$

$$K = C'_H \, b / \, S_H \tag{28}$$

Generally, the solubility of gases is higher in glassy polymers than in rubbery polymers. This is because, as said, in glassy polymers there is an additional adsorption mechanism:

$$S = So \, exp \, (-\Delta H_D/RT) \tag{29}$$

where ΔH_D = Henry type sorption enthalpy

$$b = bo \, exp \, (-\Delta H_H \, /RT) \tag{30}$$

where ΔH_H = enthalpy for sorption in Langmuir sites.

The Henry type sorption enthalpy (ΔH_D) is of the same order of magnitude of the rubbery polymers ones (ΔH_S), and ΔH_H higher than with respect to the rubbery polymers ones. These considerations are in good agreement with experimental results where it is observed that adsorption of gas in rubbery polymers is most exothermic than the corresponding ones in glassy polymers. The parameter C'_H, and so also K (being $K = C'_H b/S$), becomes zero when the temperature of the system is higher than Tg. In this case, the dependence of permeability (P) with respect to the fugacity (f), or pressure, disappears when S and D are constant. Concerning the solubility of gases (and also vapors and liquids) in glassy polymers, an excellent overview has been recently presented by Sarti (2012). The overview is mainly related to the macroscopic thermodynamic models developed to calculate the solubility in glassy polymers. To summarise this section, for both glassy and rubbery membranes, the transport properties for gases are almost similar and the relationship between temperature and the transport of small molecules is generally viewed as an activated process and obeys an Arrhenius relationship.

1.5. Gas Separation Membrane Fabrication

In gas separation, isotropic dense polymeric films or multilayered membranes with integrated skin are normally preferred. In multilayered membranes, a porous layer is used as a mechanical support for thin dense selective film (Figure 6). In order to prevent rupture and formation of defects, a highly permeable gutter layer is often used between the support and selective layer. The preparation of porous supports can be done by using traditional phase inversion processes or through more advanced techniques, enabling one to get long-range controlled structures (Gugliuzza and Drioli, 2007; Speranza et al., 2010), while the selective layer is coated onto the support via wet-dry process. The necessity to produce very thin dense layers arises from the request of higher selectivity combined with higher permeability. Lower membrane thicknesses promote large mass transfer, but increase the risk of abatement of selectivity. So, a structural compromise is necessary to achieve a good permeability /selectivity ratio. Depending on the mechanism, efficiency and productivity of a process can be enhanced by improving the skin layer affinity to specific compounds or varying the diffusive paths through it (Gugliuzza et al., 2007).

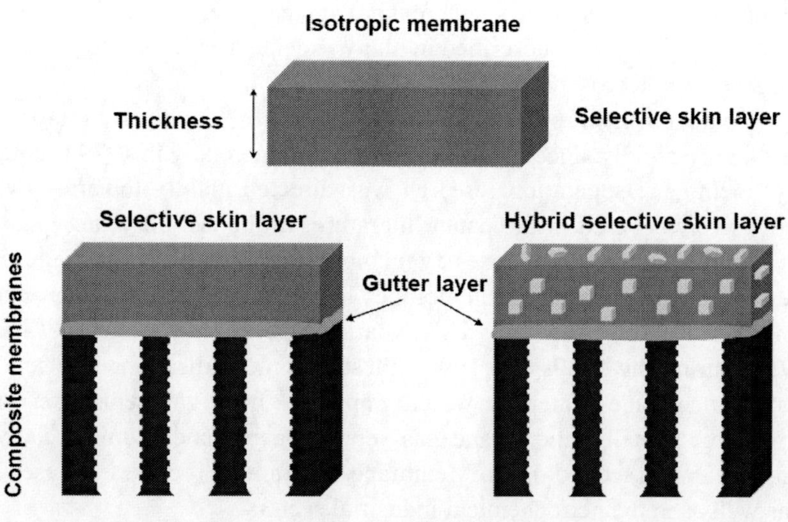

Figure 6. Conceptual scheme of a multilayered membrane.

This can also be done by combining different compounds in a unique system rather than to reduce drastically the membrane thickness (Figure 6).

In this context, the choice of materials and procedures of assembly depends on composition of gaseous streams, along with operating conditions and type of membrane operation selected. It is well-known that the majority of polymers shows low mechanical and chemical resistance, but also flexibility and easy scalability differently from what observed for inorganic membranes. This major limitation can be reduced through the design of advanced mixed matrixes. Inorganic and organic compounds can indeed be merged in a unique system, resulting in higher performance membranes. However, compatibility, uniformity, adhesion and minimum loading for maximum surface area are often regarded as critical issues, considering the tendency of nanofillers to aggregate in clusters. Solution blending (Wara et al., 1995), in situ polymerization (Patel et al., 2003) and sol-gel (Gomes et al., 2005) as well as directed assembly (Grzelczak et al., 2010) methods are frequently proposed to tailor hybrid systems. Major aggregation is often observed for polymers blended with inorganic particles, especially during solvent evaporation. In this case, the degree of interactions between polymer chains and the external surface area of the particles is dominated by van der Waals forces and hydrogen bonds.

The functionalization of nanoparticles appears a rather interesting strategy to make compatible two or more systems. Nanoparticles mixture with oligomers or polymers via sol-gel methods is often used, along with the infusion of volatile precursors of inorganic and metal-oxide nanoparticles into free volume of polymers or, more attractively, with the self-assembly of nanoparticles. In the latter case, the nanoparticles can be dragged thermo-dynamically by intermolecular interactions established at capping agent-template interface. The use of electrical, magnetic and flow fields can also direct the self-assembly of nanoparticles and colloidals in nanocomposites (Gast and Zukoski, 1989; Dijkstra, 2005; Boker et al., 2007).

1.6. Gas Separation - Polymeric Membranes for CO_2 Separation

Membrane gas separation shows different advantages over conventional processes. These aspects have been recently well described in many excellent reviews (see, for example, Maier, 1998; Stern, 2002; and Koros 2002).

The gas separation using polymeric membranes is a relatively simple process and so it has expanded in use rapidly since it was introduced commercially in 1979 (Scott, 1995).

Initially, membrane separation research was directed mainly towards reverse osmosis, which is extensively covered in the patent literature. In the last 30 years, great progress has been observed in the separation of gases by membrane technology. In fact, the first successful commercialisation of gas separation membranes for hydrogen recovery was achieved in 1977 by Monsanto/Permea followed later by Cynara, Separex and Generon (Spillman, 1989; Koros, 1991). During the 1980s and 1990s, these commercialisations also led to substantial innovations in membrane materials which improved both gas separation efficiency and membrane lifetime. Finally, membrane gas separation becomes commercially competitive with existing separation technologies. Membrane separation processes are used today in bulk chemistry as well as in the petrochemical industrial sectors.

Many companies have introduced their equipment on the market. The range of main industrial applications covers the supply of pure gases (for example He, N_2, O_2 from air), the separation of acid gases (CO, H_2S), the separation of CO_2 and H_2O from natural gas, and the separation and purification of H_2 in the petrochemical and chemical industries. Different reviews and books can be found in the literature describing these aspects (Koros, 1991, Osada and Nakagawa, 1992; Paul and Yampolskii, 1994, Baker, 2002; Basu et al., 2004), whereas a detailed and extensive description of such equipment could be found, for example, in Scott (1995).

Owing to lack of space, in this chapter the authors will concentrate only on CO_2 polymeric membrane separations. Another reason for this being that the overabundance of CO_2 in the atmosphere is one of the main contributors to global climate change. In fact, it is responsible for about 64% of the greenhouse effect and global warming (Zhou et al., 2009). This has led to scientific interest in the development of new methods for reducing the amount of CO_2 emitted into the atmosphere from industries or power plants.

Conventionally, as is well known, there are three families of processes for CO_2 capture: sorption processes (absorption in solvents or adsorption on solid sorbents), cryogenic processes and membrane processes. Among these processes, polymeric membrane processes could offer the potentiality of relatively higher capture temperature by eliminating the intermediate cooling and reheating steps, necessary, for example, in the amine-based absorption process. Moreover, membranes could offer several advantages such as low capital investment, compactness, ease of installation and operation, simplicity, flexibility and ability to perform separations at low energy penalties (Kohol and Nielsen, 1997; Basu et al., 2004; Abertz et al., 2006).

In CO_2 separation field, polymeric membranes may be used, for example, in the treatment of CO_2-rich streams, exiting from inorganic membrane reactors, previously used for hydrogen production/separation by means of the water gas shift reaction, as schematized in Figure 7.

The most important properties of polymeric membranes are summarized in Table 5.

Among the polymeric membranes with reasonable CO_2 permeability and selectivity, much attention is today paid on the following polymers:

- *Polyimides,* which are generally synthesized by the reaction of a diamine with a diahydride (PMDA, 6FDA and BTDA) in an aprotic solvent to form a polyamic acid. This polymer then undergoes a polycondensation reaction (either thermally or chemically) to form the desired polyimide. Polyimides combine excellent thermal and chemical stability with a wide range of CO_2 permeabilities.
- *Polycarbonates,* which are commonly synthesized by reaction between a diol and phosgene under a variety of conditions. The majority of the polycarbonates membranes tends to have a P_{CO2} under 40 Barrer (where: 1 Barrer = 10^{-10} cm^3 (STP) cm/cm^2 s cmHg) and $\alpha_{CO2/N2}$ in the range 15-25. One notable exception to this is the polycarbonate TMHFPC, which possesses a P_{CO2} of 111 Barrer and $\alpha_{CO2/N2}$ = 15 (Costello and Koros, 1994).
- *Polyarylates*, which are synthesized by reaction of a diol, such as bisphenol A, with an acid chloride. The reaction can be performed under a variety of conditions such as interfacial or a stepwise polycondensation. The majority of polyarylates membranes shows P_{CO2} < 25 Barrer; however, a small number shows P_{CO2} as high as 85 (Pixton

and Paul, 1995), such as the sterically bulky polyarylate, TBHFBPA/tBIA, which exhibits the highest P_{CO2} with $\alpha_{CO2/N2} = 19$ (Kumar, 2009).

- *Polypyrrolones,* which are conventionally synthesized by reaction of dianhydride and a tetra-amine functionalized monomer (Yang et al., 2005). Polypyrrolones are structurally similar to polyimides, although they are considerably more rigid (Liu et al., 2006). They are characterized by high thermal and chemical resistance. Nevertheless, owing to their high rigidity, the polypyrrolones can act as a molecular sieve.
- *Polysulfones,* which are generally synthesized by condensation reaction between bisphenol and dihalogenated diphenylsulfone. The most extensively studied polysulfone is the one formed using bisphenol A (Barbari and Datwani, 1995). Other polysulfones are structurally related to this polymer. In fact, by substitution of bisphenol A with a different diol, large numbers of PSF derivatives have been synthesized, which display a wide range of CO_2 permeabilities and selectivities.

Table 5. Main properties of CO_2 polymeric separation membranes

High CO_2 permeabilities correspondes to low CO_2/N_2 selectivities	Trade-off between selectivity and permeability: membranes with high permeability has low selectivity and vice versa.
Thermal and chemical effects	Long term stability at elevated T is requested This is important for maintaining the robustness of the membrane-based system.
Plasticization effects	Membranes must be resistant to plasticization phenomena, because they change the polymer configuration by increasing the free volume of mixture. The polymer membrane loses its selectivity due because the polymer matrix swells due to sorption of one species, accelerating the permeation of the other.
Cost effective	The cheaply manufactures into different membrane modules is necessary.

1.7. Performance of CO_2 Membrane Systems

Gas separation using membranes is a highly attractive energy efficient technique, especially for carbon dioxide capture (Basu et al., 2004; Abertz et al., 2006).

In 1992, studying the application of polymeric membranes for recovering CO_2 from a flue gas of a power plant, Feron et al. (1992) and Van Der Sluijs et al. (1992) showed that up to 76% of CO_2 removal is achievable and, moreover, that the economic competitiveness of the process depends on the membrane used and, in particular, on its selectivity to the gas transport. Without going into detail about these aspects a deeper discussion can be found, for example, in (Maier, 1998; Stern, 2002; Paul and Yampol'skii, 1994).

As already said in the Introduction section, membranes can be separated into two types: porous and nonporous (or dense) membranes. Porous membranes separate gases based on molecular size by small pores in the membrane. More common for gas separation, membranes used (for example) in natural gas applications are nonporous or asymmetric membranes. These membranes separate on the basis of solubility and diffusivity. For both porous and nonporous membranes, there are many possible separation mechanisms, but only five of them are considered the most important for gas separation through membranes: Knudsen diffusion, molecular sieving, surface diffusion, capillary condensation and solution-diffusion separation (Spillman, 1989; Paul and Yampolskii, 1994; Fritzsche and Kurz, 1990; Gallucci and Basile, 2009) and among them, molecular sieving and solution-diffusion mechanisms are the main mechanisms for nearly all gas separating membranes.

In dense membranes, the gas transport is based on a solution-diffusion mechanism and results in a selective transport of gases and in their separation. In porous membranes, different mechanisms take place. Knudson separation is based on gas molecules passing through membrane pores small enough to prevent bulk diffusion: the separation is based on the difference in the mean path of the gas molecules due to collisions with the pore walls, which is related to the molecular weight. Specifically, Knudsen selectivity for any gas pair is equal to the inverse ratio of the square root of their molecular weight. For example, for CO_2/N_2 separation, Knudsen diffusion predicts a selectivity of 0.8. In fact, 1/(SQRT(44/28)) = 0.8. For this reason, this type of diffusion process does not provide sufficient separation in most instances, although the technique was successfully used in the large-scale separation of uranium isotopes (with ceramic membranes).

Molecular sieving separates gas mixtures on the basis of size exclusion and so it is potentially useful in separating molecules of different sizes. The diffusion of smaller gases happens at a much faster rate than larger gas molecules and thus $\alpha_{CO2/N2}$ is greater than 1, as CO_2 has a smaller kinetic diameter than N_2. In surface diffusion, molecules adsorb to the surface of the pore walls of membranes (Bredesen et al., 2004). The level of interaction between the adsorbed gases and pore surface determines the rate of surface diffusion. Thus, molecules diffuse along the surface of the pore walls and separation is achieved mainly by the difference in the degree of the interactions for the individual gases. High selectivity can be achieved in cases for which preferential adsorption on one component occurs. Capillary condensation can be considered an extension of surface diffusion: it happens when the vapour pressure becomes low and adsorbed gas undergoes partial condensation within the pores. This condensed component diffuses more rapidly through the pore than gases, generating the separation of the condensable gas.

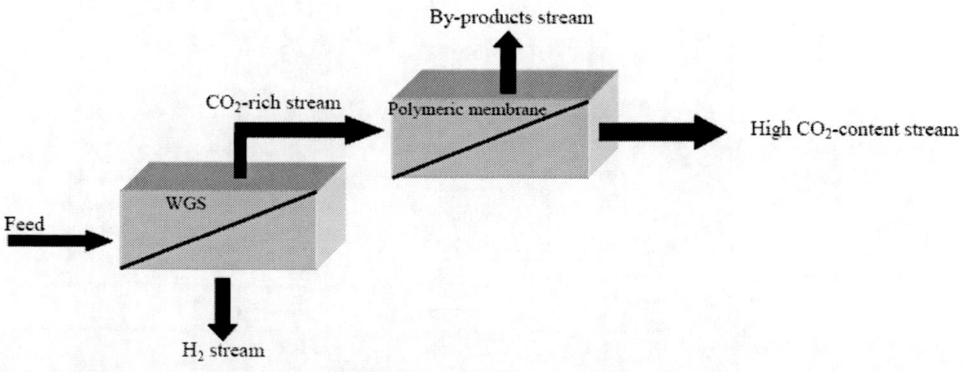

Figure 7. Scheme of treatment of a CO_2-rich stream, exiting from Pd-based MR.

The polymeric membranes currently used in most commercial applications are solution-diffusion ones. In fact, very high selectivity can be obtained in dense polymer membranes where the transport mechanism is based on the solution and diffusion of the different gases and vapours of the mixture within the membrane phase. It consists of three steps: a) sorption of the mixture components from the feed according to their solubility coefficient in the polymer matrix; b) diffusion of the gases adsorbed within the membrane phase according to their concentration gradients; c) desorption of these components from the outer face of the membrane in the permeate phase.

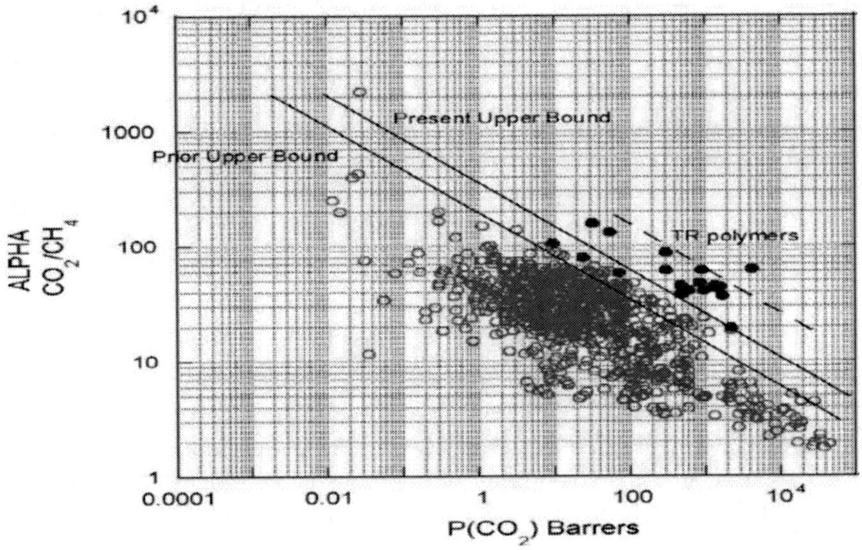

Figure 8. Robeson upper bound for the CO2 and CH4 gases (Robeson, 2008); (with permission of Elsevier).

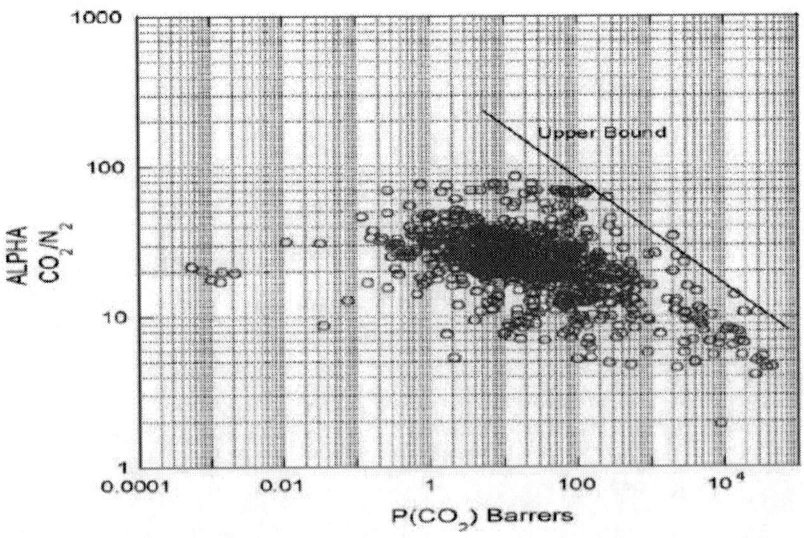

Figure 9. Upper bound correlation for CO2/N2 separation (Robeson, 2008); (with permission of Elsevier).

As already said, it is interesting to observe that there is a trade-off between selectivity and permeability: membranes with a high selectivity show a low permeability, and vice versa. Figure 8, in which the ideal $\alpha_{CO2/N2}$ versus fast component P_{CO2} of polymeric membranes is reported, well illustrates the situation for a range of glassy and rubbery membranes involved in CO_2/CH_4 separations (Robeson, 1991; 2008). Compared to other separations, such as O_2-N_2 and CO_2-CH_4 mixtures, CO_2-N_2 mixture appears to be an easer separation (Scholes et al., 2008). Figure 9 shows the same trend for the gases CO_2 and N_2. In Table 6, some experimental data points close to the present empirical upper bound for CO_2/CH_4 separation are shown. Also in this case it is evident the trade-off between the permeability of CO_2 and the selectivity of CO_2/CH_4.

Since the first paper published by Robeson in 1991, only a few examples of polymeric membranes which exceed the upper bound have been published, and overcoming it is the focus of many recently awarded patents in polymeric membranes. In fact, achieving both high carbon dioxide permeability and high selectivity is highly desirable.

It should be also added that exceeding the Robeson limit should not be considered a rigid rule. In fact, for example, the polybenzimidazole membrane exceeds the Robeson upper bound for H_2/CO_2 selectivity versus H_2 permeability in the range of temperature 100-400 °C (Berchotold, 2006). Moreover, recently, Koros and Mahajan (2000) have suggested the possibility of exceeding the upper bound by using the so-called mixed-matrix membranes.

Apart from permeability and selectivity, other membrane properties are also very important, such as their thermal, chemical and plasticization resistance, as well as the aging affects for ensuring continual performance over long time periods. Moreover, cost effectiveness to manufacture as standard membrane modules is also important. Considerable experimental research has been addressed to meeting these aims. An extensive review describing original both polymeric and inorganic membrane patents was recently published by Scholes et al. (2008) with particular attention paid to carbon dioxide separation through polymeric membrane systems for flue gas applications.

This review is particularly interesting because it focuses on recent novel approaches in polymeric membranes that achieve separation performance above Robeson's upper bound and therefore are possibly more commercially competitive than present membrane gas separation technologies.

For improving the performance of polymeric gas separation membranes, also other kind of membranes are under investigation, especially for the separation of CO_2, such as:

- Mixed-matrix membranes: this new type of membrane, being able to combine together the advantages of polymeric membranes and inorganic ones, seems to allow a superior performance.
- Hybrid membranes, which offer some advantages with respect to porous inorganic ones. In this case, the surface of a porous inorganic support material is chemically modified in order to have a good affinity with CO_2. For this reason, these membranes should be named surface-modified inorganic membranes thus avoiding confusion with mixed-matrix membranes.
- Membrane contactors, which are today considered one of the best ways to control CO_2 emissions using membrane technology. Membrane gas absorption technology combines membrane separation and chemical absorption technologies.

Table 6. Experimental data points close to empirical upper bound for CO_2/CH_4 separation (adopted from Robeson, 2008)

Polymer	P_{CO2} (Barrer)	$\alpha_{CO2/CH4}$
PVSH doped polyaniline	0.029	2,200
Polypyrrole 6FDA/PMDA (25/75)-TAB	3.13	140
Polyimide TADATO/DSDA (1/1)-DDBT	45	60
Poly(diphenyl acetylene) 3a	110	47.8
Polyimide 6FDA-TMPDA/DAT (1:1)	130.2	38.9
Polyimide 6FDA-TMPDA/DAT (3:1)	187.6	33.9
Polyimide PI-5	190	33.9
Poly(diphenyl acetylene) 3e	290	31.5
Poly(diphenyl acetylene) 3f	330	27.5
Polyimide 6FDA-TMPDA	555.7	22.7
Polyimide 6FDA-durene	677.8	20.18
6FDA-based polyimide (8)	958	24
PIM-7 1	100	17.7
PIM-1 2	300	18.4
PIM-1 2	19,000	4.42

1.8. Membrane Modules

In order to use a membrane in a separation process it must be packed in a proper device, a so-called *membrane module*. The requirements that a membrane module must meet are: low production costs, high packing density, low energy consumption, and good control of concentration polarization. In the practical applications of a separation processes membrane, the modules are quite different: stirred batch cells or dead-end filter cells are used in small-scale laboratory applications; whereas two types of configurations are used in large-scale industrial applications: flat (plate-and-frame, spiral wound and disc tube modules) and tubular (tubular, capillary and hollow fibre modules). The membrane modules used in gas separations are only spiral wound, capillary and hollow fibre (some figures are reported in Chapter 6).

The design of these modules is quite different as well as their mode of operation, production costs, and the energy requirements (mainly owing to pressure drop inside the membrane module). In Table 1 of Chapter 6 the principal characteristics of the most used membrane modules are reported.

It should be said that there is no membrane module able to solve all the problems. Vice versa, the commercial membrane modules available today are specifically designed for a specific membrane process application. In the next sections some characteristics of the modules used in gas separation are given.

1.8.1. Spiral-Wound Module

The spiral-wound module is a variation of the basic plate-and-frame concept, where the feed gas mixture, pressurized in the module and forced to pass through the membrane surface, selectively permeates the membranes forming the permeate gas mixture. The configuration spiral-wound module is formed by a membrane envelope of spacers and a membrane wound around a porous tube. The feed gas mixture is sent in the axial direction through the feed channels across the membrane surface. The permeate gas mixture is collected in the central porous tube. Owing to the pressure drop, small spiral wound modules consist of only one envelope (membrane area 1 - 2 m^2). Commercial spiral wound modules have a multi-leaf arrangement with a larger membrane surface and an extended permeate path (membrane characteristics: area 3 - 60 m^2, length 1 m, diameter 10 - 60 cm). Generally, various elements (2 to 6) are connected in series inside a single pressure vessel.

1.8.2. Capillary Membrane Module

The capillary membrane module is formed by a large number of membranes with a capillary dimension. The membranes are placed in a shell tube module and arranged in parallel. The feed gas mixture, sent inside (or outside) the capillary membranes, permeates selectively through the membranes and the permeate gas mixture is then collected in the outside shell tube (or inside the membrane bore). The capillary membrane module provides a very high membrane packing density.

1.8.3. Hollow-Fibre Membrane Modules

In the hollow fibre membranes (*o.d.* 50 - 100 μm), prepared using the spinning process, the selective layer is on the outside of the fibres. These fibres are installed as a bundle of several thousand fibres. The hollow fibre modules are formed of two basic geometries: the shell-side feed design and the bore side feed.

In the first case, the system is shell side pressurised: the feed permeates through the fibre wall and is collected at the open fibre ends. These modules operate at pressures in excess of 100 bar, because this design is easy to realise and, moreover, very large membrane areas can be used. For this reason, the fibres usually have small diameters (typically 50 μm *i.d.* and 100 - 200 μm *o.d.*) and thick walls. In the second case, the fibres are open at both ends and the feed fluid circulates through the bore of the fibres. To minimise the pressure drop inside the fibres, the diameters are usually larger than those of the fine fibres used in the previous case.

Hollow fibre membranes have the highest packing density of all module types available on the market today. Their production is very cost effective. In fact, expensive, sophisticated and very high speed automated spinning machine, fibre handling and module fabrication equipment is required to produce these modules.

1.8.4. Membrane Modules Comparison

Many factors must be considered in choosing the most suitable module, such as the specific separation to be performed, manufacturing cost, concentration polarisation and fouling (if any), packing density, ease of cleaning , permeate side pressure drop, high pressure operation and limitation of type of material. Several types of modules can be used according to the membrane processes considered.

Concerning the packing density of the commercially available polymeric membrane modules, the following trend is observed: hollow fibre > capillary > spiral wound > plate and frame > tubular. The single plate geometry shows a very low packing density, whereas single tubes typically range between 35-280 m^2/m^3. The packing density of most of the inorganic membrane modules of the shell-and-tube type is in the same range as the organic type: 150-300 m^2/m^3. In contrast to organic membranes, no commercial inorganic membranes are yet currently available in the high packing density such as spiral wound and hollow fibres. Nevertheless, quite recently, hollow fibre ceramic and carbon membranes were prepared. The packing density of these membranes is in the range 500-9000 m^2/m^3 with an area of standard module in the range 50-150 m^2. Multi-channel monolith module (with a single element in a module) presents a packing density in the range 300 – 540 m^2/m^3. As the number of elements increases in a multi-channel monolith module, the packing density tends to decrease.

It is difficult to quantify correctly the cost of a module because the same module design varies widely according to the application considered. Generally, the hollow fibre modules are cheaper than the others even if these modules must be produced for very high-volume applications in order to justify the expense in developing and building the spinning and module fabrication equipment.

Membrane modules with high packing density, *i.e.* hollow fibre or spiral-wound ones, are used for gas separation. In this context, the Monsanto hollow fibre module, equipped with polysulphone-silicone membranes, and also spiral wound modules (Separex, Grace), equipped with cellulose acetate membranes, started to enjoy great success in the late 1980s . Also Dow introduced a fine cellulose acetate hollow fibre module (packing density of 50.000 m^2/m^3) for gas permeation which separates very low transmembrane pressure differences and feed operating pressures around 8-9 bar. In these membranes, the flow pattern in the module is not counter-current as in the Monsanto module, but cross-flow (as in reverse osmosis hollow fibre modules), with the feed flowing radially through the bundle from inside to outside. It is important to stress that the design is advantageous with respect to feed side pressure losses.

A very important factor for the fabrication of a specific module design is the membrane material. During the last few decades, the technology to produce high performance ultrathin membranes in high packing membrane density has constantly improved. Nowadays, membranes with an active layer of 0.05 μm are produced. As a result of this improvement, concentration polarisation and fouling (not relevant for gas separation) are the two major factors determining system performance. Future research will be addressed at the improvement of fluid flow, to overcome (or at least to reduce) the problem of concentration polarisation, easily cleaned modules and novel membrane materials.

CONCLUSION

The importance of polymeric membranes for a sustainable gas separation is well evidenced in this chapter. At first, the principles of gas separation through polymeric membranes are introduced followed by a detailed description of the two main parameters used to characterize them: the permeability and the selectivity. Related equations are presented for both the single and gas mixtures. Also, their variation with respect to pressure and temperature is showed. A discussion about the various transport mechanisms through dense polymeric membranes in both the rubbery and glassy polymers well illustrates the potentiality of using the gas separation membrane systems. A short section is also dedicated to the membrane fabrication, with particular emphasis to the nanoparticles mixtures. A particular focus is on polymers used for CO_2 membrane separation: polyimides, polycarbonate polyarylates polypyrrolones and polysulfones. For this aspect, a section related to the performance of membrane systems is also presented, referring to the Robeson diagrams. The final part of the chapter is dedicated to the membrane modules. A short description about the spiral-wound module, the capillary membrane modules and the hollow-fibre membrane module used for gas separation is given.

REFERENCES

Abertz V., Brinkmann T., Dijkstra M. (2006), "Developments in membrane research: from material via process design to industrial application", *Adv. Eng. Mater.* 8, 328-358.

Adhikari S., Fernand S. (2006), "Hydrogen Membrane Separation Techniques", *Ind. Eng. Chem. Res.*, 45, 875-881.

Baker R.W. (2004), *Membrane Technology and Applications,* England, J. WileyandSons.

Baker R.W. (2002), "Future direction of membrane gas separation technology", *Ind. Eng. Chem. Res.*, 41 1393

Barbari T.A., Datwani S.S. (1995), "Separation properties of polysulfones membranes treated with molecular bromine", *J. Membrane Sci.* 107, 263-266.

Basu A., Akhtar J., Rahman M., Islam M. (2004), "A Review of separation of gases using membrane systems", *Petrol. Sci. Technol.*, 22 1343-1368.

Berchotold K.A. (2006), "Novel polymeric-metallic composite membrane for CO_2 separation at elevated temperatures", Conf. on American Filtration and Separation Society Fall Topical Conference, Pittsburgh (PA).

Boker A., He J., Emrick T., Russell T.P. (2007), "Self-Assembly of Nanoparticles at Interfaces", *Soft. Matter.*, 3, 1231–1248.

Bredesen R., Jordal K., Bolland O. (2004), "High-temperature membranes in power generation with CO_2 capture", *Chem. Eng. Proc.,* 43, 1129-1158.

Catalytica® (1998), *Catalytic membrane reactors: concepts and applications,* Catalytica Study N. 4187 MR.

Costello L.M., Koros W.J. (1994), "Effect of structure on the temperature dependence of gas transport and sorption in a series of polycarbonates", *J. Polymer Sci. Part B: Polymer Phys.*, 32, 701-713.

Dijkstra M. (2005), "Phase Diagram of Dipolar Hard and Soft Spheres: Manipulation of Colloidal Crystal Structures by an External Field", *Phys. Rev. Lett.*, 94, 138303-138306.

Falconer J.L., Noble R.D., Sperry D.P. (1995), "Catalytic Membrane Reactors", in *Membrane Separations Technology: Principles and Applications,* R.D. Noble and S.A. Stern (Eds.), Elsevier, 669-712.

Feron P., Jansen A., Klaassen R. (1992), *En. Convers. Mgmt.*, 33, 421.

Fritzsche A., Kurz J. (1990), 'The separation of gases by membranes', in: *Handbook of industrial membrane technology,* William Andrew Publishing, 559-593.

Gallucci F., Basile A. (2009), ‘Pd-based membranes synthesis and their application in membrane reactors’, Ch. 1 in: *Hydrogen Energy: Methods of Production, Safety and Costs,* S.V. Gorley (Ed.), Nova Science. Pub., New York.

Gast A.P., Zukoski C.F. (1989), “Electrorheological Fluids as Colloidal Suspensions”, *Adv. Coll. Inter. Sci.*, 30, 153–202.

Gomes D., Nunes S.P., Peinemann K.V. (2005), “Membranes for gas separation based on poly(1-trimetylsilyl-1-propyne)-silica nanocomposites”, *J. Membrane Sci.*, 246, 13–25.

Grzelczak M., Vermant J., Furst E.M., Liz-Marzan L.M. (2010), “Directed Self-Assembly of Nanoparticles”, ACSNANO, 4(7), 3591–3605.

Gugliuzza, A., Drioli, E. (2007), “PVDF and HYFLON AD membranes: Ideal interfaces for contactor applications”, *J. Membr. Sci.* 300, 51–62.

Gugliuzza A.., De Luca G., Tocci E., De Lorenzo L., Drioli E. (2007), “Intermolecular interactions as controlling factor for water sorption into polymer membranes”, *J. Phys. Chem. B*, 111, 8868-8878.

Hwang S.-T., Kammermayer K. (1995), *Membranes in Separation*, New York, Wiley Interscience.

Javaid A. (2005), “Membranes for solubility-based gas separation applications”, *Chem. Eng. J.*, 112, 219–226.

Kapoor A., Yang R.T., Wong C. (1989), “Surface diffusion”, *Catal. Rev.*, 31, 129-214.

Khulbe K.C., Feng C.Y., Matsuura T. (2007), “Synthetic polymeric membranes, characterization by atomic force microscopy”, *Springer* ISBN:3540739939, p. 216.

Knozinger H., Ratnasamy P. (1978), “Catalytic aluminas: surface models and characterization of surface sites”, *Catal. Rev. Sci. Eng.*, 17, 31-70.

Kohol A.L., Nielsen R. (1997), “Gas purification”, Gulf Publishing, Houston (TX), 5th ed.

Koros W (1991), ‘Gas Separation’, in: *Membrane Separation Systems - Recent Developments and Future Directions,* R.W. Baker (ed), William Andrew Publishing.

Koros W (2002), “Gas separation membranes: needs for combined materials science and processing approaches”, *Macromol. Symp.,* 188, 13.

Koros W, Mahajan R (2000), “Pushing the limits on permeabilities for large scale gas separation: which strategies”, *J. Membrane Sci.*, 175, 18.

Koros W.J., Ma Y.H., Shimidzu T. (1996), “Terminology for membranes and membrane processes”, *J. Membrane Sci.*, 120, 149-159.

Kumar P. (2009), “Inclusion chemistry in periodic mesoporous hosts: growth of quantum-confined materials and gas separation membranes”, http://etd.ohiolink.edu/send-pdf.cgi/KUMAR%20PARVEEN.pdf?ucin1186772727.

Lee K.H., Hwang S.T. (1986), “Transport of condensible vapors through a microporous vycor glass membrane”, *J. Coll. Int. Sci.,* 110, 544-555.

Liu J.G., Zhao X.J., Yang H.X., Li H.S., Fan L., Yang S.Y. (2006), “Novel high temperature polypyrrolones with asymmetric biphenyl moieties in the main chain: Synthesis and characterization”, *Macr. Chem. Phys.*, 207, 1272-1277.

Maier G. (1998), “Gas separation with polymer membranes”, *Angew. Chem. Int. Ed.*, 37, 2960.

Mallevialle J., Odendaal P.E., Wiesner M.R. (1988), *Water Treatment Membrane Processes,* McGraw-Hill Publishers, New York.

Mulder M. (1996), *Basic Principles of Membrane Technology, Kluwer Academic: Dordrecht*, 564.

Osada Y., Nakagawa T. (1992), *Membrane Science and Technology,* Marcel Dekker Inc., New York.

Patel N.P., Zielinski J.M., Samseth J., Spontak R.J. (2004), "Effects of pressure and nanoparticle functionality on CO_2-selective nanocomposites derived from cross-linked poly(ethylene glycol)", *Macromol. Chem. Phys.*, 205, 2409–2419.

Paul D, Yampolskii YP (1994), *Polymeric gas separation membranes,* CRC Press, Boca Raton.

Pixton M.R., Paul D.R. (1995), "Gas transport properties of polyarylates: substituent size and symmetry effects", *Macromolecules*, 28, 8277-8286.

Robeson L.M. (2008), "The upper bound revisited", *J. Membrane Sci.*, 320, 390-400.

Robeson L.M. (1991), "Correlation of separation factor versus permeability for polymeric membranes", *J. Membrane. Sci.*, 62, 165.

Saracco G., Specchia V. (1994), "Catalytic inorganic membrane reactors: present experience and future opportunities", *Catal. Rev. Sci. Eng.*, 36, 305-384.

Sarti G. (2012),"Predicting the solubility of gases, vapors and liquids in glassy polymers: an overview", Book of Abstracts of the workshop on Nanocomposite and Nanostructured Polymeric membranes for gas and Vapour Separations, May 15-18, Cetraro (CS, Italy), pp. 26-28.

Scholes C.A., Kentish S.E., Stevens G.W. (2008), "Carbon Dioxide Separation through Polymeric Membrane Systems for Flue Gas Applications", *Recent Patents on Chem. Eng.*, 1, 52-66.

Scott K. (1995), *Handbook of industrial membranes,* Elsevier, p. 276.

Speranza V., Trotta F., Drioli E., Gugliuzza A. (2010), "High-definition polymeric membranes: construction of 3D lithographed channel arrays through controlling natural building blocks dynamics", *ACS Appl. Mater. Interfaces*, 2(2), 459-466.

Sperry D.P., Falconer J.L., Noble R.D., "Methanol-hydrogen separation by capillary condensation in inorganic membranes", *J. Membrane Sci.*, 60 (1991) 185-193.

Spillman R (1989), "Economics of gas separation by membranes", *Chem. Eng. Prog.*, 85, 41.

Stern S. (2002), "Polymers for gas separation: the next decade", *J. Membrane Sci.*, 94, 1.

Ulhorn R.J.R., Keizer K., Burggraaf A.J. (1992), "Gas transport and separation with ceramic membranes. Part I. Multilayer diffusion and capillary condensation", *J. Membrane Sci.*, 66, 259-269.

Van Der Sluijs J., Hendriks C., Blok K. (1992), *En.Convers. Mgmt.*, 33, 429.

Van Veen H.M., Bracht M., Hamoen E., Alderliesten P.T. (1996), "Feasibility of the application of porous inorganic gas separation membranes in some large-scale chemical processes", in *Fundamentals of inorganic membrane science and technology,* A.J. Burggraaf and L. Cot (Eds.), Elsevier, 14, 641-681.

Vansant E.F., Dewolf R. (1990), "Gas Separation Technology", Proc. Int. Symp. on *Gas Separation Technology*, Antwerp (Belgium), Elsevier.

von Wroblewski S (1879), "Ueber die nature der absorption der gase durch flussigkeiten unter holen drucken", *Ann. Physik u. Chem.*, 8, 29-52.

Wara N.M., Francis L.F., Velamakanni B.V. (1995), 'Addition of alumina to cellulose acetate membranes', *J. Membrane Sci.*, 104, 43–49.

Xia Y., Lu Y., Kamata K., Gates B., Yin Y. (2003), "Macroporous materials containing three-dimensionally periodic structures", in Chemistry of Nanostructured Materials, P.Yang (Ed.), World Scientific, 69-100.

Yang K., Liu Z., Mao M., Zhang X., Zhao C., Nishi N. (2005), "Molecularly imprinted polyethersulfone microspheres for the binding and recognition of bisphenol A", *Anal. Chim. Acta*, 546 30-36.

Zhou Q., Chan C.W., Tontiwachiwuthikul P. (2009), "A monitoring and diagnostic expert system for carbon dioxide capture", *Exp. Systems Appl.*, 36, 1621-1631.

In: Membrane Processes for Sustainable Growth
Editors: Angelo Basile and Alfredo Cassano
ISBN: 978-1-62618-446-6

Chapter 2

MEMBRANE CONTACTOR TECHNOLOGY: FUNDAMENTALS, INTERFACES AND PROSPECTS

Annarosa Gugliuzza* and Angelo Basile
Research Institute on Membrane Technology-
National Research Council (ITM-CNR), Rende (CS), Italy

ABSTRACT

An overview of principles, interfaces and applications is provided with membrane contactor prospects. Basic concepts, material membranes along with manufacturing procedures and key membrane contactor operations are described, giving emphasis to advantages and shortcomings. The potential areas of applications for some advanced membrane contactors such as membrane emulsification, membrane distillation, osmotic membrane distillation and membrane crystallization are also discussed.

Keywords: membrane contactors; membrane emulsification; membrane distillation; osmotic membrane distillation; membrane crystallization; interfaces; water; emulsion; crystals; wettability; membrane pores

LIST OF SYMBOLS

A	membrane surface area
a	numerical empirical coefficient
b	numerical empirical coefficient
c	numerical empirical coefficient
C	concentration
C_{b1}	solute concentration in the bulk solution
C_f	salt bulk concentration in the feed bulk solution

* Corresponding author: a.gugliuzza@itm.cnr.it

C_{fm}	salt concentration at the hot membrane surface
C_{m1}	solute concentration at the feed solution/membrane interface
d_d	droplet diameter
d_p	membrane pore diameter
e	numerical empirical coefficient
f	function of the geometry of the system
f_s	fraction of the solid on the surface
f_v	fraction of the vapor on the surface
G	growth rate of crystals
H	total heat transfer coefficient
h_f	heat transfer coefficient in feed layer
h_{mg}	heat transfer through vapor and gas
h_{mv}	heat transfer coefficient for the liquid vaporization (latent heat)
h_p	heat transfer coefficient in permeate layer
i	generic component
id	inner diameter
J	total mass flux
J_D	water vapor flux due to diffusion of vapor through the pores
J_i	total flux of the component i
J_K	Knudsen flux
J_p	Poiseuille flux
J_w	total flux of the water vapor w
k	numerical empirical coefficient
K	mass transfer coefficient
k_b	constant of nucleation rate
K_d	membrane permeability
K_D	diffusion parameter (*eq. 14*)
K_f	mass transfer coefficient of feed side
K_g	growth rate constant
K_K	Knudsen parameter
K_M	mass transfer coefficient of the membrane
k_m	membrane conductivity
Kn	Knudsen number
K_p	mass transfer coefficient of permeate side
K_{pois}	Poiseuille parameter
M_d	mass flow rate of the dispersed phase
m_T	concentration of crystals in the magma
Nu	Nusselt number
od	outer diameter
p_{av}	mean vapor pressure in the membrane
P_c	critical pressure
$P_{c,in}$	the pressures at inlet of the membrane module
$P_{c,out}$	the pressures outlet of the membrane module
P_d	pressure of the dispersed phase outside the membrane
p_f	pressure at feed sides of the membrane
p_b	breakthrough pressure

p_p	pressure at permeate side of the membrane
Pr	Prandtl number
Q	total heat transfer
Q_f	heat transfer from feed side to membrane surface
Q_m	heat transfer across the membrane
Q_{mc}	heat conduction through the membrane
Q_{mg}	heat transfers through the vapor and gas filled pores
Q_{mv}	heat transferred across the membrane due to liquid vaporization
Q_p	heat transfer from the membrane to the permeate side
r	pore radius of the membrane
R	rugosity factor
Re	Reynolds number
R_n	nucleation rate
S	wettability
Sc	Schmidt number
Sh	Sherwood number
T	temperature
T_f	temperature of the feed side
T_{fm}	membrane surface temperature feed side
T_p	temperature of the permeate side
T_{pm}	membrane surface temperature permeate side

Greek Symbols

α	separation factor
γ	surface free tension
γ_{SV}	surface free tension between solid and vapor phase
γ_{SL}	surface free tension between solid and liquid phase
γ_{LV}	surface free tension between liquid and vapor phase
δ	membrane thickness
Δ	variation
ε	porosity of the membrane
ζ	concentration polarization coefficient
θ	contact angle
θ_w	contact angle according to Wenzel's equation
θ_{CB}	contact angle according to Cassie-Baxter's equation
μ	dispersed phase viscosity
ξ	numerical parameter
ρ	liquid density
ρ_d	dispersed phase density
τ	temperature polarization coefficient
ω	rotation rate

Acronyms

AGMD	air gap membrane distillation
DCMD	direct contact membrane distillation
DIPS	diffusion induced phase separation
G	gas
L	liquid
LEP	liquid entry-pressure
MC	membrane contactors
MCr	membrane crystallization
MD	membrane distillation
ME	membrane emulsification
OMD	osmotic membrane distillation
PEEK	polyetheretherketone
PP	polypropylene
PTFE	polytetrafluoroethylene
PVDF	polyvinylidenefluoride
RO	reverse osmosis
SGMD	sweep gas membrane distillation
SPG	Shirasu Porous Glass
TIPS	thermally induced phase separation
UF	ultrafiltration
VIPS	vapor induced phase separation
VMD	vacuum membrane distillation
VOC	volatile organic compounds

2. Introduction

Membrane Contactors (MCs) are considered with a great potential for applications in eco-friendly processes, such as water desalination, waste water treatment, food and beverage production, control of CO_2 emissions, bioseparations, drug delivery, crystallization, fermentation, along with VOC removal, semiconductors and oil extraction (Stanojević et al., 2003; Charcosset, 2009a; Gugliuzza and Basile, 2013; Banat and Simand, 1994, 1996). MCs are part of the emerging membrane technology, which can be used in place of or in combination with more conventional systems such as chemical and physical absorption, solid adsorption, carbon molecular sieve adsorption, cryogenic distillation or packed columns and tray columns (Drioli et al, 2006; Yan et al., 2007).

This advanced technology takes numerous advantages over conventional techniques, including reduction in size with packing efficiencies up to 30 times greater than traditional column, linear scale-up and operational flexibility, along with high mass transfer rate due to the highest interfacial area and easy integration (Drioli et al., 1999). In addition, typical events such as flooding at high flow rates or unloading at low flow rates as well as emulsions are not observable during MC operations.

MCs are designed to interface two streams such as liquid/liquid, liquid/gas, and gas/gas by using a membrane as a physical interface. Mass transfer is thus allowed from one phase to another under a specific driving force, also when a comparable density between the two streams is exhibited. The dispersion of one phase within another is prevented as structural and chemical features are well matched in the membrane. However, some drawbacks must also be considered, including the limited life-time, (bio)fouling phenomena and breakthrough pressure, which may sometime limit the operating pressures conditions. Undoubtedly, the successful outcome of a MC operation depends on the choice of the membrane along with the engineering of the process. In this chapter, basic concepts, membrane-types, configuration and transport mechanisms are discussed for introducing the potential of this advanced technology in highlighted applicative areas of the life science.

2.1. Membranes Contactors: Basic Concepts and Membrane Features

In MC devices hydrophobic and hydrophilic membranes work as physical interfaces between two aqueous or organic phases (Figure 1); the dispersion of one phase within another is prevented (Gabelman and Hwang, 1999). Gas/liquid, liquid/liquid or liquid/gas/liquid phases bring into contact at the entrance of the membrane pores and chemical species can diffuse from one side to the other with a separation factor (α), which depends on the distribution coefficient of the component in the two phases (Kosaraju and Sirkar, 2007). The diffusion of the species is induced by a driving-force, such as a gradient of concentration, temperature, pressure or partial pressure.

High-performance processes can be achieved as high interfacial area per unit of volume, high volumetric mass transfer rate and controlled mutual membrane-liquid interactions are well matched. From a morphological point of view this implies the use of membranes with controlled structural features, including pore size of 0.02 to 0.2 um, regular pore shape, narrow pore distribution, along with overall porosity (ε) of 75 to 90%.

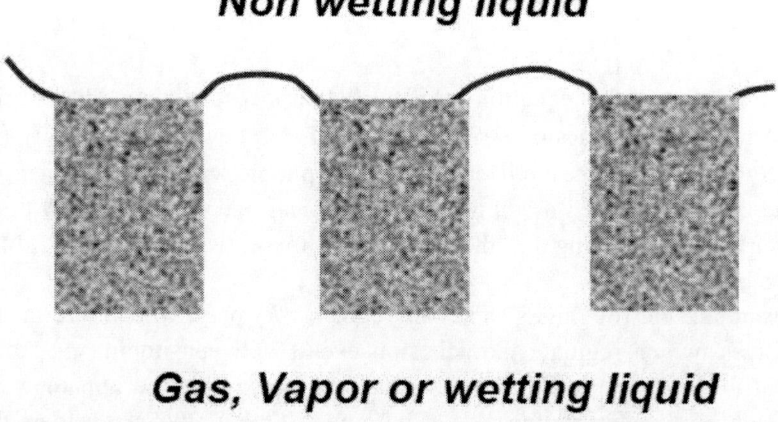

Figure 1. Scheme of MC interfacing two phases at the entrance of the membrane pores.

A suitable balance between pore size and overall porosity must be achieved in order to prevent viscous liquid intrusion inside the membrane pores and, hence, additional resistance to the transport. Also, membrane-liquid affinity can critically control the wetting of the membrane pores.

In this respect, the Young-Laplace equation (1) enables one to evaluate the suitability of a membrane for membrane contactor operations:

$$p_b = \frac{-2\gamma\cos\theta}{r} \tag{1}$$

This equation correlates the breakthrough pressure (p_b), which is the pressure exceeding the pressure of the liquid phase during the operation, to the surface free tension (γ) between the two phases and the contact angle (θ) between the membrane and the liquid. It is inversely proportional to the maximum pore radius (r).

As hydrophobic membranes are operated, the pressure of the aqueous phase is usually higher than the pressure of the gas/organic phase, but is lower than the breakthrough pressure.

For hydrophilic membranes with pores filled by the liquid, the pressure is lower than that of the gas/organic phase, resulting in a value below the breakthrough pressure.

As just mentioned, the concept of wettability is also relevant to establish the ability of the liquid to enter in the pores. Wettability means liquid spreading on the membrane surface and it is usually expressed as:

$$S = \gamma_{SV} - (\gamma_{SL} + \gamma_{LV}) \tag{2}$$

where S is the degree of wetting, γ_{SV} the surface free tension between solid and vapor phase, γ_{SL} the surface free tension between solid and liquid phase, γ_{LV} the surface free tension between liquid and vapor phase. An experimental and direct measurement of the wettability is the apparent contact angle, expressed by the Young's equation:

$$\cos(\theta) = \frac{\gamma_{SV} - \gamma_{SL}}{\gamma_{LV}} \tag{3}$$

Given the validity of this equation for ideal flat surfaces, values of $\theta<90°$ give a positive wettability ($S>0$), whilst values of $\theta>90°$ correspond to a negative wettability ($S<0$). Values of $\theta>150°$ ($S<<0$) can be measured on super-hydrophobic surfaces. However, the apparent contact angle can be regarded as the result of chemical and morphological elements of the membrane surface. Depending on the balance of these two parameters, different liquid behaviors can be observed.

As an example, the roughness factor or rugosity (R) plays a key role in the control of interfacial forces, which regulate the adhesion events between liquid and surface. It is an irregularity of the surface profile and can cause an increase in the apparent contact angle, producing heightened waterproofness. Wenzel model (Quéré, 2002) considers the roughness as a sort of grooves filled by the liquid and describes the effect of the rugosity on the contact angle according to this relation:

$$\cos(\theta_w) = R\left[\frac{\gamma_{SV} - \gamma_{SL}}{\gamma_{LV}}\right] \tag{4}$$

If the grooves are filled by air, the Cassie–Baxter's model (Quéré, 2002) describes the behavior of the liquid droplet by using equation 5:

$$\cos(\theta_{CB}) = f_s \cdot \cos(\theta_a) + f_v \cdot \cos(\theta_v) \tag{5}$$

where θ_a is the angle measured on the flat surface, whereas f_s and f_v are the fractions of the solid and vapor on the surface, respectively. In these circumstances, the adhesion line between liquid and substrate is significantly reduced.

Considering all these aspects, it is possible to balance chemical and structural membrane features in order to get the best compromise between all factors necessary to a successful outcome of the processes. Despite that pore sizes lower than 0.2 μm are preferred for membranes contactors, superhydrophobic membranes *(θ~150°), Liquid Entry Pressure of water (LEPw= 2.5) bar* with larger pore size (2.0 μm) have also been demonstrated to meet the basics of MC technology due to a good compromise between chemical and morphological relationships (Gugliuzza and Drioli, 2007, 2009). Related wettability shortcomings can be avoided by using coated membranes. Their use in place of porous membranes takes advantage to prevent liquid intrusion inside the pore allowing operations in a broader range of pressure. The use of highly permeable coating prevents additional resistance to the transport (Nymeijer et al., 2004). The performance of a membrane is the result of well-addressed selection of materials, manufacturing and assembly procedures, as discussed hereafter.

2.2. Membrane Materials

The choice of materials depends on the chemical composition of the stream as well as on the kind of operation selected (Koonaphapdeelert et al., 2009; Feng et al., 2004; Luo et al., 2009; Gugliuzza et al. 2009a; Prince et al. 2012). Materials for hydrophobic membranes are usually polytetrafluoroethylene (PTFE), polypropylene (PP) and polyvinylidenefluoride (PVDF) (Gugliuzza and Drioli, 2007; Gugliuzza et al., 2009a, 2010). In some cases, other polymers such as polyethersulphone (PES) and polyetherketones (PEEKs) could be very suitable for the fabrication of membranes with such a structure to be competitive or superior to traditional hydrophobic polymer (Gugliuzza et al., 2008; Speranza et al., 2010). Cellulose acetate along with blends of hydrophobic polymers and hydrophilic pore formers are also used as hydrophilic membrane materials (Bey et al., 2011; Yang, et al., 2007). Inorganic or nanocomposite membranes are also considered (Susanto, 2011; Peng et al., 2005). The hydrophilic or hydrophobic nature of the membrane along with the stream composition determines the position of the interface phases.

2.3. Fabrication of Interfaces for Membrane Contactors

Different techniques can be used to fabricate porous polymeric membranes for MC operations, even if not all are suitable to get desired and interplayed structural and chemical properties. Hereafter, some of the most used and competitive manufacturing approaches will briefly be described in relation to the requirements of MC technology (Iversen, 1997).

2.4.1. Phase Separation Methods

The phase separation is based on the principle of precipitation of the polymer, which can be induced by exchange of solvent-non solvent (Diffusion induced phase separation - DIPS process), variation of temperature (Thermally induced phase separation - TIPS process), or exposition to vapor (Vapor induced phase separation - VIPS) (Gu et al., 2006; Cheng, 1999; Gugliuzza et al., 2006; Gugliuzza and Drioli, 2009; Gugliuzza et al., 2009a; Gryta and Barancewicz 2010; Baker, 2004).

All methodologies are flexible, feasible, scalable and energy-saving. Depending on experimental conditions different types of structures, i.e. lacy-like, sponge-like, finger-like and particulate-like morphology, can be achieved via directed thermodynamics and kinetics. Parameters such as concentration and chemical composition of the casting solution (Deshmukh and Li, 1998), harshness of the coagulation bath (Young et al., 1999; Lin et al., 2006; Wang et al, 2011), temperature and relative humidity (Cheng, 1999; Gugliuzza and Drioli, 2007; Gugliuzza et al., 2006), and solubility parameters of the materials and additives (Bottino et al., 1991) decide the arrangements of the polymer regions wherein various typologies of pores can be detected.

Quick precipitations bring to a densification of the membrane skin, whilst salts or hydrophilic additives are often used in order to produce pores through the membrane surface with the objective to enhance the mass transfer through the membranes (Lü et al., 2011; Feng et al., 2004). This gives rise, however, loss of hydrophobicity and heightened spreading of polar liquids on the membrane surface with a concrete risk of liquid intrusion inside the pores (Tomaszewska, 1996; Chen and Hong, 2002). One of the most studied polymers for the preparation of membranes for MCs is the semicrystalline PVDF. This kind of polymer offers the opportunity to get different morphology depending on the velocity of demixing (Gugliuzza and Basile, 2013). Solid-liquid mechanism (delayed demixing) could be better than liquid-liquid phase separation (quick demixing), because particulate-like skin-less structures are formed with such an increased surface roughness to result in an enhanced waterproofness (Gugliuzza and Drioli, 2007, 2009; Cheng, 1999).

An evolution of the traditional phase inversion via DIPS mechanism is based on the use of a supercritical fluid in place of a traditional non solvent (Cardea et al., 2010; Huang et al., 2007), Aerogel-membranes with performance comparable or superior to that achieved by conventional techniques can be fabricated (Reverchon and Cardea, 2006; Cardea et al., 2006; Cardea et al., 2009).

The advantage is of achieving sponge-like structures without using pore formers or post-treatments, thereby preserving intrinsic characteristics of the materials used.

Recently, some authors have also proposed to combine the traditional phase inversion with the Micro Moulding Technique (Vogelaar and Wessling, 2002). This approach consists of covering a lithographed master with a polymeric solution and inducing the polymer precipitation via traditional DIPS, TIPS, and VIPS approaches. The polymeric layers are, hence, depleted from micro-master and membranes with ordered and periodic porous surfaces are realized. The major difficulty of this approach is to get nano-master moulds with high aspect ratio pins.

2.4.2. Track-Etching Technique

Membranes with circular and almost cylindrical pores can be prepared through track-etching methods, consisting of two steps: formation of tracks on the surface by collision with charged particle in a nuclear reactor and treatment in an etching bath containing harsh solution (Apel, 2001). The pores density is usually determined by the residence time in the reactors, while the pore size is decided by the residence time in the corrosive bath. These membranes exhibit low pore density and sometimes the pores are not perfectly cylindrical through the overall thickness of the films. Films with thickness of 10 to 15 μm are available, while thicker polymer films are not yet available on commercial scale.

2.4.3. Stretching Technique

Stretched membranes are largely used in membrane contactor operations. They are prepared from semicrystalline polymer such as polyethylene or poly-tetrafluoro-ethylene, which are extruded at close to their melting point and, then, rapidly drawn-down, thereby alienating the crystallites (http://www.gore-tex.com). After annealing and extrusion, the film is stretched perpendicularly to the direction of drawing in order to cause partial fracture and thus generate pores with diameter of 0.2 to 20 μm. Porosity of 90% are achieved and flat and capillary geometries can easily be achieved.

2.4.4. Self-Assembly Techniques

Differently from the traditional phase inversion, self-assembly procedures allow to get high control of the degree of structural order, resulting in a good agreement with the requirements of MC technology. The use of building blocks such as colloidal crystals enables self-assembly and rearrangements in packed micro-spheres, forming rigid and interconnected structures following annealing. After infiltration of a hydrogel, the latex beads are removed by chemical or thermal treatment yielding regular structures (Kotov et al., 2004).

The limitation of this technique is currently the control of 3D morphologies at length scales below the micrometres along with the limited degree of interconnectivity.

Other building blocks such as surfactants, microsegregation of block copolymers, or water droplets can be used to form three-dimensionally ordered structures (Imhof and Pine, 1997; Kresge et al., 1992; Srinivasarao et al., 2001; Gugliuzza et al., 2009b; Speranza et al 2010). The use of water droplets as pore builders has, for example, been demonstrated to be particularly attractive for the formation of membranes with very high periodicity and degree of order through 3D space. Furthermore, the droplets take the advantage over other building block of being sacrificial non-toxic templates largely available in nature and easily removable. No additional steps are requested to prepare membranes with ideal structural features.

The process can be resumed as follows: natural condensation of water droplets on cold surfaces; self-assembly in semi-crystalline lattices; full propagation of the lattice through the polymer solution; formation of well-ordered 2D or 3D air bubble arrays, where the bubbles are the pores of the membrane perfectly monodisperse and packed in honeycomb geometries.

Membranes prepared according to this approach showed better performance than commercial modules, resulting in higher water vapor flux at lower operating conditions. Enhanced resistance to the liquid intrusion and long operational time has been experimentally confirmed (Gugliuzza et al., 2008, 2009b; Speranza et al., 2010). Despite that the control of thermodynamics and kinetics can sometimes be complex, the scale up of this manufacturing approach is expected to give propulsion to the development of next generation membranes for membrane contactors.

2.4. Membrane Module Assembly

The membrane module configuration is depending on process related factors such as high rates of both heat and mass transfer, high flow rates for both feed and permeate, high packing density-membrane surface area to module volume ratio, low pressure drop, high evaporation efficiency and high thermal stability and heat recovery, along with economics.

Plate and frame, spiral wound, tubular, and hollow fiber modules are usually used (Gaeta, 2003; Teoh et al., 2008; Khayet et al., 2003; Koschilowski et al., 2003; Kumar et al., 2002; El-Zanati and El-Khatib, 2007; Winter et al., 2011), even if hollow fiber modules are usually preferred because exhibit higher packing density (1500-3000 m^2/m^3), low capital and operating costs (Kumar et al., 2002; Yan et al., 2007). On the contrary, tubular module-types exhibit low fouling tendency, good cleaning. Plate and frame, as well as spiral wound modules exhibit middle characteristics.

In plate and frame as well as spiral wound modules, parallel membrane sheets are assembled with interposed spacers for preventing mechanical abrasions, whilst in tubular and hollow fiber modules cylindrical membranes are collected in a tubular-shaped housing extending along an axis between two opposing ends and defining an inner chamber. A very important concern is with the geometry of the modules, since the optimization of the mass transfer depend on it in agreement with the relation proposed by Gekas and Hallström (1987):

$$Sh \approx \mathrm{Re}^a \, Sc^b \, f \qquad (6)$$

where *Sh* is the Sherwood number, *Re* the Reynolds number, *Sc* the Schmidt number, *f* a function of the geometry of the system, and *a* and *b* numerical empirical coefficients to be appropriately selected for the given hydrodynamic conditions.

2.5. Advanced Membrane Contactor Operations and Related Prospects

Compared with conventional techniques MCs have brought in the last decades important advances in various strategic fields of the agro-food, biomedical and environmental science. The purpose of this section is to give a description of four distinctive membrane contactor-based technologies, highlighting for each one the principles, membrane-types, and areas of application. The focus will essentially be on membrane emulsification (ME), membrane distillation (MD), osmotic membrane distillation (OMD), and membrane crystallization (MCr).

2.5.1. Membrane Emulsification (ME)

A system where two or more immiscible phases are dispersed within others is traditionally called emulsion (Piacentini et al., 2011). Depending on the nature of the dispersed phase and the number of phases, simple or complex emulsions can be formed (Figure 2). Oil-in-water is traditionally indicated as an o/w emulsion, whilst water-in-oil is indicated as a w/o emulsion.

More complex systems such as water-in-oil-in-water are indicated as w/o/w emulsions. Simple emulsions use surfactants and the management of a few parameters is requested to control their stability. Instead, triple, quadruple and multi emulsions are the result of emulsions of emulsions and their stability is particularly sensing to thermodynamics and susceptible to rupture (Mathioulakis et al., 2007). However, compared to stirring methods, the preparation of w/o/w or o/w/o emulsions by membrane contactor technology is demonstrated to yield more homogeneous and uniform systems with ability to delivery drugs and nutrients in a more controlled and slower manner (Okochi and Nakano, 1997; Dowding et al., 2001; Supsakulchai et al., 2002).

Membrane emulsification (ME) is indeed considered an advanced branch of MC technology. It represents a valid alternative to more traditional techniques such as high-pressure homogenizers, ultrasound homogenizers and rotor/stator systems, colloid mills or toothed disc dispersing machines (Charcosset et al., 2004; Giorno et al., 2003), because it takes numerous advantages over conventional mechanical approaches in terms of lower shear stress, pressure and energy (Altenbach-Rehm et al., 2002). The use of regular porous membranes enables one to decide droplet size and uniformity of the droplet distribution without requiring turbulent droplet break-up.

Inorganic and polymer track-etching membranes are often used. Shirasu Porous Glass (SPG) membranes (Nakashima et al., 1991) based on $CaO–Al_2O_3–B_2O_3–SiO_2$ with uniform cylindrical pores ranging in pore size from, a wide spectrum of available 0.05 to 30 μm and overall porosity of 50–60%, ceramic aluminium oxide (α-Al_2O_3) membranes, α-alumina- and zirconia coated along with polytetrafluoroethylene (PTFE) and polypropylene membranes are among the most used. However, a great potential is also for high-definition membranes prepared by breath figure method (Speranza et al., 2010).

These types of membranes meet morphological and chemical requirements of ME in terms of regular circular pores, monodisperse pore distribution and well-balanced wetting properties. In membrane emulsification the dispersed phase is pressed through the membrane pores under an applied pressure, thereby promoting the formation of droplets that detach in a tangentially circulating continuous phase along the membrane surface as a determined size is reached (Figure 3).

In order to prevent coalescence, surfactants are usually dispersed in the continuous phase, thereby forming a stabilising layer around each single droplet. Depending on the desired emulsion, membranes with appropriate wettability must be chosen.

As a rule, oleophilic/hydrophobic membranes are selected for dispersing water in oil, whereas membranes with hydrophilic/oleophobic properties are preferred for the dispersion of oil in water. This means that the dispersed phase would not wet the membrane pores (Joscelyne and Trägårdh, 2000). A schematic representation of a small-scale membrane emulsifier for the preparation of o/w emulsions is depicted in Figure 4.

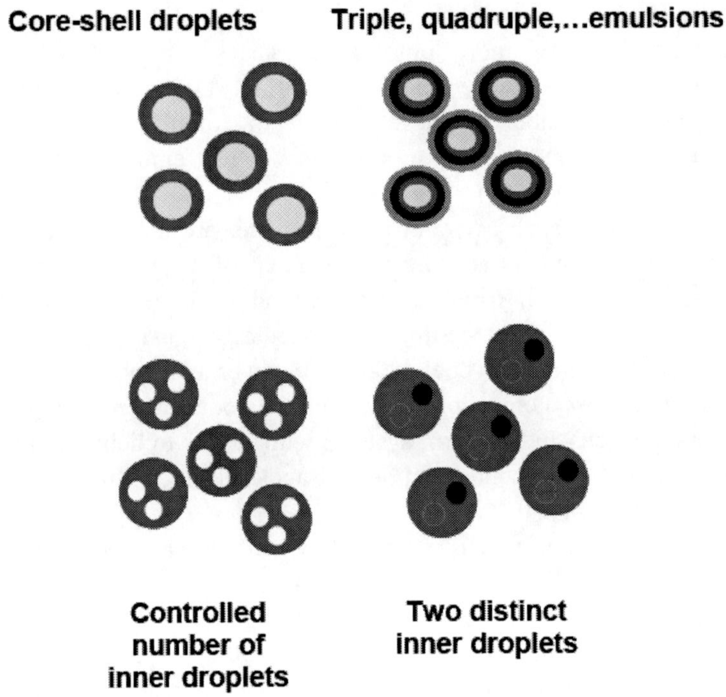

Figure 2. Representation of different kind of simple and complex droplet emulsions.

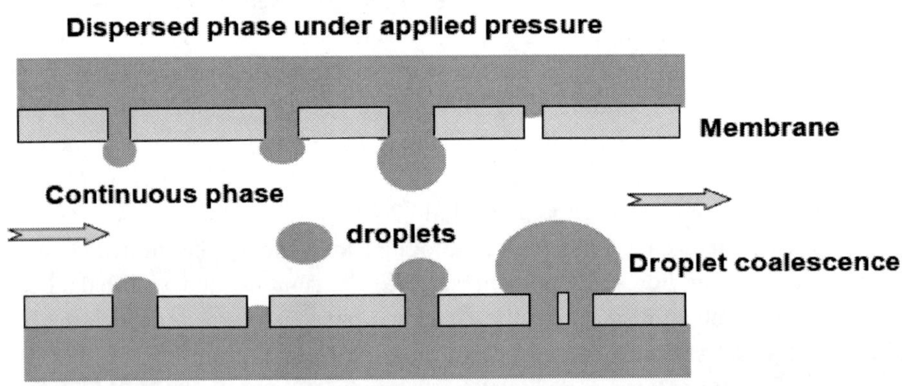

Figure 3. Production of droplet emulsions by using a membrane emulsifier.

The device assembly includes a tubular membrane, a feed vessel containing an aqueous solution, a pump and a pressurized oil tank. The oil phase is the dispersed phase to be pumped through the membrane pores, while the aqueous solution is circulated through the lumen of the membranes. Process and membrane parameters represent again key issues for successful membrane emulsification operations (Oh et al., 2010). Pressure, velocity of the circulating phase along with temperature, pH and surfactants are parameters to be well-addressed during a membrane emulsification process.

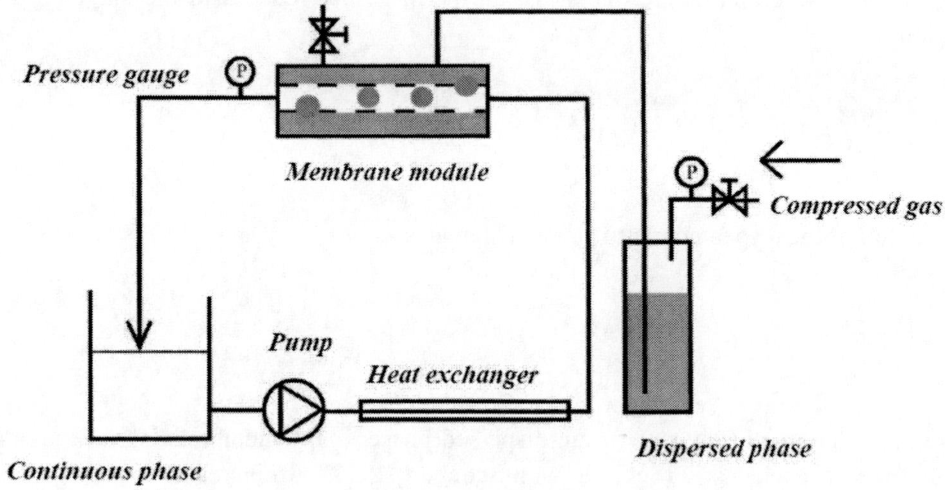

Figure 4. Schematic representation of membrane emulsification operation.

The trans-membrane pressure is defined as:

$$\Delta P = P_d - \frac{(P_{c,in} + P_{c,out})}{2} \tag{7}$$

where P_d is the pressure of the dispersed phase outside the membrane; $P_{c,in}$ and $P_{c,out}$ the pressures at inlet and outlet of the membrane module, respectively. Assuming cylindrical pores, the ΔP necessary to run the dispersed phase in the continuous one is given by the capillary pressure expressed as

$$P_c = \frac{4\gamma \cos\theta}{\bar{d}_p} \tag{8}$$

where P_c is the critical pressure, γ the o/w interfacial tension, θ the contact angle of the oil droplet on the membrane surface in contact with the continuous phase and d_p the mean pore diameter. The calculated value is often lower than real one, considering that the tortuosity of the real pores as well as morphological and chemical factors control the wetting events. The relation that allows the calculation of flux (J_d) and permeability of the dispersed phase in agreement with the Darcy's law:

$$J = \frac{K_d \Delta P}{\mu \delta} \tag{9}$$

where K_d is the membrane permeability, δ the membrane thickness, and μ the dispersed phase viscosity. Deviation from *Darcy*'s law is, however, observed as the pores are not completely opened. Assuming *n* cylindrical and uniform membrane pores of radius *r*, and pore length almost equal to the membrane thickness, the permeability reduces to the *Hagen–Poiseuille* equation:

$$K_d = \frac{nr^2}{8\pi} \tag{10}$$

and the flux of the dispersed phase can be calculates as

$$J = \frac{M_d}{\rho_d A} \tag{11}$$

where M_d is the mass flow rate of the dispersed phase, A the membrane surface area and ρ_d the dispersed phase density (Schröder and Schubert, 1999). An increase in temperature gives rise to a general decrease in the solution viscosity, resulting in changes of the the average emulsion droplet size (Katoh et al., 1997a, 1997b)

The pH could seriously change the surface charge, affecting the wetting properties and consequently altering the emulsified droplet size and distribution. But, a variation of the droplet size can also be observed with the velocity of the continuous phase.

A decrease in size is observed with raising wall shear stress until the droplet reaches a value over which no dependence on the velocity is observed (Schröder and Schubert, 1997; Schröder et al., 1997). Structural and chemical features of the membranes are undoubtedly other important aspects to be considered in a membrane emulsification process. One the most important features is the pore size, the droplet dimension being strictly dependent on it according to this relation:

$$\bar{d}_d = c\bar{d}_p \tag{12}$$

where d_d is the droplet diameter, d_p the membrane pore diameter and c a constant with values of 2 to 10 or 3 to 50 depending on the kind of membrane used (Charchosset et al., 2004).

Membranes with monodisperse pore distribution are requested to get droplet emulsions with very uniform dispersion, even if the porosity parameter is another critical factor. There is, in fact, a minimum critical distance between the adjacent pores to prevent coalescence when the droplets detach. A distance of 1.5% has been calculated on membranes with pore size of 5.0 μm (Abrahamse et al., 2001). Fewer pores per surface area affect however the quality of the emulsion. Thus, a compromise between number of pores per surface area and distance between neighbour pores must be achieved.

Growth rate and detachment of the droplets can be regarded as the result of coupling effects (Figure 5), which have been the subject of various experimental and theoretical studies (Schröder et al.,1998); Peng and Williams 1998; Rayner and Tragårdh, 2002;De Luca et al., 2004).

It is particularly interesting to examine the contribution of various forces involved in these processes, including the interfacial tension force ($F\gamma$); the static pressure difference force (FSP); the drag force (FD); the dynamic lift force (FL); the buoyancy force (FB); the inertial force (FI).

In this respect, Fγ is related to the adhesion of the dispersed phase around the pore edges; FSP is generated by the pressure difference between the dispersed phase and the continuous phase at the membrane surface; FD derives from the continuous phase running past the droplet parallel to the membrane surface; FL is due to the asymmetric velocity profile of the continuous phase near the droplet; FB is generated by the density difference between the continuous phase and the dispersed phase and, finally FI is given from the dispersed phase, which moves through the capillary increasing the droplet volume (Charcosset et al., 2004).

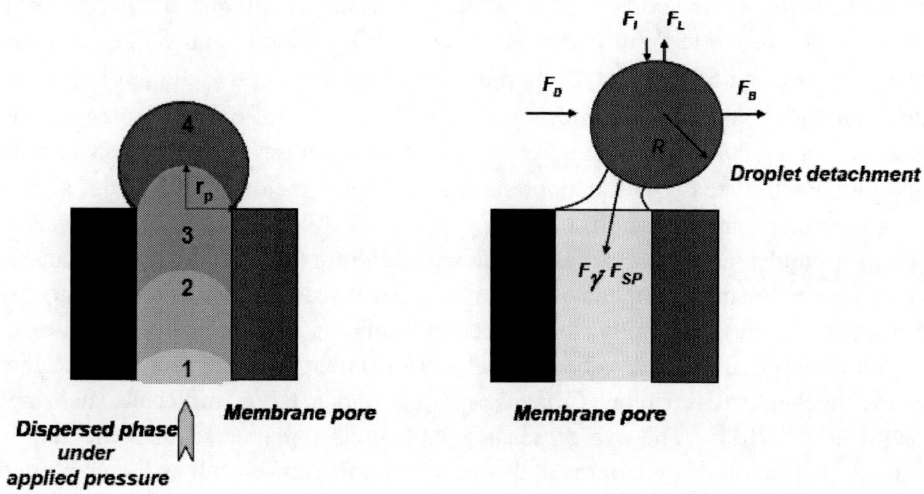

Figure 5. Growth rate and detachment of the droplet from membrane pore.

All these forces can take different directions during growth of the droplets, affecting the final size. Despite that some papers are reported in literature, this matter requires still a more quantified description of the various events through comparative experimental results and theoretical data. Given the great potential of cross-flow membrane emulsification on industrial scale, the overcome of some limitations such as, for example, the low flow rate through membranes appears to be rather urgent for scaling up the emulsion production on commercial scale (Gijsbertsen-Abrahamse et al., 2004). Scherze et al. (1999) referred about oil fluxes ranging from 5 to 15 $dm^3m^{-2}h^{-1}$ with a 0.2 μm (pore size) membrane and from 10 to 45 $dm^3m^{-2}h^{-1}$ with a 0.5 μm (pore size) membrane for the preparation of o/w emulsions. Higher fluxes (50-250 $dm^3m^{-2}h^{-1}$) were achieved by using α-alumina membranes with pore of 0.2 μm (Joscelyne and Tragårdh, 1999). With glass membranes having pore size of 1 μm a flux of 2300 $dm^3m^{-2}h^{-1}$ has been obtained for the preparation of w/o emulsions (Katoh et al., 1996). However, these fluxes are still too low for industrial applications. This depends on lower density of opened pores, but also on membranes thickness too large. In order to overcome these frequent drawbacks, a repeated premix membrane emulsification has been proposed, resulting in an average oil flux of 20,000 $dm^3m^{-2}h^{-1}$ (Suzuki et al., 1998; Park et al., 2001; Ribeiro, et al., 2005).

As just mentioned, the preparation of simple and complex emulsions regards with various segments of the food, biomedical and chemical industry. Artificial milks, dressings, toppings, cream liqueurs, ice-cream, along with fats, mayonnaise and sauces are some examples of food formulations where the control and stabilization of simple or complex emulsions are strongly required (Charcosset, 2009a). Emulsions along with micro-encapsulations can also be used as vehicles of active ingredients such as additive, preservative, flavours or probiotics (Song et al., 2003; Taylor et al., 2005). In microencapsultions, biopolymers such sugars, starches, gums, proteins, synthetics, dextrins and alginates are usually used as solid micro carriers of drugs or nutrients, but also liposomes vesicles have also been proposed as carriers of proteins and enzymes, vitamins, minerals and stabilizers against degradation and oxidation (Mui, et al. 2003). Their preparation by membrane extrusion could takes advantages from membrane emulsification technology for application on larger scale productions. In a recent review, Charcosset (2009b) refers about some attempts to develop membrane emulsification in batch and semi-continuous mode (Williams et al., 1998), while an industrial plant of culinary cream was designed by using SPG, α-Al_2O_3 membrane and microsieves, yielding droplets with size of 1 to 3 μm and production volumes of 20 m^3h^{-1} with 30% of disperse phase (Gijsbertsen-Abrahamse et al., 2004). Fluxes higher than 100 kg $m^{-2}h^{-1}$ with submicron droplets were obtained on industrial-scale with membranes having pore size of 0.1 μm under a shear stress of 135 Pa (Joscelyne and Tragårdh, 1999).

Complex emulsions w/o/w can also find application in biomedical area as transporters of molecular-targets for the treatment of cancer or other diseases (Higashi et al., 1999; Higashi and Setoguchi, 2000). Recently, biomolecular stimulus responsive water-in-oil-in-water (w/o/w) multiple emulsions have been developed in order to promote a specific interaction between a biochemical receptor (*Concanavalin A*) and a target molecule such as glucose (Piacentini et al., 2011). The use of sensing emulsions is expected to open new horizons towards the integration of membrane and sensor technologies as well as the combined use of microfluidics engineering appears to be relevant to the fields of biology and chemistry, considering the advances in the manufacturing of "made-to-measure" droplets (Vladisavljević et al., 2012). Double, triple and multiple emulsions encapsulating different types of active compounds can be prepared in a single step facilitating thus quicker diagnostics and high-throughput screening for drug discovery.

2.5.2. Membrane Distillation (MD)

Membrane distillation (MD) is a branch of the MC technology, which may play a key role in the production of high purity water from seawater, brackish water or industry wastewater (Lawson and Lloyd, 1997; Boi et al., 2005; El-Bourawi et al., 2007; Gryta et al., 2006a). Particularly attractive is its potential use in the concentration of aqueous solutions from salts or inorganic acids, sugar, fruit juice, streams, along with the breaking of the azeotropic mixtures (Garcia-Payo et al., 2000; Nene et al., 2002). Despite that MD has a long history (Findely, 1967; Weyl, 1967), its industrial development has been somewhat slow due to the difficulty to get low-cost processes in terms of low operating temperature and long-life hydrophobic membranes (Susanto, 2011; Wang et al., 2009).

In the last years, the concrete possibility to associate membrane process with renewable /alternative energies, i.e. solar, water and wind, has however allowed to reassessed MD processes for seawater and brackish water desalination (Charcosset, 2009a; Mathioulakis et al., 2007), especially if benefits such as compactness, low energy consumes, and low operational pressure are considered. Figure 6 displays the basics of this technique. Two aqueous solutions are separated by a hydrophobic membrane. The diffusion of water vapor occurs through the membrane under a difference of temperature applied across feed and permeate side, resulting in a thermally-driven MD process.

Table 1. Resume of MD configuration: advantages and limitations

Configuration	Short description	Advantages	Limitations	Literature
DCMD	Both the streams are in direct contact with the membrane. Vapor diffuses through the pores of the membrane and condenses directly into a solution flowing on the permeate side (distillate)	Separation of water vapor as the major permeate component	Evaporation and condensation surfaces are too close and high conductive heat flux is produced along with mass flux	Winter et al., 2011
AGMD	Only the feed solution is directly in contact with the membrane: an air gap is formed between the hydrophobic membrane and a cool condensing surface	No wetting at permeate side; Reduction of energy loss through the membrane	Fluxes lower than MD modes	He et al., 2011
SGMD	A flowing gas is used to sweep the vapor at the permeate side of the membrane, maintaining the gradient necessary for transport	Reduced resistance to the transport.	Additional energy for separation, because the permeate vapor is diluted by the sweep gas	Lawson and Lloyd, 1997
VMD	Evaporation is induced by applying the vacuum at the permeate side, where low pressure is kept constant	Negligible conductive heat loss	Degradation of the hydrophobicity of the microporous membrane: possible wetting events under vacuum	Wang et al., 2009

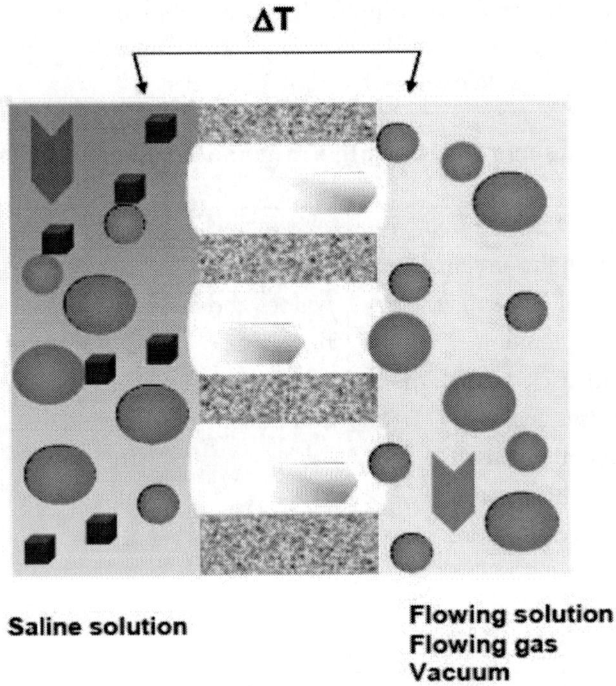

Figure 6. Scheme of a membrane distillation operation.

Usually, the feed stream is slightly warmed in order to promote evaporation of water, which is diffused thorough the membrane and, then, condensed by cooling at permeate side. The difference of temperature between inlet and outlet requires the use of an evaporator and a condenser, respectively, even if the feed temperature is lower than water boiling temperature, making just attractive the use of solar energy, geothermal energy or waste heat. In this context, the role of the membrane is decisive, considering the necessity to prevent liquid intrusion inside the pores (LEPw), to promote large mass transfer of water, to keep low thermal conductivity, along with high resistance to fouling (Mathioulakis et al., 2007).

Hydrophobic (θ=100-140°) and super-hydrophobic (θ=150-180°) membranes with high degree of porosity (ε>90%) and pore size ranging from 0.2 to 1.0 μm are often preferred. Membrane thickness of 60-250 μm can be considered, provided that no thermally conductive interfaces are operated. A constant difference of temperature across the membrane is indeed necessary for assuring long operational time. MD operations can have four types of configurations: direct contact membrane distillation (DCMD), air gap membrane distillation (AGMD), sweep gas membrane distillation (SGMD), and vacuum membrane distillation (VMD). The main features of these configurations are briefly described in Table 1.

Concerning the transport mechanisms, feed and cold solution temperatures, along with velocities of both the hot and cold streams are factors that can influence the water vapor transport. The water mass transfer J_w (kg/m^2·h), is described according to a simplified mathematical relation:

$$J_w = K_M \Delta p_m \tag{13}$$

where Δp_m is the vapor pressure difference ($p_f - p_p$), with $p_f > p_p$, between the warm and cold membrane sides (Pa), K_M the membrane distillation coefficient (kg/m^2·h·Pa). This last parameter depends strongly on the membrane morphological parameters such as porosity, mean pore size, thickness, tortuosity and wettability, along with flow parameters of the aqueous solution. The mechanism, which directs the vapor transfer, can be regarded as the result of a combination of three different types of transport: a) Knudsen diffusion; b) Poiseuille flow; c) Molecular diffusion.

In this respect, the Knudsen number (*Kn*), defined as the mean free path of the gas/pore diameter, is a useful parameter to identify the predominating transport mechanism. Values > 1 indicate that the transport mechanism follows Kundsen diffusion; values between 0.01 and 1 suggest a combination of Knudsen and molecular diffusion; values << 0.01 are typical of a molecular diffusion (Wang et al., 2009). Another parameter is the resistance to the mass transport, expressed as 1/K. It is the sum of the inverse of various contributions: $1/K_f$ (feed side), $1/K_m$ (membrane) and $1/K_p$ (permeate side). Inside the membrane pores, the resistance arises from the collision of diffusing molecules with others or with membrane, whereas the real limiting step is the diffusion through the boundary layer close to the membrane surfaces. As hydrophilic membranes are used, the dependence on the resistance to the mass transfer inside the membrane pore can assume a relevant importance.

The mass transfer must be examined together with the heat transfer (Q_m), because the temperature difference across the membrane causes also a thermal gradient so-called *Soret effect.*

Indeed, the heat transfer rate convection from the hot solution to the membrane surface is related to the mass flux and *vice versa*. From mathematical point of view, the heat transfer can be described as the sum of different contributions:

$$Q_m = Q_{mv} + Q_{mc} + Q_{mg} = (h_{mv} + k_m/\delta + h_{mg})\cdot(T_{fm} - T_{pm}) \quad (14)$$

In this relation each contribution describes a distinct phenomenon at the single stage.

At the membrane surface the liquid is vaporized and the heat is transferred across the membrane according to this relation:

$$Q_{mv} = h_{mv}\cdot(T_{fm}\text{-}T_{pm}) \quad (15)$$

where, h_{mv} is the heat transfer coefficient for the liquid vaporization (latent heat), T_{fm} the temperature at the membrane surface, T_{pm} the temperature at permeate layer.

Depending on thermal conductivity, membrane thickness and porosity, the heat is transferred through the solid membrane material by heat conduction:

$$Q_{mc} = (k_m/\delta)\cdot(T_{fm} - T_{pm}) \quad (16)$$

where k_m is the membrane conductivity, δ the membrane thickness.

The heat is also transferred through the vapor and gas filled pores according to this relation:

$$Q_{mg} = h_{mg} \cdot (T_{fm} - T_{pm}) \tag{17}$$

where h_{mg} is the heat transfer through vapor and gas.

When the vapor condenses, the transfer of heat to permeate side is described according to

$$Q_p = h_p \cdot (T_{pm} - T_p) \tag{18}$$

where h_p is the heat transfer coefficient through the permeate boundary layer.

The relation resuming the transfer of heat through a membrane under a driving difference of temperature is hence reported:

$$Q = H \cdot (T_f - T_p) \tag{19}$$

where H is the overall heat transfer coefficient, T_f the temperature at feed side and T_p the temperature at permeate side.

The coefficient H is again expressed as the sum of resistance in series:

$$H = [1/h_f + 1/(h_{mv} + k_m/\delta + h_{mg}) + 1/h_p] \tag{20}$$

where h_f is the heat transfer coefficient in feed, k_m the membrane conductivity, and δ the membrane thickness..

An empirical parameter frequently used to describe the heat transfer through the boundary layer is the Nusselt number (Nu) expressed as:

$$Nu = \mathrm{a}\, Re^b\, Pr^c \tag{21}$$

where a, b and c are empirical constants that depend on laminar or turbulent flow regime.

The temperature polarization is a frequent shortcoming which can take place during MD operations, especially when the temperature near the membrane wall is lower than the main flow temperature, due to heat absorption from aqueous solution: $T_{fm} < T_f$ and $T_{pm} > T_p$.

The temperature polarization is quantified by the temperature polarization coefficient (τ), given by:

$$\tau = (T_{fm} - T_{pm})/(T_f - T_p) \tag{22}$$

where T_{fm} and T_{pm} are the temperatures at the hot and cold membrane surfaces, respectively, and T_f and T_p the temperatures in the feed and permeate bulk solutions, respectively.

When the τ value is proximal to unity, the MD system is well-designed. Differently, values equal to zero are expression of poor-quality MD systems. A reduction of the temperature profile at the membrane side is expected to decrease the temperature polarization (Drioli et al., 1999).

The temperature polarization can be expressed by the coefficient ζ, which depends on the concentration parameter. It is to be considered that the concentration polarization occurs as the concentration near the membrane wall is higher than the main flow concentration. This is due to the water vapor transfer across the membrane:

$$\zeta = C_{fm}/C_f \qquad (23)$$

where C_{fm} is the salt concentration at the hot membrane surface and C_f the bulk concentration in the feed bulk solution.

The solute concentration C_{m1} at the feed solution/membrane interface can, hence, be correlated to the mass flux (J) according to this relation (Martinez-Diez and Vasquez-Gonzales, 1999; Martinez-Diez and Florido-Diaz, 2001):

$$C_{m1} = C_{b1}\, exp(J/\rho\, K) \qquad (24)$$

where K is the solute mass transfer coefficient, C_{b1} the solute concentration in the bulk solution, and ρ the liquid density.

All these events can be minimized through the control of some parameters, including heat transfer in the liquid phase, the latent heat for vaporization, the hydrophobic character of the membrane (Wang et al., 2009; Khayet et al., 2003).

Hydrodynamic conditions along with properties of the fluids can affect the heat transfer, whilst the temperature polarization and heat conduction can affect the latent heat necessary to vaporize the water to transport through the membranes. An increase in the membrane thickness could limit the heat conduction, even if the risk of a reduction of mass transfer would be considered. Summarizing, a just compromise would be found in order to assure longer operational time.

Table 2. List of benefits and application of MD technology

Type	Degree
Operating pressure	*Low*
Operating temperature	*Low*
Vapor space	*Low*
Membrane mechanical strength	*Reduced*
Fouling	*Absent*
Rejection level	*Very high*
Water purity	*High*
Application	Literature
Production of ultrapure water, wastewater treatment and water purification	Lawson and Lloyd, 1997), (Gryta et al., 2006), (Boi et al., 2005), (Macedonio and Drioli, 2008, Gugliuzza et al. 2010).
Desalination of brackish sea water and highly saline water	(Gryta et al, 2000), (Martinez-Diez and Florido-Diaz, 2001).
Concentration of non-volatile solutes: salt, sugar, fruit juices, blood; waste water treatment solution concentration; crystallization	(Nene et al, 2002 ; Al-Asheh et al., 2006, Zhao et al, 2008).
Deprivation and callback of volatile solute in solution	(Banat and Simandl, 1996, El-Bourawi et al., 2007, Ding et al., 2006) .

Compared with conventional thermal and other membrane processes, MD operations offer some benefits for various applications, as resumed in Table 2. MD is expected to give great propulsion to the development of integrated systems for supplying drinkable water, in view of the increasingly demand that is estimated to be 4 to 4,000 km^3/y (Meindersma et al., 2006; Charcosset, 2009a), concentration of syrups and juices (Al-Asheh et al., 2006; Zhao et al., 2008), but also for wastewater treatment (Phattaranawik et al., 2008; Ding et al., 2006; Gryta et al., 2006b).

A combination of MD process and wastewater biological process has been done by submerging a MD module in an aerobic bioreactor. The thermally-driven MD process, caused water to diffuse through the membrane. High-quality water with water TOC levels below 1 p.p.m. was achieved.

MD technology has also been proposed in combination with a fermentation membrane bioreactor, for the production of alternative liquid fuels such as ethanol with an efficiency of 0.40–0.51 g_{EtOH}/g_{sugar} and a production rate of 2.5–40 g_{EtOH}/dm^3•h against an efficiency of 0.35–0.45 g_{EtOH}/g_{sugar} a production rate of 0.8–2.0 g_{EtOH}/dm3•h. obtained with traditional batch fermentation (Gryta et al., 2000; Lewandowicz et al., 2011).

2.5.3. Osmotic Membrane Distillation (OMD)

Osmotic membrane distillation (OMD) is a variant of the thermally-driven MD, where a microporous hydrophobic membrane interfaces (PTFE or PVDF) two aqueous solutions having different solute concentrations (osmotic pressure). Water vapor is transferred through the membrane under driving-force established by a difference of concentration between two streams, which result in a different vapor pressure (Figure 7). OMD takes the advantage over MD of using lower operating temperature and pressure. This allows preserving heat sensitive components from thermal degradation and mechanical damage (Kunz et al., 1996). Briefly, in OMD process a porous hydrophobic membrane separates two different non-wetting aqueous solutions with different water activity, which corresponds to different water vapor pressures. Stripping solutions with high solubility in water, high superficial tension, low volatility and viscosity, and non-toxicity such as sucrose, NaCl, $CaCl_2$, $MgCl_2$, $MgSO_4$, K_2HPO_4, and KH_2PO_4 or glycerol and polyglycerol are often used (Kunz et al., 1996). Also in this case the Knudsen number allows the predominating mechanism to be identified. Values of $Kn < 1$ suggest Poiseuille capillary transport, giving a flux (J_p) depending on the square root of the pore radius:

$$J_P = K_{pois} \cdot r^2 \cdot p_{av} \cdot (p_f - p_p) \tag{25}$$

where K_{pois} is the Poiseuille parameter, r the radius of the membrane pore, p_{av} the mean vapor pressure in the membrane, and p_f and p_p the respective pressures at both sides (feed and permeate) of the membrane.

Values of $Kn > 1$ indicate Knudsen diffusional transport. In this case the flux (J_K) is proportional to the pore radius:

$$J_K = K_K \cdot r \cdot p_{av} \cdot (p_f - p_p) \tag{26}$$

where K_K is the Knudsen parameter, which depends on temperature, membrane porosity, molar weight of the solvent and pore length.

The diffusion of vapor through the ends of the pores could also be described in agreement with a third model:

$$J_D = K_D \cdot (p_f - p_p) \tag{27}$$

where K_D is the diffusion parameter which depends on temperature, vapor diffusion coefficient, membrane porosity, molar fraction of air, molar weight of the solvent and pore length.

Similarly to MD, the mass transfer implies also heat transfer, considering the gradient of temperature caused by the water evaporation that cools the feed side and the water vapor condensation that warms up the extract side (Thanedgunbaworn et al., 2009; Wang and Min, 2011).

These events result in concentration and temperature polarization, which cause a significant reduction of the water vapor flux through the membrane (Rawindra et al., 2008), making actually the process less competitive on commercial scale. The resolution of this limitation is today matter of numerous studies (Ravindra Babu et al., 2008; Warczok et al., 2007; Alves and Coelhoso, 2006).

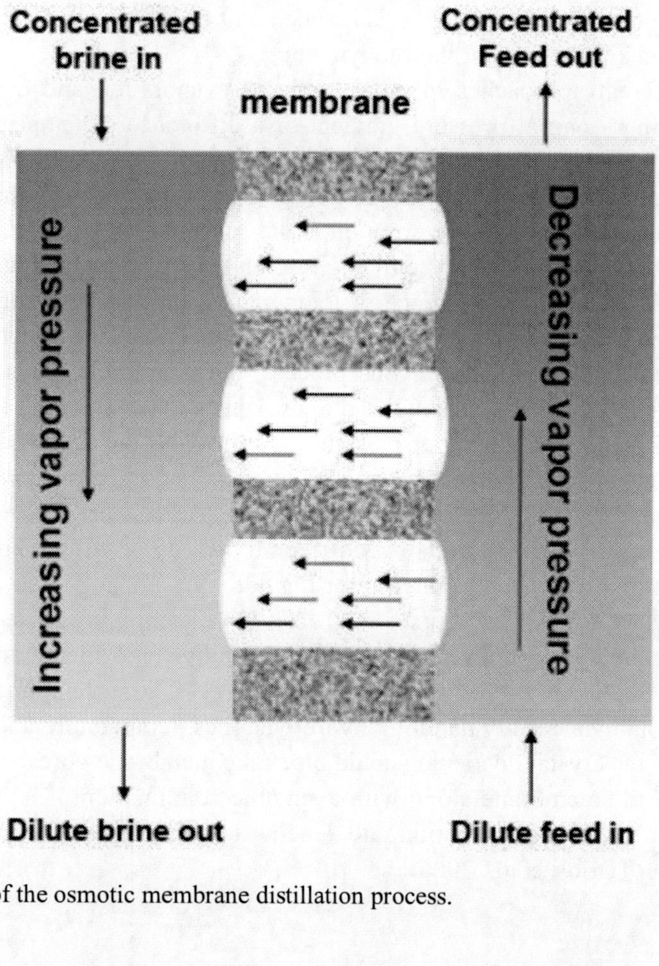

Figure 7. Scheme of the osmotic membrane distillation process.

This technique is used for the concentration of pulpy juices, sugars and milk and other liquid foodstuffs, production of ultrapure water, treatment of dye wastewater and is under evaluation for the concentration of aqueous solutions of thermally labile compounds (Ge et al, 2012; Macedonio and Drioli,2008; Alves and Coelhoso, 2006). OMD is a relatively expensive process, as compared to more conventional membrane processes, such as MD, ulltrafiltration (UF) and reverse osmosis (RO). However, this technique becomes more competitive as highly concentrated solutions of heat- or stress-sensitive products are requested. In this case the increase in capital cost should be justified. Anyway, hybrid systems arising from integration of OMD with UF and RO are expected to promote an abatment of the costs in favour to better quality concentrated streams.

2.5.4. Membrane Crystallization (MCr)

MCr has common basic concepts with MD and for some aspects with OMD (König and Weckesser 2005; Curcio et al, 2001), making the processes ease to combine and integrate. While MD is addressed at supplying clean water, MCr has the main objective to recover valuable compounds in form of crystals at retentive side (Drioli and Macedonio, 2010; Zhang et al., 2008). The technique allows the concentration of a solution via water vapor mass transfer through a microporous hydrophobic membrane (Figure 8). Under forced flow applied to membrane a degree of supersaturation is reached and nucleation and growth of crystals is induced via directed kinetics and/or thermodynamics.

There are different approaches to induce solvent evaporation and transfer through the membrane, including a partial pressure gradient or a difference of temperature between the crystallising solution (at feed side) and the stripping solution (at permeate side). When using a difference of partial pressures, an osmotic process is established between the two streams and the vapor is transferred from the dilute solution to the concentrated solution.

This approach is preferred for crystallizing thermo-sensitive compounds biomolecules. If the driving force is, instead, a difference of temperature, the vapor moves from the heated crystallizing solution to cool permeate stream. The latter is usually selected in presence of inorganic compounds such as salts. In this process, the membrane plays a key role, since heterogeneous crystallization must be induced in proximity to the surface.

The enhanced control of the vapor mass transfer through this interface allows further a better control of the distribution of the crystal nuclei. In this respect, morphological and chemical factors can be well-addressed at controlling the supersaturation events, which affect the nucleation induction, enabling one to get differently sized crystals in very short-time.

The induction time can be controlled by using additives (Zhang et al., 2006), improving the membrane surface roughness (Curcio et al., 2006) or combining surface roughness with established interfacial forces between crystallizing solution and membranes at preferential adsorption sites (Gugliuzza et al., 2006, 2009a). However, it is very important to prevent flooding phenomena by carefully handling hydrodynamics, temperature and crystallizer plant assembly, because the crystal deposition could block the membrane pores.

The choice of the membrane along with a suitable combination of laminar flux and low shear stress can influence the nucleation rate, leading to uniform distribution of crystals with good crystal lattices (Drioli et al., 2006).

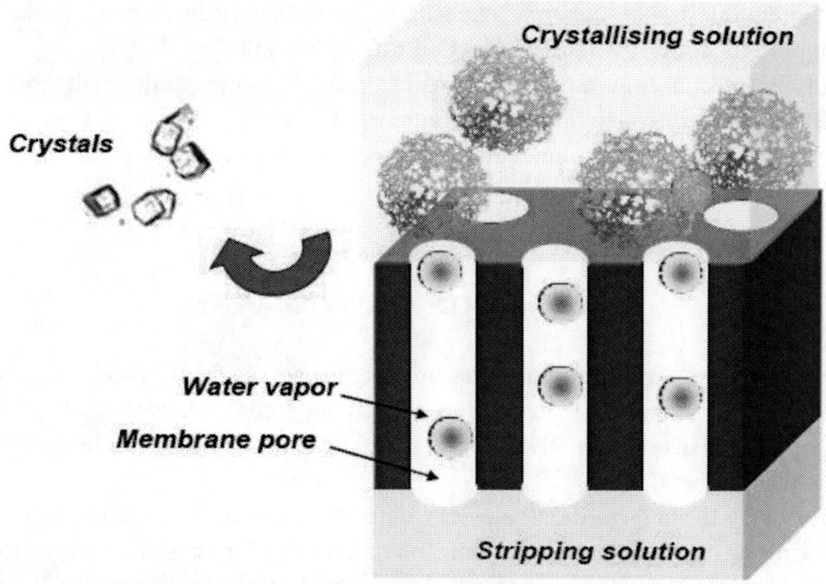

Figure 8. Representative formation of protein crystals via membrane crystallization process.

The nucleation rate, usually indicated by R_n parameter, can be described by the following equation:

$$R_n = k_b m_T^{\xi} \omega^e \Delta C^k \quad (28)$$

where m_T is the concentration of crystals in the magma, ω^e the rotation rate, k is a numerical empirical coefficient and ΔC the difference of concentration between the mother liquor and solubility value at equilibrium. k_b is the constant of the nucleation rate and its value can range from 1 to 2.5; as the collision between crystals and vessel prevail over the collisions between crystal and crystal values of ξ close to unit can be achieved. Proportionality between the growth rate (G) and concentration is given by

$$G = K_g \Delta C^g \quad (29)$$

where K_g is a growth rate constant, which depends on the temperature.

All these parameters can be varied by working on membrane features, difference of temperature between mother liquor and stripping solution, along with the concentration of the stripping solution. The membrane crystallization is considered a promising technology for the production of solid crystalline products for pharmaceuticals, chemicals and many others. High-purity crystals can easily be recovered at relatively low capital costs and eco-friendly operating conditions by using handle and compact devices.

In the logic of the sustainable process intensification, this technology is expected to give an important contribution to integrated productive cycles such as seawater desalination (Peinemann, and Nunes, 2010). The integration of membrane distillation plants with membrane crystallizers can allow large production of freshwater with significant reduction of polluting brines. In perspective, the crystallization of (bio)macromolecules based on membrane contactors on industrial scale could also support the development of new molecular targets due to the concrete opportunity to get single protein in crystal form and identify due protein–activity relationships (Di Profio et al., 2009).

Currently, the technology on small plants has provided interesting results, but a possible scale up will require necessarily higher and more rigid control of the process, given the lowest crystallization kinetics of macromolecular compounds, which are slower to diffuse owing to their more heterogeneous and complex structure.

Conclusion and Remarks

Membrane contactors technology resumes concepts such as flexibility, modularity, feasibility, along with efficient productivity-efficiency ratio. This area of the membrane technology is expected to bring advances to the chemical synthesis, engineering, molecular biology, and many others.

The possibility to integrated MC devices with other membrane operations or with more conventional plants may give an important impulsion to the process intensification strategy, promoting the opening and the growth of new markets. Many efforts are still necessary to overcome some current shortcomings, but the purpose of this chapter is also to stimulate major and more realistic contributions to the realization of new materials, concepts and engineering set-up in order to promote an evolution of the role of MC in competitive processes on commercial scale.

References

Abrahamse, A.J., van der Padt, A., Boom, R.M., Heij, W.B.C. (2001), Process fundamentals of membrane emulsification: simulation with CFD. *AIChE J* 47, 1285–1291.

Al-Asheh, S., Banat, F., Qtaishat, M., Al-Khateeb, M. (2006), Concentration of sucrose solutions via vacuum membrane distillation. *Desalination*, 195, 60-68.

Altenbach-Rehm, J., Suzuki, K., Schubert, H. (2002), Production of o/w-emulsions with narrow droplet size distribution by repeated premix membrane emulsification, in 3 i`eme Congrés Mondial de l'Emulsion, 24–27 September 2002, Lyon, France, 212.

Alves, M.D., Coelhoso, I.M. (2006), Orange juice concentration by osmotic evaporation and membrane distillation: a comparative study, *J. Food Eng.,* 57, 153-163.

Apel, P. (2001), Track etching technique in membrane technology, *Radiation Measurements* 34(1–6), 559–566.

Baker, R.W. (2004) Membrane technology and applications, (RW Baker Ed), John Wiley and Sons.

Banat, F.A., Simandl, J. (1996), Removal of benzene traces from contaminated water by vacuum membrane distillation, *Chem. Eng. Sci.*, 51, 1257-1265.

Banat, F.A., Simandl,J. (1994), Theoretical and experimental study in membrane distillation, *Desalination*, 95, 39-52.

Bey, S., Criscuoli, A., Simone, S., Figoli, A., Benamor, M., Drioli E. (2011), Hydrophilic PEEK-WC hollow fibre membrane contactors for chromium (VI) removal, *Desalination* 283, 16–24.

Boi, C., Bandini, S., Sarti, G.C. (2005), Pollutants removal from wastewaters through membrane distillation, *Desalination*, 183, 383-394.

Bottino, A., Roda, G.C., Capannelli, G., Munari, S.M. (1991), The formation of microporous polyvinylidene difluoride membrane by phase separation, *J. Membr. Sci.*, 57, 1-20.

Cardea, S., Gugliuzza, A., Schiavo Rappo, E., Aceto ,M., Drioli, E., Reverchon ,E. (2006), Generation of PEEK-WC Membranes by Supercritical Fluids, *Desalination,* 200, 58–60.

Cardea, S., Gugliuzza, A., Sessa, M., Aceto, M.C., Drioli, E., Reverchon E. (2009), Supercritical gel drying: a powerful tool for tailoring symmetric porous PVDF-HFP membranes, *ACS Applied Materials and Interfaces*, 1(1), 171-180.

Cardea ,S., Sessa ,M., Reverchon, E. (2010) Supercritical phase inversion to form drug-loaded poly(vinylidene fluoride-co-hexafluoropropylene) membranes, *Ind. and Eng. Chem. Res.*, 49(6), 2783-2789.

Charcosset, C., Limayem, I., Fessi, H. (2004), The membrane emulsification process –a review, *J. Chem. Technol. Biotechnol.*, 79, 209–218.

Charcosset, C. (2009a), A review of membrane processes and renewable energies from desalination, *Desalination, 245*, 214-231.

Charcosset, C. (2009b), Preparation of emulsions and particles by membrane emulsification for the food processing industry, *J. Food Eng.* 92, 241–249.

Chen, N., Hong, L. (2002), Surface phase morphology and composition of the casting films of PVDF–PVP blend, *Polymer*, 43, 1429-1436.

Cheng, L.P. (1999), Effects of temperature on the formation of microporous PVDF membranes by precipitation from 1-octanol/DMF/PVDF and water/DMF/PVDF systems, *Macromolecules*, 32, 6668-6674.

Cheng, L.P., Young, T.H., Fang, L., Gau, J.J. (1999), Formation of particulate microporous poly(vinylidene fluoride) membranes by isothermal immersion precipitation from the 1-octanol/dimethylformamide/poly(vinylidene fluoride) system, *Polymer,* 40 (9), 2395-2403.

Curcio, E., Criscuoli, A., Drioli, E. (2001), Membrane Crystallizers, *Ind. Eng. Chem. Res.*, 40(12), 2679-2684.

Curcio, E., Fontananova, E., Di Profio G., Drioli, E. (2006), Influence of the structural properties of poly(vinylidene fluoride) membranes on the heterogeneous nucleation rate of protein crystals. *J. Phys. Chem B,* 110, 12438-12445.

De Luca, G., Sindona, A. Giorno, L. Drioli, E. (2004), Quantitative analysis of coupling effects in cross-flow membrane emulsification, *J. Membr. Sci.* 229, 199–209.

Deshmukh, S.P., Li, K. (1998), Effect of ethanol composition in water coagulation bath on morphology of PVDF hollow fibre membranes, *J. Membr. Sci.*, 150, 75-85.

Ding, Z., Liu, L., Li, Z., Ma, R., Yang, Z. (2006), Experimental study of ammonia removal from water by membrane distillation (MD): The comparison of three configurations. *J. Membr. Sci.*, 286, 93-103.

Di Profio, G., Curcio, E., Drioli, E. (2009), A Review on membrane crystallization, Chemistry Today, 27 (2), 27-31.

Drioli, E., Criscuoli, A., Curcio, E. (2006), Membrane Contactors: Fundamentals, Applications and Potentialities, Membrane Science and Technology, Vol. 11 , Elsevier: Amsterdam.

Drioli, E., Laganà, F., Criscuoli, A., Barbieri ,G. (1999), Integrated membrane oprations indesalination processes, *Desalination*, 122, 141-145.

Drioli, E., Macedonio, F. (2010), Integrated Membrane Systems for Desalination in: Membranes for water treatment, Peinemann, K.V., Pereira Nunes, S. (Eds), Wiley-CH Verlag GmbH andCo, Weinheim, Germany.

Dowding, P.J., Goodwin, J.,Wand Vincent, B. (2001), Production of porous suspension polymer beads with a narrow size distribution using a cross-flow membrane and a continuous tubular reactor, *Colloids Surf. A,* 180, 301-309.

El-Bourawi, M.S., Khayet, M., Ma, R., Ding, Z., Zhang, X. (2007), Application of vacuum membrane distillation for ammonia removal. *J Membr Sci,* 301, 200-209.

El-Zanati, E., El-Khatib, K.M. (2007), Integrated membrane – based desalination system, *Desalination,* 205, 15–25.

Feng, C., Shi, B., Li G., Wu, Y. (2004), Preparation and properties of microporous membrane from poly(vinylidene fluoride-co-tetrafluoroethylene) (F2.4) for membrane distillation, *J. Membr. Sci.* 237, 15.

Findley, M.E. (1967), Vaporization through porous membranes, *Ind. Eng. Chem. Des. Dev.*, 6, 226–230.

Gabelman, A., Hwang, S.-T. (1999), Hollow Fiber Membrane Contactors, *J. Membr. Sci.*, 159, 61-106.

Gaeta, S.N. (2003), Membrane contactors in industrial applications, Proc. 1st Italy-Russia Workshop on membrane and membrane processes, Cetraro, Italy, 51-55.

Garcıa-Payo, M.C., Izquierdo-Gil, M.A., Fernández-Pineda, C. (2000), Air gap membrane distillation of aqueous alcohol solutions, *J. Membrane Sci.,* 169, 61–80.

Ge, Q, Wang, P, Wan, C, Chung, TS (2012), Polyelectrolyte-promoted forward osmosis-membrane distillation (FO-MD) hybrid process for dye wastewater treatment, *Environ. Sci. Technol.*, 46(11), 6236-6243.

Gekas, V., Hallström, B. (1987), Mass transfer in the membrane concentration polarization layer under turbulent cross flow: I. Critical literature review and adaptation of existing Sherwood correlations to membrane operations, *J. Membr. Sci.*, 30, 153-170.

Giorno, L., Li, N., Drioli, E. (2003). Preparation of oil-in-water emulsions using polyamide 10 kDa hollow fiber membrane, *J. Membr. Sci.*, 217, 173–180.

Gijsbertsen-Abrahamse, A.J., van der Padt, A., Boom, R.M., (2004) Status of cross-flow membrane emulsification and outlook for industrial application, *J. Membr. Sci.* 230, 149–159.

Gryta, M, Tomaszewska, M, Karakulski, K. (2006a), Wastewater treatment by membrane distillation. *Desalination,* 198, 67-73.

Gryta, M., Barancewicz, M. (2010), Influence of morphology of PVDF capillary membranes on the performance of direct contact membrane distillation, *J. Membrane Sci.,* 358(1-2), 158-167.

Gryta, M., Karakulski, K., Morawski, A. (2006b), Separation of effluents from regeneration of a cation exchanger by membrane distillation. *Desalination,* 197, 50-62.

Gryta M., Morawski, A.W., Tomaszewska, M. (2000), Ethanol production in membrane distillation bioreactor, Catal. Today, 56, 159-165.

Gu, M., Zhang, J., Wang, X., Tao, H., Ge L. (2006), Formation of poly(vinylidenefluoride) (PVDF) membranes via thermally induced phase separation, *Desalination* 192, 160.

Gugliuzza, A., Aceto, M.C., Macedonio, F., Drioli, E. (2008), Water droplets as template for next generation self-assembled poly-(etheretherketone) with Cardo membranes, *J. Phys. Chem B* 112(34), 10483-10496.

Gugliuzza, A., Drioli, E. (2007), PVDF and HYFLON AD membranes: Ideal interfaces for contactor applications, *J. Membr. Sci.* 300, 51–62.

Gugliuzza, A., Drioli, E. (2009), New performance of hydrophobic fluorinated porous membranes exhibiting particulate-like morphology, *Desalination*, 240, 14-20.

Gugliuzza, A., Ricca, F., Drioli, E. (2006), Controlled pore size, thickness and surface free energy of super-hydrophobic PVDF® and Hyflon®AD membranes, *Desalination*, 200, 26-28.

Gugliuzza, A., Speranza, V., Macedonio, F., Drioli, E. (2010) High-performance hydrophobic membranes for contactors and desalination technologies, Proc. of Advances in Science and Engineering for Brackish Water and Seawater Desalination (ECI), May 8-12, Cetraro, Italy, 122-124.

Gugliuzza, A., Aceto, M.C., Drioli, E. (2009a) Interactive functional poly(vinylidene fluoride) membranes with modulated Lysozyme affinity: a promising class of new interfaces for contactor crystallizers, *Polymer International,* 58 (12), 1452-1464.

Gugliuzza, A., Speranza, V., Trotta, F., Drioli, E., (2009b), Bio-inspired membranes with well-defined channels, *Chem. Eng. Trans*., 17, 1537-1542.

Gugliuzza, A., Basile, A. (2013) Membrane contactors: fundamentals, membrane materials and key operations in: Handbook of membrane reactors Volume II: Fundamental materials science, design and optimisation, A. Basile (Ed.), Woodhead Publishing Limited, Cambridge, UK, in press.

He, K., Hwang, H.J., Moon ,I.S. (2011), Air gap membrane distillation on the different types of membrane, *Korean J. Chem. Eng*., 28(3), 770-777.

Higashi, S., Tabata, N., Kondo, K.H., Maeda, Y., Shimizu, M., Nakashima, T., Setoguchi, T. (1999), Size of lipid microdroplets effects results of hepatic arterial chemotherapy with an anticancer agent in water-in-oil-in-water emulsion to hepatocellular carcinoma,. *J. Pharm. Exp. Ther,* 289, 816–819.

Higashi, S., Setoguchi, T. (2000), Hepatic arterial injection chemotherapy for hepatocellular carcinoma with epirubicin aqueous solution as numerous vesicles in iodinated poppyseed oil microdroplets: clinical application of water-in-oil-in water emulsion prepared using a membrane emulsification technique*, Adv. Drug Deliv. Rev*, 45, 57–64. http://www.gore-tex.com

Huang, S., Wu, G., Chen, S. (2007), Preparation of microporous poly(vinylidene fluoride) membranes via phase inversion in supercritical CO_2, *J. Membr. Sci.,* 293, 100-110.

Imhof, A., Pine, D.J. (1997), Ordered macroporous materials by emulsion templating, *Nature,* 389, 948-951.

Iversen, S.B., Bhatia, V.K., Dam-Johansen, K., Jonsson, G. (1997), Characterization of microporous membranes for use in membrane contactors, *J. Membr. Sci.* 130, 205-217.

Joscelyne, S.M., Tragårdh, G., (1999), Food emulsions using membrane emulsification: conditions for producing small droplets, *J. Food Eng.* 39:59–64.

Joscelyne, SM., Tragårdh, G. (2000), Membrane emulsification –a literature review *J. Membr. Sci.* 169, 107–117.

Katoh, R, Asano, Y, Furuya, A, Sotoyama, K, Tomita, M. (1996), Preparation of food emulsions using a membrane emulsification system. *J. Membr. Sci,* 113, 131–135.

Katoh, R., Asano, Y., Furuya, A., Sotoyama, K., Tomita, M., Okonogi, S. (1997a) Preparation of food emulsions using a membrane emulsification system 2. Conditions for preparation of W/O food emulsions using a membrane emulsification system, Nippon Shokuhin Kagaku Kaishi, 44, 44–49.

Katoh, R., Asano, Y., Furuya, A., Sotoyama, K., Tomita, M., Okonogi, S. (1997b), Preparation of food emulsions using a membrane emulsification system 3. A method of preventing dispersed droplets from crushing using a membrane emulsification system, Nippon Shokuhin Kagaku Kokaku Kaishi, 44(3), 233–237.

Khayet, M., Godino, M.P., Mengual, J.I. (2003), Theoretical and experimental studies on desalination using the sweeping gas membrane distillation method, *Desalination,* 157, 297-305.

König ,A., Weckesser, D. (2005), Membrane based evaporation crystallization, 16th *Int. Symp. on Industrial Crystallization* ISIC 2005, September 11-14, 2005, Dresden, In: VDI-Berichte 1901.2, 1171-1176.

Koonaphapdeelert, S., Zhentao, W., Li K. (2009) Carbon dioxide stripping in ceramic hollow fibre membrane contactors, *Chem. Eng. Sci.*, 64, 1– 8.

Kosaraju, P.B., Sirkar, K.K. (2007), Novel solvent-resistant hydrophilic hollow fiber membranes for efficient membrane solvent back extraction, *J. Membr. Sci.,* 288, 41–50.

Koschilowski, J., Wieghaus, M., Rommel, M. (2003), Solar thermal driven desalination plants based on membrane distillation, *Desalination,* 156, 295-304.

Kotov, N.A., Liu, L., Wang, S., Cumming, C., Eghtedari, M., Vargas, G., Motamedi, M., Nichols, J., Cortiella, J. (2004), Inverted Colloidal Crystals as Three-Dimensional Cell Scaffolds, *Langmuir,* 20 (19), 7887-7892.

Kresge, C.T., Leonowicz, M.E., Roth, W.J, Vartuli, J.C., Beck J.S. (1992), Ordered mesoporous molecular sieves synthesized by a liquid-crystal template mechanism, *Nature,* 359, 710-712.

Kumar, P.S., Hogendoorn, J.A., Feron, P.H.M., Vesteeg, G.F. (2002), New absorption liquids for removal of CO2 from diluite gas streams using membrane contactors, *Chem. Eng. Sci.,* 57, 1639-1651.

Kunz, W., Benhabiles, A., Ben-Aim, R. (1996), Osmotic evaporation through macroporous hydrophobic membranes: a survey of current research and applications", *J. Membrane Sci.,* 121, 23-36.

Lawson, K.W., Lloyd, D.R. (1997), Membrane Distillation, *J. Membrane Sci.,* 124, 1-25.

Lewandowicz, G., Białas, W., Marczewski, B., Szymanowska ,D. (2011), Application of membrane distillation for ethanol recovery during fuel ethanol production, *J. Membrane Sci.,* 375, 212–219.

Lin, D.J., Chang, H.H., Chen, T.C., Lee, Y.C., Cheng, L.P. (2006), Formation of porous poly(vinylidenefluoride) membranes with symmetric or asymmetric morphology by immersion precipitation in the water/TEP/PVDF system, *Eur. Polym. J.,* 42, 1581-1594.

Lü, L.Z., Yang, Z.S., Wang, Z.Y., Yang, Y.C., Tian, S.N. (2011), Preparation and performances of PVDF hydrophobic microporous membrane via immersion precipitation assisted with template, *Journal of Tianjin Polytechnic University,* 30(4), 6-10.

Luo, C., Huang, W., Han, Y. (2009), Formation of Two Kinds of Hexagonally Arranged Structures in ABC Triblock Copolymer Thin Films Induced by a Strongly Selective Solvent Vapor Macromol. *Rapid Commun.*, 30, 1917-1921.

Macedonio, F., Drioli, E. (2008), Pressure-driven membrane operations and membrane distillation technology integration for water purification, *Desalination,* 223, 396-409.

Martinez-Diez, L., Florido-Diaz, F. J. (2001), Desalination of brines by membrane distillation, *Desalination,* 137, 267-273.

Martinez-Diez, L., Vasquez-Gonzales, M.I. (1999), Temperature and concentration polarization in membrane distillation of aqueous salt solutions, *J. Membrane Sci.,* 156, 265-273.

Mathioulakis, E., Belessiotis, V., Delyannis, E. (2007), Desalination by using alternative energy: Review and state-of-the-art, *Desalination*, 203, 346–365.

Meindersma, G.W., Guijt , C.M., de Haan, A.B. (2006), Desalination and water recycling by air gap membrane distillation, *Desalination,* 18, 291-301.

Mui, B., Chow, L., Hope, M.J., (2003) Extrusion technique to generate liposomes of defined size. *Meth. Enzymol.* 367, 3–14.

Nakashima, T., Shimizu, M., Kukizaki, M. (1991), Membrane emulsification by microporous glass, *Key Eng. Mat.,* 61–62, 513–516.

Nene, S., Kaur ,S., Sumod, K., Joshi, B., Raghavarao, K.S.M.S. (2002), Membrane distillation for the concentration of raw cane-sugar syrup and membrane clarified sugarcane juice, *Desalination,* 147, 157-160.

Nymeijer, D.C., Folkers, B., Breebaart, I., Mulder, M.H.V., Wessling M. (2004), Selection of top layer materials for gas-liquid membrane contactors, *J. of Appl. Polym. Sci.*, 92(1), 323-334.

Oh, D.H., Balakrishnan, P., Oh, J-k., Kim, D-D., Yong, C.S., Choi, H-G. (2011), Pharmaceutical Nanotechnology Effect of process parameters on nanoemulsion droplet size and distribution in SPG membrane emulsification, *International Journal of Pharmaceutics,* 404, 191–197.

Okochi, H., Nakano, M., (1997), Comparative study of two preparation methods of w/o/w emulsions: stirring and membrane emulsification, *Chem Pharm Bull,* 45, 1323–1326.

Park, SH, Yamaguchi, T., Nakao, SI. (2001), Transport mechanism of deformable droplets in microfiltration of emulsions, *Chem. Eng. Sci.* 56, 3539–3548.

Peinemann, K-V., Pereira Nunes, S.(2010), *Membrane for Water Treatment*, (Peinemann, K-V., Pereira Nunes, S.Eds), Wiley-VCH, Weinheim, Germany.

Peng, P., Fane, A.G, Li, X. (2005), Desalination by membrane distillation adopting hydrophilic membrane, *Desalination*, 173, 45–54.

Peng, S.J., Williams, R.A. (1998) Controlled production of emulsions using a crossflow membrane. *Part Part Syst. Charact.*, 15, 21–25.

Phattaranawik, J., Fane, A.G., Pasquier, A.C.S., Bing W. (2008), A novel membrane bioreactor based on membrane distillation, *Desalination*, 223, 386–395.

Piacentini, E., Giorno, L., Figoli, A., Drioli, E. (2010) Membrane Emulsification in: *Comprehensive Membrane Science and Engineering*, Drioli, E.; Giorno, L. (Eds), Elsevier Science,

Piacentini, E., Drioli, E, Giorno, L. (2011), Preparation of stimulus responsive multiple emulsions by membrane emulsification using Con A as biochemical sensor, *Biotechnol. Bioeng.,*108(4), 913-923.

Prince, J.A., Singh, G., Rana, D., Matsuura, T., Anbharasia, V., Shanmugasundarama T.S., (2012) Preparation and characterization of highly hydrophobic poly(vinylidene fluoride) – Clay nanocomposite nanofiber membranes (PVDF–clay NNMs) for desalination using direct contact membrane distillation, *J. Membr. Sci.* 397-398, 80- 86.

Quéré, D. (2002) Rough ideas on wetting, Physica A 313, 32-46.

Ravindra Babu,, B., Rastogi N.K., Raghavarao, K.S.M.S. (2008), Concentration and temperature polarization effects during osmotic membrane distillation, *J. Membr. Sci.,* 322, 146-153.

Rayner, M., Tragårdh, G. (2002), Membrane emulsification modelling: how can we get from characterisation to design? *Desalination* 145,165-172.

Reverchon, E., Cardea, S. (2006), PVDF−HFP Membrane Formation by Supercritical CO_2 Processing: Elucidation of Formation Mechanisms, Ind. and *Eng. Chem. Res.,* 45, 8939-8945.

Ribeiro, H.S., Rico, L.G., Badolato, G.G., Schubert, H. (2005) Production of o/w emulsions containing astaxanthin by repeated premix membrane emulsification, *J. Food Sci.* 70 (2), 117-123.

Scherze, I., Marzilger, K., Muschiolik, G. (1999) Emulsification using micro porous glass (MPG): surface behaviour of milk proteins, *Colloids Surf. B* 12, 213-221.

Schröder, V., Behrend, O., Schubert, H. (1998), Effect of dynamic interfacial tension on the emulsification process using microporous, ceramic membranes. *J. Coll. Interf. Sci.* 202, 334–340.

Schröder, V., Schubert, H. (1999), Production of emulsions using microporous, ceramic membranes. *Colloids Surf. A* 152,103-109.

Schröder, V., Schubert, H. (1997), Emulsification using microporous, ceramic membranes, in: Proceedings of the First European Congress on Chemical Engineering (ECCE 1), Florence, Italy, 4–7 May 1997, 2491-2494.

Schröder, V., Wang, Z., Schubert, H. (1997), Production of oil-in-water emulsions by microporous membranes, in: Proceedings of the Third International Symposium on Progress in Membrane Science and Technology, Euromembrane 1997, University of Twente, 439.

Song, S.H., Cho, Y.H., Park, J., (2003), Microencapsulation of Lactobacillus casei YIT 9018 using a microporous glass membrane emulsification system. *J. Food Sci.* 68, 195-200.

Speranza, V., Trotta, F., Drioli, E., Gugliuzza, A. (2010), High-definition polymeric membranes: construction of 3D lithographed channel arrays through controlling natural building blocks dynamics, *ACS Appl. Mater. Interfaces,* 2(2), 459-466.

Srinivasarao, M., Collings, D., Philips, A., Patel, S. (2001), -dimensionally ordered array of air bubbles in a polymer film, *Science,* 292, 79-83.

Stanojević, M., Lazarevic, B., Radic, D. (2003), Review of membrane contactors designs and applications of different modules in industry, *FME Transactions,* 31, 91-98.

Supsakulchai, A., Ma, G.H., Nagai, M., Omi, S. (2002), Uniform titanium dioxide (TiO_2) microcapsules prepared by glass membrane emulsification with subsequent solvent evaporation, J Microencapsulation 19, 425-449.

Susanto, H. (2011), Towards practical implementations of membrane distillation, *Chem. Eng. and Proc.*, 50, 139-150.

Suzuki, K, Fujiki, I., Hagura, Y. (1998), Preparation of corn oil/water and water/corn oil emulsions using PTFE membranes, *Food Sci Technol* 4,164-167.

Taylor, T.M., Davidson, P.M., Bruce, B.D., Weiss, J., (2005) Liposomal nanocapsules in food science and agriculture. *Crit. Rev. Food Sci. Nutr.* 45, 587-605.

Teoh, M.M., Bonyadi, S., Chung, T.S. (2008), Investigation of different hollow fiber module designs for flux enhancement in the membrane distillation process, *J. Membr. Sci.,* 311, 371-379.

Thanedgunbaworn, R., Jiraratananon, R., Nguyen, M.H., Vapour Transport Mechanism in Osmotic Distillation Process, *Int. J. Food Eng.,* 5(5), (2009) 1-19.

Tomaszewska, M. (1996), Preparation and properties of flat-sheet membranes from poly(vinyldene fluoride) for membrane distillation, *Desalination,* 104, 1-11.

Vogelaar, L., Wessling, M. (2002), Fabrication of polymeric microsieves by phase separation micro moulding. In: Network Young Membrains, NYM, 5-7July, 2002, Toulouse, France.

Vladisavljević, G.T., Kobayashi, I., Nakajima, M. (2012), Production of uniform droplets using membrane, microchannel and microfluidic emulsification devices, *Microfluid Nanofluid* (2012), 13, 151–178.

Wang, L., Min, J. (2011), Modelling and analyses of membrane osmotic distillation using non-equilibrium thermodynamics, *J. Membr. Sci.*, 378, 462-470.

Wang, Z.-S., Gu, Z.-L., Feng, S.-Y., Li, Y. (2009), Applications of membrane distillation technology in energy transformation process-basis and prospect, Chinese Science Bulletin (Springer), 54, 2766-2780.

Wang, Z.Y., Li J.L., Kong, X.S., Yang, Z.S., Li C.L. (2011), Effect of octanol on wettability and permeability of PVDF porous membrane via dry-wet phase inversion, Tianjin Daxue Xuebao (Ziran Kexue yu Gongcheng Jishu Ban)/*Journal of Tianjin University Science and Technology,* 44 (7), 628-632.

Warczok, J., Gierszewska, M., Kujawski, W., Guellet, C. (2007), Application of osmotic membrane distillation for reconcentration of sugar solutions from osmotic dehydration, *Sep. and Purif. Technol.*, 57, 425-429.

Weyl, P.K. (1967), Recovery demineralized water from saline waters, *United States Patent* 3,340,186.

Williams, R.A., Peng, S.J., Wheeler, D.A., Morley, N.C., Taylor, D., Whalley, M., Houldsworth, D.W. (1998), Controlled production of emulsions using a crossflow membrane, Part II: Industrial scale manufacture, *Chem. Eng. Res. Des.* 76, 902–910.

Winter, D., Koschikowski, J., Duever, D. (2011), Spiral wound modules for membrane distillation: modelling, validation and module optimization, *Proc. Int. Workshop on Membrane Distillation and Related Technologies,* Ravello (Italy), Oct. 9-12, 58-59.

Yan, S.-P., Fang, M.-X., Zhang, W.-F., Wang, S.-Y, Xu, Z.-K., Luo, Z.-Y., Cen, K.-F. (2007). Experimental study on the separation of CO2 from flue gas using hollow fibre membrane contactors without wetting, *Fuel Proc. Techn.*, 88(5), 501-511.

Yang, X., Cao, Y-M., Wang, R., Yuan, Q. (2007) Study on highly hydrophiliccellulose hollow fiber membranecontactors for thiol sulfur removal, *J. Membr. Sci,* 305 (1-2), 247–256.

Young, T.H, Cheng, L.P., Lin, D.J., Fane, L., Chuang, W.Y. (1999), Mechanisms of PVDF membrane formation by immersion-precipitation in soft (1-octanol) and harsh (water) nonsolvents, *Polymer* 40, 5315-5323.

Zhang, X., El-Bourawi, M.S., Wei, K., Tao, F., Ma, R. (2006), Precipitants and additives for membrane crystallization of lysozyme, *Biotechnol. J.,* 1(11), 1302-1311.

Zhang, X., Zhang, P., Wei, K., Wang, Y., Ma, R. (2008), The study of continuous membrane crystallization on lysozyme, *Desalination,* 219, 101-117.

Zhao, Z.P., Ma, F.W., Liu, W.F., Liu, D.-Z. (2008), Concentration of ginseng extracts aqueous solution by vacuum membrane distillation. 1. Effects of operating conditions, *Desalination*, 234, 152-157.

In: Membrane Processes for Sustainable Growth
Editors: Angelo Basile and Alfredo Cassano
ISBN: 978-1-62618-446-6

Chapter 3

PERVAPORATION AS SUSTAINABLE PROCESS

***Alberto Figoli*[1]*, *Pieter Vandezande*[2], *Francesco Galiano*[1] *and Sina Zereshki*[3]**
[1]Institute on Membrane Technology,
ITM-CNR, c/o University of Calabria, Rende (CS), Italy
[2]Flemish Institute for Technological Research (VITO),
Unit Separation and Conversion Technology (SCT), Boeretang, Belgium
[3]Chemical Engineering Department,
Kermanshah University of Technology, Kermanshah, Iran

ABSTRACT

Nowadays, pervaporation (PV) can be considered as a membrane separation process that can satisfy the claims of the modern society. Low energy consumption and no solvents requirements make PV a "green" process to be considered alternative to the traditional ones. In this chapter, it is reported the history of PV, starting from 1917, when the word "pervaporation" was firstly used, to most recent years during which many breakthroughs in this field have been done. Then, basic aspects of the separation PV process as well as the applications, classified as solvent dehydration, organic/water separation and organic/organic separation, are discussed in details.

Keywords: pervaporation (PV); solvent dehydration; organic/water separation; organic /organic separation

1. INTRODUCTION

During the last few decades, the increased world-wide competitiveness has been forcing the chemical, pharmaceutical, food and other industries to make their current production processes more efficient, economical and sustainable. Consequently, the reorganization of

* Corresponding Author: e-mail: a-figoli@itm.cnr.it

existing process designs and the development of new processes, with the possible integration of new technologies, is of growing importance to industry.Membrane technologies have emerged as an additional category of separation processes to the well-established thermal separation processes. The advantages of membrane separations include high selectivity, low energy consumption, moderate cost to performance ratio, and compact and modular design.

In particular, pervaporation (PV) has gained acceptance in industry as an effective process tool for the separation and recovery of liquid mixtures. It offers potential solutions in a wide range of applications, but is currently best identified with the well established dehydration of organic solvents (Nagai, 2010; Fleming and Slater, 1992).

PV involves the partial vaporization of a liquid-liquid mixture through a selective membrane which can be either a nonporous polymeric or a nanoporous inorganic (ceramic/zeolite) membrane.

The process is thus characterized by the imposition of a membrane barrier between a liquid feed phase and a gaseous permeate phase, with mass transfer occurring selectively across the barrier to the vapour side. The vaporous permeate is subsequently condensed to obtain a liquid product. Because of the unique phenomenon of phase change and the permselective evaporation of the molecules diffusing through the membrane, the process is termed "pervaporation", being a contraction of "per-meation" and "e-vaporation". Since different species permeate through the membrane at different rates, a substance at low concentration in the feed stream can be highly enriched in the permeate. The feed can thus be separated with the efficacy of separation being determined by the physicochemical nature of the membrane and the affinityof the targets compounds for it.

The choice of the membrane material in PV is critical and depends on the temperature and composition of the feed mixture, the separation/purification target and the desired performance. A broad range of membrane materials can be used, both organic (polymeric), inorganic (ceramic, zeolite) and composite (inorganic particles dispersed in a polymeric matrix) in nature. The key to commercialization of PV has been the development of thin film composite membranes, consisting of an ultrathin, selective polymer coating on top of a porous support. Besides the differences in membrane material, PV membranes can be classified according to the character of the separation.

To minimize the heat of vaporization and thus the energy required to separate the liquids, the membrane is usually chosen such that the minor fraction is removed. Therefore, three main application types are usually distinguished and discussed in this chapter, i.e. dehydration of organic solvents using hydrophilic membranes, removal of organics from dilute aqueous solutions using hydrophobic membranes, and organic-organic separations using target-organophilic membranes.

In general, PV can be employed when the target compounds to be removed are (semi-) volatile, of low molecular weight (typically < 200 Da) and present at relatively low concentrations (typically < 10%). In such cases, the most appropriate driving force for transport through a non-porous or nanoporous membrane to take place is a gradient in the chemical potential or activity of the target compounds between the feed and permeate sides of the membrane.

2. History

PV was first mentioned by Philip Adolph Kober in 1917 in the course of some dialysis experiments (Kober, 1995). Kober reported the evaporation of water from an albumin/toluene solution through a tightly closed collodion bag at room temperature, indicating the presence of an extremely fine porosity allowing vapour molecules to pass freely and preventing liquid to pass. Until the late 1950s, no major progress was made in the field of PV, although few researchers published their findings on pervaporative dehydration (Heissler et al., 1956; Farber, 1935). In 1958, Binning et al., assigned to the American Oil Company (Amoco), reported the use of PV for the separation of a ternary isopropanol-ethanol-water azeotrope (Fleming and Slater, 1992), being the start of several decades of dedicated research. From 1959 to 1961, more than 10 patents on PV were issued to Amoco (Binning et al., 1958), demonstrating the preliminary interest of industry in PV. The capacity of PV in separating organic-organic mixtures was demonstrated by 1965, but commercial developments did not proceed, primarily due to the lack of a clear market demand. Traditional separation technologies including distillation, extraction and adsorption were deemed sufficient at that time. Moreover, the membranes being utilized usually lacked the required selectivity and permeability to make PV economically attractive. In fact, in those years there was much greater interest in other membrane processes being developed, particularly reverse osmosis and ultrafiltration.

The energy crisis of the 1970s, however, refocused interest in separation technologies offering a high potential for energy savings. PV was widely investigated in the seventies, especially in Europe, because of its demonstrated ability to dewater mixtures for fuel utilization.The first industrial PV based dehydration units using polymeric membranes, producing high ethanol purities and ling azeotropic distillation, were put into service in 1983 in Brazil by GFT (Germany) (Brüschke, 2003). The primary drivers for the GFT process included the reduced energy demand, the low overall system costs, the fact that no additive was required for separation, and the possibility to achieve superior separations with no limitations being imposed by the thermodynamics of the system. In the late 1970s and continuing into the early 1980s, other integrated distillation-PV plants for ethanol dehydration were built all over the world, mostly with moderate capacity (Fleming and Slater, 1992). As the cost of the membrane modules gradually decreased and selectivity increased, the achievable ethanol purity increased (up to 99.85%) and the integrated process gained wider industrial acceptance. In the mid-1980s, multiple membranes were developed, all based on novel proprietary composite polymer technology allowing for the production of economical, chemically and thermally stable membranes in a flat plate geometry. Besides a vast number of ethanol dehydration units, commercial PV plants for solvent recovery, removal of organics from wastewaters and dealcoholization of wine and beverages, all using polymeric membranes of the thin film composite type, were installed in the 1980s. In contrast, the first large-scale dehydration unit using NaAzeolite membranes was only operational almost 20 years later, in 2001, in Japan (Morigami et al., 2001).

Despite the promising achievements in the 1970-80s and the large number of published papers on membrane and process development, the progress in PV technology during the last three decades stagnated somehow and dehydration of organic solvents still remains the most important industrial application (Jonquières et al., 2002).

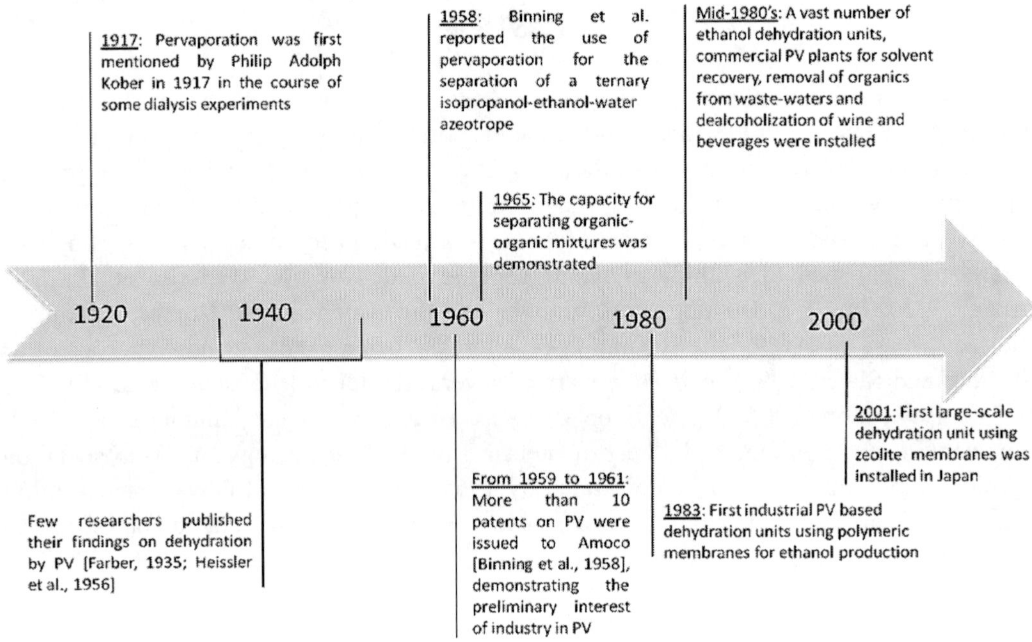

Figure 1. Resume of history of PV.

However, in the current context of increasing energy costs, more stringent environmental legislation and growing environmental awareness, increasing competition from low cost economies, mild, non-thermal and intrinsically energy efficient separation technologies such as PV are regaining more and more industrial interest. Due to disposal norms that are becoming stricter, more and more companies are nowadays being confronted with increasing wastewater treatment costs, to such extent even that their production capacity is being limited by the availability of sufficient treatment capacity.

Due to the lack of cost-efficient separation technologies, many waste and side streams are simply incinerated instead. Furthermore, as large volumes of these dilute waste waters typically containing a few weight percentages of organics are being generated, significant volumes of useful and expensive chemicals are being lost. PV could offer a potential solution here as it allows to remove the organics below the disposal threshold and enables reuse of the enriched permeate in the process. Besides conventional applications as an end-of-pipe technology PV could also play a significant role in future biorefineries, for instance in the removal of inhibitory compounds from fermentation broths or the recovery of high added value bio-chemicals from liquid biomass derived streams.

3. Theory

PV differs from other membrane processes in that the membrane constitutes a barrier between the feed in the liquid phase and the permeate in the gas phase. The driving force applied across the membrane creates a chemical potential gradient in the liquid phase, and the relative flow of the different components is determined by the selectivity of the membrane. In PV, the driving force for transport is generally recognized as a gradient in partial vapour pressure between the liquid feed and the vaporous permeate (Nagai, 2010; Mulder, 1996). In practice, the permeate side vapour pressure is usually lowered by establishing a vacuum. The driving force can be further enhanced by lowering the permeate pressure or increasing the feed temperature.Usually, a relatively small vacuum pump is sufficient to maintain the driving force and perpetually drive the process as the condensing permeate continually creates a sufficient vacuum and lowers the temperature at the permeate side (Fleming and Slater, 1992). In the industrial practice, one aims at switching off the vacuum pump once the appropriate vacuum level has been established in the downstream circuit in order to minimize energy costs, and ideally let the condensation unit(s) maintain the vacuum alone. Alternatively, PV can also be driven by the partial vapour pressure difference generated between a liquid feed and an inert sweep gas at the permeate side (Figure 2). Besides the energy to maintain a vacuum or apply a sweep gas at the permeate side, the operational energy required in PV also comprises the enthalpy of vaporization of the permeating molecules.

Similar to reverse osmosis and gas separation membranes, polymeric PV membranes are nonporous, and the transport across the membrane barrier is usually described with the solution-diffusion mechanism (Wijmans and Baker, 1995). This model was formulated in the 19th century, based on the work of Graham (1829) (Graham, 1866), Mitchell (1831) (Mitchell, 1995) and von Wroblewski (von Wroblewski, 1879). According to the solution-diffusion model, transport can be described with a three step mechanism (Nagai, 2010; Fleming and Slater, 1992; Mulder, 1996). Firstly, solutes partition and selectively sorb from the bulk feed liquid into the membrane material according to the thermodynamic equilibrium at the liquid-membrane interface. Subsequently, the permeant molecules diffuse across the membrane owing to the concentration gradient. At the permeate side of the membrane, a vacuum or carrier gas promotes continuous desorption of the vaporous permeate molecules, maintaining in this way a concentration gradient and hence a continuous transmembrane flux. In contrast to purely sorption controlled processes, both sorption and diffusion control the process performance in PV.

The key feature in PV is the mass transfer of the target components through the membrane and the liquid-vapour phase transition occurring simultaneously. Basically, the permeability (*P*) of a PV membrane can be described in terms of the solubility (*S*) and the diffusivity (*D*) of the target components:

$$P = S \times D \quad (1)$$

The solubility is a thermodynamic parameter providing information on the amount of penetrant sorbed by the membrane under equilibrium conditions. The diffusivity, on the other hand, is a kinetic parameter indicating the transport rate of the penetrant through the membrane. As indicated above, the driving force in PV is the chemical potential (μ_i) across the membrane, which is expressed as:

$$d\mu_i = RTd\ln(\gamma_i n_i) + v_i dp \quad (2)$$

where γ_i is the activity coefficient, n_i the mole fraction ($\gamma_i n_i = a_i$), $a_i$is the activity, v_i the partial molar volume, p the pressure of component i, R the universal gas constant and T the absolute temperature.

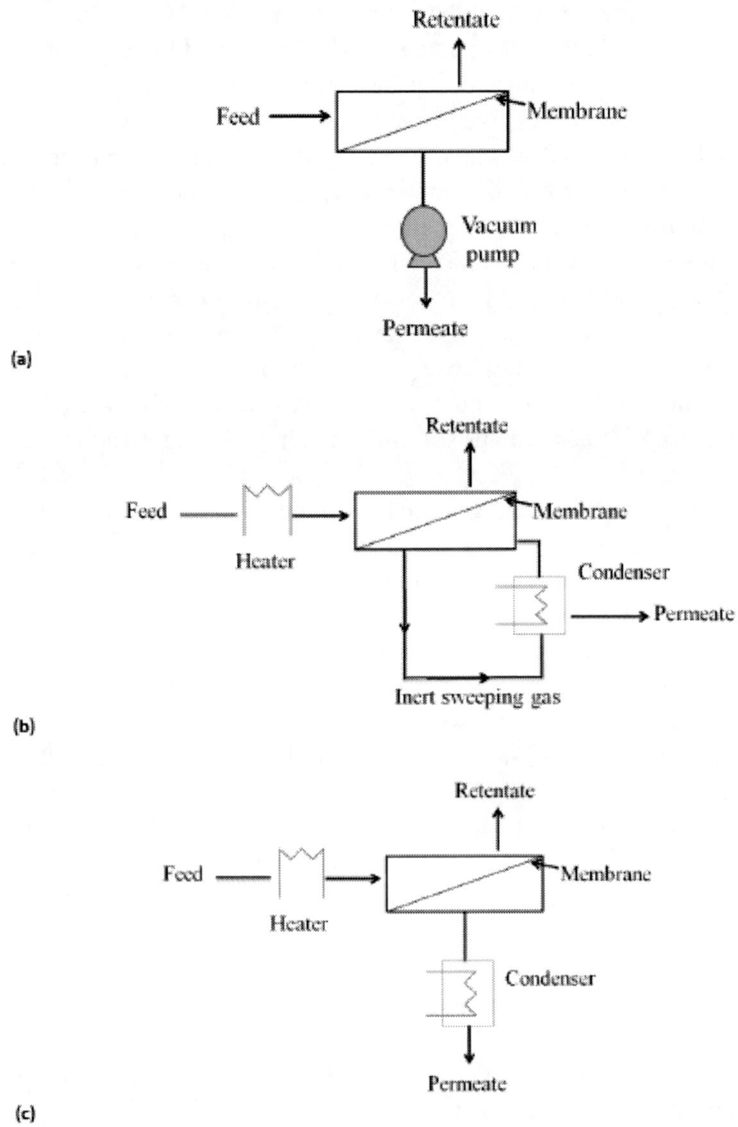

Figure 2. PV driving force by: a) vacuum, b) sweeping gas and c) temperature gradient (adapted from Baker, 2004).

The flux, J_i, of the target component i can be described by an equation similar to Fick's first law:

$$J_i = -L_i \frac{d\mu_i}{dx} \quad (3)$$

where $d\mu_i/dx$ represents the chemical potential gradient of the target component i across the membrane and L_i a coefficient which describes the proportionality of the flux with the chemical potential.

Prior to defining the model of permeation in PV, some assumptions must be made. Firstly, the fluid and vapour on either side of the membrane are expected to be in equilibrium with the membrane material at the interface. Secondly, the solution-diffusion model assumes that the pressure within the membrane is uniform and that the chemical potential gradient across the membrane is expressed only as a concentration gradient (Wijmans and Baker, 1995). Figure 3 represents a schematic diagram of the driving force gradients according to the solution-diffusion model.

At the feed side, two incompressible phases exist, i.e. the membrane and the liquid feed mixture, of which the volumes do not change appreciably with pressure. Integrating Equation 2 with respect to concentration and pressure then gives:

$$\mu_i = \mu_i^0 + RTln(\gamma_i n_i) + v_i(p - p_i^0) \tag{4}$$

where μ_i^0 is the chemical potential of the pure component i at a reference pressure p_i^0. For the equilibrium conditions at the feed side liquid/membrane interface (*0*), the following equation can be written:

$$\overbrace{\mu_i^0 + RTln(\gamma_{i_0}^L n_{i_0}) + v_i(p_0 - p_{i_{sat}})}^{\text{Liquid feed}} = \overbrace{\mu_i^0 + RTln\left(\gamma_{i_{0(m)}} n_{i_{0\,(m)}}\right) + v_i(p_0 - p_{i_{sat}})}^{\text{Membrane}} \tag{5}$$

in which the subscripts 0 and $0(m)$ denote the liquid feed phase and the liquid in the membrane at point 0 (Figure 3), respectively, and the superscript L stands for liquid.

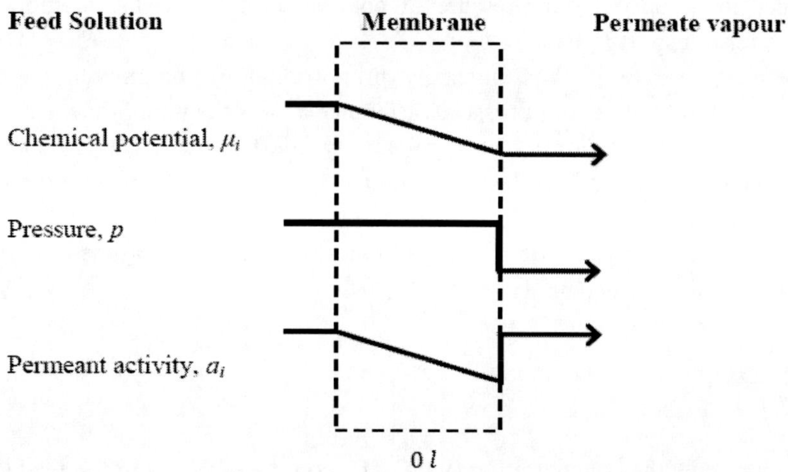

Figure 3. Driving force gradients for permeation through a nonporous membrane according to the solution-diffusion model (adapted from Wijmans and Baker, 1995).

At the permeate side vapour/membrane interface, on the other hand, one side is a compressible gas of which the molar volume changes with pressure. By integrating Equation 2 and applying the ideal gas law, the following expression is obtained:

$$\mu_i = \mu_i^0 + RTln(\gamma_i n_i) + RT\ln\left(\frac{p}{p_i^0}\right) \tag{6}$$

To concede to the equilibrium conditions at the permeate side vapour/membrane interface, the following equation can be derived:

$$\overbrace{\mu_i^0 + RTln(\gamma_{i_l}^G n_{i_l}) + RT\ln\left(\frac{p_l}{p_{i_{sat}}}\right)}^{\text{Gaseous permeate}} = \overbrace{\mu_i^0 + RTln\left(\gamma_{i_{0(m)}} n_{i_{0(m)}}\right) + v_i(p_0 - p_{i_{sat}})}^{\text{Membrane}} \tag{7}$$

At the permeate side interface l (Figure 3), the membrane phase is at the feed pressure as assumed by the solution-diffusion model, while the permeate phase is at a lower permeate pressure (vacuum). Since in an isothermal system the chemical potential consists of a pressure term and an activity term, as expressed by Equation 2, the activity in the permeate phase must be higher than the activity in the membrane phase to fulfil the equilibrium conditions (Figure 3).

Taking into account all these equilibrium conditions implied by the solution-diffusion model, the component flux J_i(kg/m^2.h) can be derived as:

$$J_i = \frac{P_i^G}{\theta}(p_{i_0} - p_{i_l}) \tag{8}$$

where P_i^G represents the permeability coefficient, θ the membrane thickness and $(p_{i_0} - p_{i_l})$ the difference in partial vapour pressure of component i across the membrane, being the driving force for transport in PV.

The permeation rate is thus inversely proportional to the membrane thickness and proportional to the partial vapour pressure difference across the membrane.

In contrast to gas separation, the permeability coefficient in PV is dependent on both composition and temperature due to the pronounced interaction between organic liquids and the membrane polymer.

The separation efficiency of PV membranes is usually expressed by means of the separation factor α_{ij}, defined as:

$$\alpha_{ij} = \frac{c_i^P / c_j^P}{c_i^F / c_j^F} \tag{9}$$

in which c represents the concentration (wt%) of a component i or j in the feed (superscript F) or permeate (superscript P). The corresponding enrichment factor is defined as:

$$\beta_i = \frac{c_i^P}{c_i^F} \tag{10}$$

The problem with reporting performance data of pervaporation membranes in terms of fluxes (Equation 8) and separation (Equation 9) or enrichment factors (Equation 10) is that these values are not only a function of the intrinsic properties of the membrane, but also depend on the operating conditions, i.e. feed composition, permeate pressure and feed temperature.This makes comparison of pervaporation data sets obtained under different conditions difficult. Moreover, it can lead to confusion in understanding the real trend of intrinsic membrane properties as the operating conditions change. Therefore, Richard Baker and co-workers recently suggested to report pervaporation data as driving force normalized properties, i.e. membrane permeabilities P_i^G, defined by Equation 8, and membrane selectivities α'_{ij}, defined as the ratio of the permeabilities of components i and j through the membrane:

$$\alpha'_{ij} = \frac{P_i^G}{P_j^G} \tag{11}$$

The use of permeabilities and selectivities thus allows decoupling of the variations in flux and separation factor due to vapour-liquid equilibrium effects from true variations in the permeation properties of the membrane (Baker et al., 2010).

Compared with conventional pressure-driven membrane processes such as ultrafiltration and reverse osmosis, fluxes in PV are generally low, typically below 20 $kg/m^2.h$. However, separation factors can be extremely high, often exceeding 1000. Full recovery of water or organics from the feed is usually difficult as the driving force for transport will become very low at low feed concentrations of the penetrant.

While concentration-polarization is usually not an issue, temperature polarization is typical for PV. By definition, PV requires the volatilization of a portion of the liquid feed and the enthalpy of vaporization must be consumed by the feed. For a permeant such as water, the resulting temperature drop can be substantial and a large thermal gradient is generally established across the membrane at the liquid side with continual heat loss to the permeate. As the flux is strongly proportional with the temperature, interstage heaters are often employed to reheat the feed mixture between the membrane modules, thereby compensating for this temperature drop.

4. Pervaporation versus Distillation

PV is often seen as a valuable alternative to or complementary technique to conventional distillation. While there are certainly similarities, i.e. the occurrence of a vapour-liquid phase transition and the integration of mass and heat transfer, there are some notable features that make PV considerably more energy-efficient than distillation (Nagai, 2010; Shao and Huang 2007; Fleming and Slater, 1992). In distillation, the liquid and vapour phases are in equilibrium and mass transfer between both phases takes place in both directions with the more volatile component being enriched in the vapour phase. PV, however, is a non-equilibrium process where transport occurs from the liquid feed to the vaporous permeate side with penetrants being vaporized as they "feel" the vacuum.

PV is not governed by the vapour-liquid equilibrium of the mixture to be separated, as is the case in distillation, but by the selectivity of the membrane. The latter is determined by the solubility of the permeants in the membrane and their diffusivity through the membrane, as explained above. Since highly selective membranes are used, the separation in PV is thus mainly based on the affinity of the target molecules for the membrane material, whereas distillation is entirely governed by differences in boiling point or vapour pressure.

Thanks to the selectivity introduced by the membrane, only the permeating molecules need to be vaporized, while in distillation the entire feed is subjected to repeated vaporization-condensation cycles. Although the membrane selectivity is theoretically gained at the expense of lower overall fluxes, in practice this can be compensated by a larger membrane area as commercial membrane costs are competitively low. The membrane is generally chosen in such a way that the minor component in the feed mixture selectively permeates, thereby minimizing the required latent heat of vaporization. PV is therefore especially effective and economically attractive for dilute solutions containing traces or minor concentrations (usually below 10%) of the target compound.

Furthermore, as PV can be carried out at mild temperatures, it offers undeniable advantages over thermal separation processes in case of feed mixtures containing heat-sensitive (bio) molecules. Finally, PV is performed in the absence of any additives (e.g. entrainers), avoiding the risk on product contamination and the need for extra recovery columns such as in extractive or azeotropic distillation.

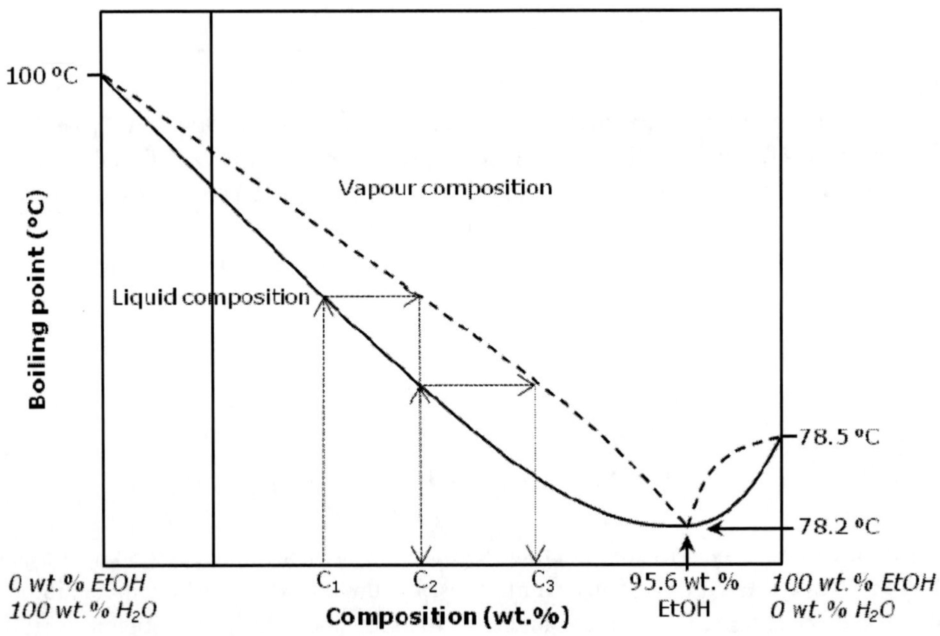

Figure 4. Boiling point as a function of the composition for a binary ethanol/water mixture (positive azeotrope).

Thanks to these distinct features, PV is particularly attractive in the separation of azeotropic or close-boiling point mixtures, or in those cases where higher temperatures would affect the activity, texture or color of the target compounds, all situations where conventional distillation fails, or is at best energetically ineffective or very expensive. An azeotropic mixture consists of at least two liquids of which the compositions of the vapour and liquid phases are identical.

Therefore, azeotropic mixtures cannot be separated by conventional distillation. The boiling point of the azeotropic mixture can be either higher (negative azeotrope) or lower (positive azeotrope) than its individual components (Hadler et al., 2009). A well-known example of a positive azeotrope is the binary mixture containing 95.6 wt% ethanol and 4.4 wt% water (Figure 4).

When a mixture with composition C_1, left from the azeotropic composition, would be distilled, it will boil at a temperature given by the liquid composition curve and produce a vapour with composition C_2. Reheating of the condensed vapour C_2 will result in a new vapour with composition C_3. By repeating this boiling-condensing-reboiling sequence, the compositions of the vapour and liquid will finally end up in the azeotropic point. This means that pure ethanol cannot be obtained by conventional distillation. Pure water is recovered in the boiling flask as the distillation residue, while the composition of the distillate corresponds to the azeotrope. Heating a mixture with a composition at the right-hand side of the azeotropic mixture, on the other hand, will result in pure ethanol in the boiling flask and the same azeotropic mixture in the distillate.

More sophisticated distillation techniques, such as pressure swing, vacuum or extractive distillation could be used to purify the mixture beyond the azeotropic point, but all of them require significantly more energy and/or rely on additives (e.g. entrainer) that alter the relative volatilities of the liquids to be separated, and must be recovered afterwards. As discussed above, PV is not governed by the thermodynamics of the liquid-vapour equilibrium and as it relies on the selectivity of the membrane, it can break azeotropes at mild temperatures and without addition of entrainers. Indeed, since the permeate composition, hence the selectivity, is determined by the mass transport through the membrane, vapour-liquid equilibria with their associated azeotropic effects are irrelevant. In many industrial bio-ethanol plants, the distillate of the final distillation column, having the azeotropic composition shown in Figure 4, is nowadays effectively dewatered up to 99.9 wt%, using hydrophilic PV membranes.

5. APPLICATIONS

5.1. Solvent Dehydration

Dehydration of organic solvents, i.e. removing traces of water from solvents in order to obtain higher purities, is the most widespread PV application, accounting for the majority of the plants built to date (Jonquières et al., 2002). In the chemical process industry, there are literally unlimited numbers of organic streams that become contaminated with small (< 10%) amounts of water. In most cases, removal of this water to ppm levels prior to the next processing step is economically desirable. PV has made major contributions to these applications and is now a preferred process option for dehydration of organic solvents. The used hydrophilic membranes are designed such that they have a high affinity for water, either by dipole-dipole interactions, hydrogen bonding or ion-dipole interactions (Bolto et al., 2011; Chapman et al. (a), 2008; Semenova et al., 1997). A vast amount of research has been performed on the development of polyvinyl alcohol (PVA) based membranes, due to PVA's pronounced abrasion resistance, elongation, tensile strength, flexibility, chemical stability and excellent water-permselective properties (Nagai, 2010; Shao and Huang, 2007). It is therefore not surprising that the first commercial hydrophilic polymeric membranes manufactured by SulzerChemtech (Switzerland), formerly known as GFT, were thin film composite membranes with a PVA top layer of typically 0.1 μm. For the inorganic membranes which can operate at much higher temperatures, such material monopoly does not exist at present. Tubular hydrophilic membranes based on NaA zeolites are available from, amongst others, Mitsui (Japan) and Fraunhofer IKTS (Germany), while microporous silica membranes can be purchased from Pervatech (the Netherlands). Recently, the Energy Centre of the Netherlands (ECN) developed a new hydrophilic membrane, HybSi®, which merges organic and inorganic fragments in a silica based amorphous material (Castricum et al., 2011). This novel modified silica membrane with superior stability under acidic and hydrothermal conditions is now also manufactured and marketed by Pervatech. As the water flux is strongly dependent on the temperature, the superior thermal stability of the inorganic membranes, allowing operation at higher temperatures, in many cases compensates for their higher cost. Feed water contents of 0.1 to 10 wt% can be economically treated with PV (Fleming and Slater, 1992). Greater amounts of water, however are best removed with extraction or other bulk techniques, while at lower concentrations, adsorption may become competitive. Generally, water levels down to about 20 ppm are attainable. However, to approach 1 ppm or less, much more membrane surface and in some cases reduced permeate pressures are required as the driving forces will become very low.

Dehydration of solvents is even more studied than water treatment. The main reason of such a long history should be the unfortunate priority of financial considerations over environmental considerations. Dehydrated solvent is commercially more valuable than the low purity one. Furthermore, water treatment and environmental aspects had almost no importance for the old industry. Another reason may be the abundance of hydrophilic polymers, mainly PVA and cellulose based polymers, in former times.

PV is a claimed technique for water recovery from organic solvents and organic mixtures. Hydrophilic membranes are used for solvent dehydration. In fact, by means of interactions between water and membrane material the permeance of water through the membrane is favoured. In many applications the removal of water from organic solvents is not possible using other common techniques such as distillation or liquid-liquid extraction. Providing the latent heat of water evaporation in distillation, for instance, requires a high energy demand. Dehydration of solvents that form an azeotrope with water is another challenge in distillation. Both these difficulties are less pronounced when the water amount is very low though. Azeotropic or extractive distillation and pressure swing adsorption are common solutions, however they are associated with other issues such as the recovery of the entrainer and potential contamination of the product streams.

PV can be used to break the azeotrope with high selectivity and low energy requirement. Dehydration of alcohols can be considered as the main field of application of PV in solvent dehydration, followed by dehydration of other organics such as ethylene glycol, THF and acetic acid (Chapman et al. (a), 2008). Ethylene glycol, for instance, is miscible with water in all proportions. It is produced by hydration of ethylene oxide obtained from an ethylene feed, and needs to be dehydrated to allow use as an antifreeze agent for example. The pervaporative dehydration of alcohols, particularly ethanol, has been extensively studied.

Ethanol is both well soluble in water and forms an azeotrope at 4.4 wt% water, as described above. In order to be useful as gasoline octane enhancer, the water content of bioethanol must be below 1 wt%. Therefore hybrid PV-distillation plants have become industrially common. Isopropanol (IPA) which is widely used in the electronics industry where its purity is crucial for the quality of the products, is another important solvent that is dehydrated by PV at industrial scale.

Many polymers have been studied during the years for the manufacturing of solvent dehydration membranes. Among these materials, especially PVA, chitosan and sodium alginate have been extensively investigated due to their pronounced water selectivity (Svang-Ariyaskul et al., 2006; Huang et al., 1999; Ghazali et al., 1997).

Membranes prepared with these materials, however, showed low mechanical resistance and poor chemical stability. Thus, in order to enhance their performance in solvent dehydration, different strategies such as crosslinking, blending and grafting were employed (Varghese et al., 2010).

PVA is one of the most studied polymers used, as already previously mentioned, in pervaporative dehydration due to its hydrophilic nature, chemical resistivity and wide availability. However, there are some problems associated to PVA due to its pronounced swelling in water, resulting in a loss of selectivity and stability. In order to overcome these problems, PVA was blended or crosslinked (Burshe et al., 1998). Vauclair et al. (1997) prepared blended PVA/poly(acrylic acid) (PAA) membranes for the dehydration of fusel oil, a by-product of bio-ethanol distillation containing five alcohols and water. They found an improvement in water permeation flux, mechanical stability and resistance in comparison to pure PVA membranes. Yuan et al. (2011) tested PVA coated poly(acrylonitrile) (PAN) hollow fibers in the dehydration of ethyl acetate while PVA/ polyethersulfone (PES) composite membranes were used for ethanol dehydration by Hong et al. (2011).

Chitosan is the deacetylated form of chitin and represents a very attractive polymer due to its abundance in nature, biocompatibility and hydrophilicity (Dogan and Hilmioglu, 2010). Chitosan mainly found application in the dehydration of water-alcohol mixtures. In its chemical structure, the reactive amino and hydroxyl groups can interact with water molecules by hydrogen bonds. Often, chitosan was blended with other polymers resulting in a more selective membrane. By blending two polymers, it is possible to combine the respective advantages of both polymers (Svang-Ariyaskul et al., 2006). Zhang et al. (2009), for instance, studied the pervaporative dehydration of ethyl acetate/ethanol/water mixtures using chitosan /poly (vinyl pyrrolidone) (PVP) membranes crosslinked by glutaraldehyde (GA).

With increasing PVP loading into the membrane matrix (from 0 to 25 wt%), a decrease in total flux was observed (from 1000 $kg/m^2.h$ to 700 $kg/m^2.h$). On the contrary, as expected, the separation factor increased (from 250 to 950) with increasing PVP content. This behaviour was explained by the fact that PVP caused a variation of the membrane density, thereby increasing the free volume and hence increasing the diffusion of all components in the feed mixture.

The role of the crosslinker GA was also evaluated. Increasing its concentration resulted in a reduction of free volume which determined a lower permeation flux and a higher selectivity.

Table 1. Literature data of solvent dehydration

Membrane material	Thickness (μm)	Feed mixture	Water feed concentration (wt%)	T. (°C)	Total flux (kg/m²h)	α	Reference
PVA-PEC (50 wt%)	55	Water/Isopropanol	10	60	1.35	1002	Zhao et al., 2009
PVA-PAN	-	Water/Ethyl acetate	2	40	0.0345	7270	Yuan et al., 2011
PVA (16,67 wt%)-PES	180	Water/Ethanol	10	80	0.29	86.1	Hong et al., 2011
PVA-SA-PS	1.8 (active layer)	Water/n-butanol	10	45	0.585	606	Dong et al., 2006
PVA-SA-PS	1.8 (active layer)	Water/Ethanol	10	45	0.384	384	Dong et al., 2006
PVA-SA-PS	1.8 (active layer)	Water/Isopropanol	10	45	0.414	1727	Dong et al., 2006
PVA-SA-PS	1.8 (active layer)	Water/tert-butanol	10	45	0.37	725	Dong et al., 2006
PVA-Iron oxide (10 wt%)	58	Water/Acetonitrile	13.64	35	0.03	406.14	Mandal et al., 2011
PAni	100	Water/Tetrahydrofuran	4	55	0.622	34	Chapman et al. (b),2008
CS/HEC-CA	110	Water/Ethanol	5	60	0.22	16606	Jiraratananon et al., 2002
CS-Nylon	70	Water/1,4-dioxane	18	40	0.075	900	Smitha et al., 2006
CS-H-ZSM-5	25	Water/Ethanol	10	80	0.23	152.82	Sun et al., 2008
CS-TiO_2 (6wt%)	-	Water/Ethanol	10	80	0.34	196	Yang et al., 2009
CS-PTFE	1.26 (PTFE layer)	Water/Isopropanol	30	70	1.73	775	Liu et al., 2006

Thin film composite chitosan/PTFE membranes were prepared by Liu et al. (2007) for the dehydration of isopropanol. In this case, a chitosan solution was cast on modified PTFE support. The main advantage of using composite membranes lies in the thinner dense top layer which allows to significantly increase the permeation flux by decreasing the permeation resistance. The membranes were tested on a 70 wt % isopropanol/water solution at different temperatures, and a flux of 1730 g/m^2.h and separation factor of 775 were observed.

Furthermore, the durability of the composite chitosan/PTFE membranes was evaluated by immersing the membrane in a 70 wt% aqueous isopropanol solution at 25°C. It was found that the chitosan/PTFE membrane maintained its PV separation ability for 45 days, while the pure chitosan membrane only showed a stable performance during two days.

The use of mixed matrix membranes, produced by incorporating inorganic filler materials in polymeric matrices, has been extensively reported in literature (Kickelbick, 2003). Mandal et al. (2011), for instance, applied iron oxide filled PVA membranes for the dehydration of aqueous acetonitrile solutions. The membranes were also benchmarked against the commercial, PVA based PERVAP 2202 membrane. The PVA-Fe oxide membranes showed higherthickness normalized permeation fluxes but lower selectivities then the commercial membrane.

Numerous studies have been carried out using alginate based membranes. The hydrophilicity of the polymer and its wide availability made it a good candidate for the production of PV membranes for alcohol dehydration (Kalyani et al., 2008; Huang et al., 1999). In order to increase the mechanical strength and stability of the polymer, sodium alginate has been blended with other polymers such as chitosan (Kanti et al., 2004), used as coating material on PAN sublayers (Wang et al. (b), 2001; Wang, 2000), or filled with zeolites (Bhat and Aminabhavi, 2007).

In Table 1 are reported the literature data of the main solvent dehydration applications by PV using different membrane materials. Inorganic membranes were also used for PV dehydration. They offer chemical and thermal stability without being prone to swelling. In particular, ceramic silica and NaA zeolite membranes have been widely used for alcohol and solvent dehydration (Verkerk et al., 2001, Van Gemert and Cuperus, 1995, Van Hoof et al., 2006; Yu et al., 2011).

5.2. Organic/Water Separations

Hydrophobic membranes rely on their ability to separate organic compounds from dilute aqueous mixtures. These membranes typically find application in waste water treatment to remove volatile organic compounds (VOCs), solvent recovery from washing waters of gas scrubbers, aroma recovery from plant extracts, partial dealcoholization of wine and other beverages, and *in situ* bio-alcohol recovery from fermentation broths. Hydrophobic PV membranes and processes only became commercially available around 1990. The currently available composite materials that have been developed for these applications usually have good chemical properties but limited separation capabilities.

At present no real breakthrough of hydrophobic PV in industry has been achieved due to the combination of a complicated process design, the relatively high investment costs, the lack of highly selective and permeable membranes, and the limited number of commercial membrane suppliers. The relatively low attainable separation factors in hydrophobic PV, as compared to pervaporative solvent dehydration, can be explained by the fact that, despite the often very high sorption selectivity of the membranes, water diffuses easily due to the simple fact that it is a smaller molecule than the targeted organics and thus experiences less sterical hindrance during diffusion. The overall permselectivity of hydrophobic PV is usually lower than in strictly sorption controlled processes, although it is still favourable compared to evaporation.

The increasing interest in the production of bio-alcohols as alternative for or additive to fossil fuels and as platform molecules for the synthesis of bio-chemicals, has led to a large number of papers on fermentation processes. Butanol, for instance, is an important commodity chemical used as a solvent and as a precursor for chemical synthesis. In addition, it holds tremendous promise as a second-generation biofuel (García et al., 2011; Vane, 2008). It has a considerably higher combustion value than ethanol and its chemical properties allow blending in fuels more readily than ethanol. The production of n-butanol with solvent-producing strains of *Clostridium*, commonly known as ABE (acetone-butanol-ethanol) fermentation, is one of the largest biotechnological processes ever developed. Due to the inability of the fermentation process to compete with the booming petrochemical industry, it was however almost completely discontinued in the 1950s and 1960s.

One of the main bottlenecks of fermentative bio-alcohol production is the product toxicity and associated cell inhibition, low product accumulation, low space-time-yields and high purification costs. Micro-organisms usually experience strong ethanol inhibition above 5-8 wt% while strong butanol inhibition occurs above about 1 wt% (Oudshoorn et al., 2009; Vane, 2005; Ezeji et al., 2004). Besides the optimization of the fermentation itself, e.g. by improving the strain stability and chemical/thermal resistance of the micro-organisms (either bacteria or yeasts), a facile and efficient recovery of the bio-alcohols from the broths would greatly enhance the product yield, and thus boost the overall techno-economical feasibility and industrial applicability of fermentative biofuel production processes. While most current industrial bio-alcohol plants rely on energy-intensive distillation techniques for the recovery of the produced alcohols after fermentation, membrane processes working at ambient conditions offer the possibility to remove the alcohols from the broth during the fermentation itself. Significant economic gains can be anticipated when the efficiency of the alcohol recovery step would be increased. Integration of a suitable *in-situ* alcohol removal technique with the fermentation is an attractive strategy to overcome this bottleneck.

Based on energy requirements of the product capture, PV has been identified as one of the most promising unit operations for *in situ* ethanol and butanol recovery (Oudshoorn et al., 2010; Oudshoorn et al., 2009; Vane, 2005). Several authors experimentally demonstrated that by coupling the fermenter to a hydrophobic PV unit, the inhibitory solvents can be continuously removed, thus avoiding that their concentrations become inhibitory, alleviating the product toxicity and leading to higher bio-alcohol yields (Van Hecke et al., 2012; Vane, 2005). Although many hydrophobic membrane materials have been studied for the recovery of ethanol and butanol from dilute fermentation broths or model mixtures, the materials with good alcohol-permselective properties are rather scarce and it must be recognized that their PV performances do not yet meet the demands of industrialization at this stage (Peng et al., 2011).

Although other polymeric materials have been considered, most developments have focused on the well-known silicone rubbers (Shao and Huang, 2007; Peng et al., 2003). Poly (dimethylsiloxane) (PDMS) based membranes have found wide use and are commercially available from SulzerChemtech, Pervatech, SepraTek (South Korea), Membrane Technology and Research, Inc (USA) and other suppliers. These membranes consist of a thin selective layer of about 1-2 μm thickness on top of a porous support, and can be provided in flat sheet, tubular and hollow fiber geometry.

Thanks to its pronounced hydrophobicity and convenient processability, PDMS has been known as the most representative membrane material for the removal of alcohols from aqueous solutions at low alcohol concentrations. For unmodified PDMS membranes, the permeate fluxes reported in literature are generally between 1 and 1000 $g.m^{-2}.h^{-1}$, while ethanol-water separation factors are below approximately 10. The butanol-water separation factors reported in literature range from approximately 40 to 60 (Peng et al., 2011; Vane, 2005). This wide range of performance values arises from a variety of factors, including the source of the polymer starting materials, the used method of film casting, the crosslink density, the thickness of the selective layer, the support material (if any) and the test conditions.

Much efforts have been devoted to searching for polymeric materials with better alcohol-water separation performance than PDMS, however only seldom materials with better separation performance have been reported. To enhance their PV performance, PDMS membranes have been modified in several ways, e.g. by optimizing the support layer, blending or copolymerization with other polymers or surface treatment (Peng et al., 2011; Vane, 2005). Besides PDMS, its chemically modified derivatives with reduced partial water flux such as poly(octylmethylsiloxane (POMS) (She and Hwang, 2006) and other hydrophobic, elastomeric materials such as polyether-polyamide block copolymers (PEBAX) (Le et al., 2011; Hao et al., 2009) have been studied.

Poly (1-trimethylsilyl-1-propine) (PTMSP), a hydrophobic glassy polymer with extremely high free volume fraction, has been recently employed to prepare alcohol-selective thin film composite membranes with very high fluxes and ethanol/water separation factors ranging from 9 to about 26 at 50°C, which is far beyond the correponding performance of PDMS membranes under similar conditions. Moreover, the flux of these supported PTMSP membranes could be further enhanced, up to about 10 $kg.m^{-2}.h^{-1}$, by incorporation of hydrophobic silica particles in the polymer matrix (Claes et al., 2012). Despite many unique and outstanding properties,and the significant attention the polymer received over the last two decades, PTMSP did so far not find a commercial application as a membrane material, mainly because of its tendency toward physical and/or chemical aging, resulting in a significant decline of flux and separation factor over time (Volkov and Kujawski, 2011; Vane, 2005). Performance instabilityof PTMSP membranes has been reported in literature during butanol recovery from real fermentation broths (Fadeev et al., 2003; Fadeev et al., 2000). This phenomenon was tentatively ascribed to internal fouling of fermentation by-products inside the free volume. To overcome this issue and increase the physical stability of the polymer, PTMSP has been modified in order to yield a more physically stable material, for instance by grafting fluoroalkyl groups (Nagase et al., 1991), chemical crosslinking (Kelman et al., 2008) or using different polymerisation catalysts (Volkov et al., 2009).

While no organic membrane has yet succeeded to challenge PDMS membranes as the benchmark hydrophobic PV membrane material, inorganic membranes based on hydrophobic zeolites have shown both considerably higher ethanol-water separation factors and fluxes than PDMS membranes (Wee et al., 2008; Vane, 2005;). Zeolites are hydrated aluminosilicates with uniform crystalline structures and molecular-sized pores. Small molecules can adsorb in and diffuse through these pores. Bacause of the small pore size, the diffusion rates of molecules with different sizes and adsorption strengths inside the zeolite crystals may differ. Separation is thus based on differences in adsorption and diffusion (Wee et al., 2008; Bowen et al., 2004). Compared to polymer membranes, zeolite membranes, prepared by depositing polycrystalline layers on porous supports, show higher fluxes and higher separation factors. Additionally, they have a low swelling tendency, high chemical resistance and thermal stability. Modernite framework inverted (MFI) zeolites, particularly silicalite-1 and ZSM-5, have been studied most frequently in the context of alcohol/water separations. Silicalite-1has been reported to deliver ethanol-water separation factors from 12 to about 100 with a typical value around 40 (Peng et al., 2011; Vane, 2005). Additionally, fluxes observed with silicalite membranes meet or exceed those of the thinnest PDMS membranes. Ikegami and co-workers, for instance, prepared silicalite-1 membranes by hydrothermal synthesis and reported ethanol-water separation factors of 60 in the separation of a 5 vol% ethanol/water mixture at 30°C, in combination with fluxes up to 680 $g.m^{-2}.h^{-1}$(Ikegami et al., 1999). Due to selective sorption of alcohols, the ethanol concentration in the permeate was found higher than the vapour-liquid equilibrium, irrespective of the feed composition (Sano et al., 1994). Despite many efforts to produce silicalite based modules on a larger scale, defect-free, commercial scale manufacturing of zeolite membranes remains difficult. Nevertheless, Fraunhofer IKTS recently released a tubular ZSM-5 zeolite membrane, being the first hydrophobic inorganic PV membrane on the market.

In an attempt to overcome the drawbacks of both polymeric and inorganic materials and exploit their synergies, many authors have been studying mixed matrix membranes in which inorganic nanofillers are dispersed in a polymer matrix. Since such membranes would be fabricated in almost the same way as standard polymeric membranes, the cost per unit area of mixed matrix membrane modules should also be similar, and several times lower than that of zeolite modules. The inorganic filler particles can be either nonporous as is for instance the case with the PTMSP-silica membranes discussed above, or porous hydrophobic zeolite fillers may be used. Since it was first demonstrated that incorporation of hydrophobic zeolites can significantly enhance the ethanol-permselectivity (TeHenneppe et al., 1987), zeolite-filled PDMS membranes have gained increasing attention in literature. Separation factors in the range 7 - 59 have been reported using silicalite-1, the most hydrophobic zeolite, as a filler (Peng et al., 2011; Vane, 2005). This range overlaps with the ranges reported for PDMS and silicalite-1 membranes. The PV performance of these PDMS-silicalite-1 mixed matrix membranes is highly dependent on the preparation conditions, such as the siloxane chain length, the concentration of the crosslinking agent, the density of the reactive groups, the solvent type, the zeolite type and loading, the mixing method and the support type. It is clear that the fillers should have nanodimensions and be well dispersed, for instance by sonication, to permit defect free top layers. Moreover, the surface chemistry of the particles needs to be altered to ensure that they are embedded well in the PDMS matrix. Zeolite particle loadings as high as 77 wt% have been reported, but some systems may be limited to a maximum of 30 wt% or lower due to particle agglomeration and defect formation (Peng et al., 2011). To enhance the practicality of zeolite membranes for *in situ* alcohol recovery, the deterioration in PV performance caused by accumulation of by-products during fermentation must be resolved. To counteract this decline in flux and separation factor, Ikegami and co-workers have coated the top layer of a silicalite membrane with silicone rubber, thereby protecting the zeolite surface and "filling" defects, such as non-zeolitic pores, in the silicalite-1 layer

(Ikegami et al., 2004; Ikegami et al., 2003). Stable PV performances were obtained by entirely covering the silicalite membrane, but the flux was strongly limited by the thickness of the silicone rubber sheet. Overall, the separation factors of the PDMS-coated silicalite-1 membranes ranged from 47 to 125 (Ikegami et al., 2003).

Another potential technique for the removal of alcohols from dilute aqueous solutions is the use of liquid PV membranes. Such membranes are commonly prepared by immobilization of a liquid with high boiling point in the pores of a hydrophobic substrate so as to form a nanoporous coating on the membrane surface in contact with the broth. Thongsukmak and Sirkar (2007) studied the removal of ethanol using polypropylene hollow fibers with trioctylamine as the immobilized liquid and having a nanoporous fluorosilicone coating.

The reported butanol-water and ethanol-water separation factors were very high, up to respectively 275 and 80 at 54°C, but fluxes were very low due to the relatively thick trioctylamine layer in the porous substrate wall of the coated fibers. Izák and co-workers immobilized the ionic liquid tetrapropylammoniumtetracyanoborate in a relatively open PDMS membrane (pore size of 60 nm) and demonstrated the stable and efficient continuous removal of butanol from an Acetone butanol ethanol (ABE) broth using a solvent producing *Clostridium acetobutylicum* culture (Izák et al., 2008).

A comparison of the literature studies on pervaporative recovery of bio-alcohols from fermentation broths is difficult due to the large number of variables such as the type, strain and loading of the used micro-organisms, the source, pretreatment and concentrations of the substrate and nutrients, the temperature and pH and their respective history, the oxygen concentration, the fermenter design and level of mixing, and the membrane type, permeate pressure and condensation strategy. Despite this complexity, it is a general observation that the fermenter productivity increases upon application of hydrophobic PV if the micro-organism used exhibits inhibition by the solvents produced (Vane, 2005). Moreover, integration of PV with the fermenter enables the use of more concentrated substrate solutions, further increasing the productivity of the fermenter (Van Hecke et al., 2012).

Pervaporative alcohol removal is understandably most cost-effective at the maximum possible feed temperature because of the positive effect of temperature on the driving force. Therefore, if heat-sensitive compounds such as mesophilic micro-organisms are present in the feed, the PV system will either have to be operated at a temperature below the operational optimum, the sensitive compounds will need to be removed prior to exposing the broth to the PV membranes, or a certain loss of the sensitive compounds will need to be accepted (Vane, 2005). In most biofuel applications, sensitive particulate materials can be removed with either a filtration device, for instance microfiltration, or a centrifugation unit. The presence of a solid-liquid separation device before the PV unit not only removes temperature-sensitive broth components, thereby enabling PV at elevated temperatures, it also protects the PV modules from fouling by solids.

The recovered biosolids can obviously be returned to the fermenter to increase its productivity by increasing the concentration of viable cells. An economic trade-off obviously exists between on the one hand the increased efficiency of the PV process by enabling higher temperature operation combined with the ensuing, increased fermenter productivity, and on the other hand the increased complexity of the system by addition of a pre-PV separation device. Alternative to the introduction of a separation device for the protection and removal of heat-sensitive micro-organisms, thermotolerant or thermophilic cultures may be used.

These micro-organisms could be exposed to elevated temperatures in a PV system and be recycled to the fermenter with minor loss in viability (Vane, 2005).

Literature contains several contradictory observations regarding the impact, sometimes positive but in most cases negative, of fermentation broth components on the performance of a variety of PV membranes and modules (Vane, 2005).

Only a few of these studies attempted a systematic study to unravel the mechanism of this performance decline and identify the problematic broth constituents. However, it is clear that the membrane performance can be significantly affected by fermentation by-products and other broth components, even in small concentrations, which limits the commercial applicability of PV for *in situ* bio-alcohol recovery.

The impact of the different broth constituents on the performance of the membrane is likely to be dependent on the temperature as sorption, diffusion, evaporation and precipitation are all temperature dependent phenomena (Vane, 2005).

Organic acids such as acetic acid, butyric acid and succinic acid have been identified as negatively impacting the performance of silicalite-1 membranes (Ikegami et al., 2003). This might be ascribed to a competitive, but reversible sorption process in which the organic acids are preferentially sorbed relative to the alcohols, thereby blocking the alcohol transport and making the membrane more hydrophilic. However, an irreversible interaction between the acids and reactive sites in the zeolite may also occur.

Sugars and salts were found to increase the membrane flux and separation factor due to their effect on the thermodynamic behavior of alcohols (Lipnizki et al., 2004). Simply adding more salt or sugar to the broth will increase the activity coefficient of the alcohols such that they prefer to partition out of the water phase, into the membrane (salting-out effect). Gaykawad et al. (2012) recently reported the effect of yeast cellular polymers which might be released in the broth upon cell lysis, on the performance of commercial PDMS and POMS membranes. Among the different yeast-originating polymeric compounds (proteins, lipids, phospholipids, RNA and polysaccharides), lipids were identified as the most detrimental components, causing a 50% and 33% flux decline in PDMS and POMS membranes, respectively. This fouling was irreversible and tentatively ascribed to hydrophobic interactions between the lipids and the membrane material, leading to strong lipid adsorption on the membrane surface. This confirms the observations of Offeman and Ludvik (2011) and of Fadeev et al. (2000) who reported the detrimental effect of fatty acids on the performance of ZSM-5 filled PDMSand PTMSP membranes, respectively. In case of PTMSP membranes, internal fouling was suggested as the major cause of the decline in flux and separation factor over time (Volkov and Kujawski, 2011; Fadeev et al., 2003). It was found that the free volume cavities of a fouled PTMSP membrane were occupied with highly sorbing and low volatility fermentation by-products such as glycerol and butanediol (Fadeev et al., 2003). Given the uncertainties and lack of concrete explanations for the observed performance decline, additional experimental studies are needed to further elucidate the impact of broth components on the performance of PV membranes. As foulants are being identified and the phenomena being understood, engineering controls and cleaning protocols can be developed to minimize their impact.

A summary of the main results of ethanol/water separation with different membrane materials are reported in Table 2.

Table 2. Literature data of ethanol/water separation

Membrane material	Membrane type	Thickness (μm)	Ethanol feed concentration (wt%)	T. (°C)	Total flux (kg/m².h)	$\alpha_{EtOH/water}$	Reference
PDMS	TFC	5	1,5	66	-	14.0	Mori and Inaba, 1990
PDMS	TFC, supported liquid membrane	25	4	25	-	8.8-12.6	Kashiwagaet al., 1988
PDMS	TFC, MTR PDMS membrane	20	6	25	-	7.7	Schmidt et al., 1997
PDMS	unsupported	120	1.5	66	-	10.4	Mori and Inaba, 1990
PDMS	unsupported, GE 615 PDMS	200-400	6	35	-	9.0	Moermans et al., 2000
PDMS	unsupported	160	16.5	40		8	Takegami et al., 1992
PDMS	TFC, Pervatech PDMS	2	5	50	2,6	6.7	Claes et al., 2012
PDMS-silicalite-1	TFC, PERVAP 4060 (Sulzer)	2	5	50	1,3	7.1	Claes et al., 2012
PDMS-silicalite-1 (77%)	unsupported, mixed matrix	-	7	22	-	59	Jia et al., 1992.
PDMS-silicalite-1 (65%)	unsupported, mixed matrix	-	5	50	-	43.1	Vane et al., 2008
PDMS-ZSM-5 (50%)	TFC, mixed matrix	-	-	30	0.460 L/m².h	14	Adnajevid et al., 1997
PDMS-USY (50%)	TFC, mixed matrix	-	-	30	0.610 L/m².h	16.1	Adnajevid et al., 1997
PTMSP-silica (25%)	TFC	2.8	5	50	9.5	18.3	Claes et al., 2012
PTMSP	TFC	-	5	50	0.28	24	Claes et al., 2012
PTMSP	unsupported	14-20	6	30	0.3	20	Fadeev et al., 2003
PTMSP	unsupported	14-25	6	30	0.3-0.5	15.1-19.9	Volkov et al., 2004
PTMSP	unsupported	20-70	3-6	25	0.08-0.15	21-22	Schmidt et al., 1997
PTMSP	unsupported	15-175	10	75	0.15-1.05	4-11	González-Marcos et al., 2004
PTMSP/PDMS	unsupported	32	5	30	0.02-0.8	1-18	Kang et al., 1994
Silicalite-1	Zeolite	-	5	60	-	up to 106	Lin et al., 2001
Silicalite-1	Zeolite	-	4	30-60	-	60	Sano et al., 1994
Silicalite-1 with PDMS coating	TFC	-	4	30	-	47-125	Matsuda et al., 2002

5.3. Organic-Organic Separations

The last category of PV membranes, the so-called target-organophilic membranes, is applied in the separation of mixtures of two or more organic liquids. Since decades, the industrial separation of organic compounds is traditionally performed using distillation processes. Distillation is a robust technology that is well suited and adapted for such applications so that it is rather difficult to imagine an empirical substituting method. However, the high energy consumption is considered as an inherent disadvantage of distillation. Several successful improvements and optimizations have been made in order to make distillation more efficient, particularly as customizations to special cases. But the inherently high energy consumption and the related environmental considerations are instigating more and more companies to consider non-thermal separation technologies as an alternative for or complementary technique to distillation. Among the different membrane processes, PV is usually considered as the most appropriate one for organic separations. Although there are some industrial PV based organic separation processes operational for years, it is not actually a competitive process for distillation by now. Although the separation of organic solvent mixtures has been widely investigated in the early years of PV research almost half a century ago, and despite the huge industrial potential, it still represents the least developed field of PV. Historically, the earliest interest in PV was focused on the separation of benzene/cyclohexane mixtures (and James, 1958). These two solvents form an azeotrope at approximately 50% benzene, a mixture that cannot be separated by a conventional distillation column. However, PV allows to further enrich this mixture, thereby producing a vaporous permeate containing more than 95% benzene.

This example illustrates the advantages of PV over simple distillation for separating azeotropes and close boiling mixtures. A large number of studies have been conducted on the pervaporative separation of benzene/cyclohexane mixtures, as reviewed by Villaluenga and Tabe-Mohammadi (2000). The lack of membranes and modules able to withstand aggressive solvent mixtures under the operating conditions of PV is the key problem hindering the application of PV to organic mixtures. Commercial target-organophilic PV membranes are scarces as the separation layer of the membranes needs to be tuned for almost each distinct application. Nevertheless, several commercial PV plants have been developed and installed. Sulzer Chemtech reported two cases where organo-selective membranes were used in industrial applications, i.e. the separation of methanol from a mixture with methyl-*tert*-butyl ether and the separation of ethanol from a mixture with ethyl-*tert*-butyl ether. In Germany, PolyAn GmbH supplies molecular surface engineered (MSE) polymeric membranes, which can be tailor-made for specific organic-organic separations. MSE is a technique in which the pores of a porous membrane are "filled" with a functional polymer, rendering nonporous target-organophilic PV membranes.

Pervaporative organic/organic separations can be tentatively classified as shown in Figure 5: *i)* separation of polar/non-polar solvent mixtures, *ii)* separation of aromatic/aliphatic and aromatic/alicyclic mixtures, *iii)* separation of aliphatic hydrocarbons, and *iv)* separation of isomers. In Table 3 a list of the main literature data on different types organic/organic separation by PV is reported.

5.3.1. Separation of Polar/Non-Polar Solvent Mixtures

Interactions between polymeric membranes and penetrants play an important role in pervaporative liquid-liquid separations with nonporous polymeric membranes. If there is a large difference in chemical nature between the mixture components, a satisfactory separation will usually be obtained by using polymeric membranes which interact preferentially with one of the mixture components, preferably the minor fraction. This is the case with polar/non-polar solvent mixtures. A number of investigations have been reported in literature were PV was mainly applied in the separation of mixtures containing aliphatic alcohols and hydrocarbons.

In these studies, it was observed that generally polar molecules permeate faster through polar membranes than non-polar molecules, and vice versa. Polar polymer membranes such as those based on PVA, Nafion and cellulose acetate were found to be selectively permeable for alcohols over hydrocarbons. Aliphatic and aromatic hydrocarbons, on the other hand, always permeated preferentially with respect to alcohols when more hydrophobic membranes, based on for instance polyethylene, polypropylene and natural rubber, were used.

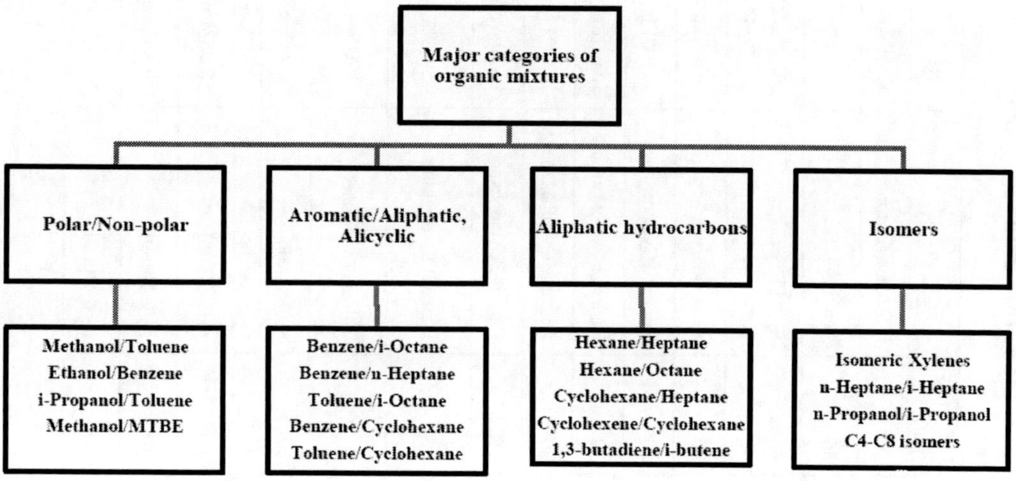

Figure 5. Classification of organic-organic PV separation. (Adapted from Smitha et al., 2004 and Park, 1993).

Some of the most frequently studied pervaporative polar/non-polar separations are methanol/aliphatic hydrocarbon, methanol/aromatic hydrocarbon (Zhou et al., 1996), methanol/methyl tert-butyl ether (Zereshki et al.(a), 2011), methanol/dimethyl carbonate, ethanol/aliphatic hydrocarbon, ethanol/aromatic hydrocarbon, isopropanol/hydrocarbon, butanol/toluene, aromatic alcohol/benzene, acetic acid/aromatic hydrocarbon, acetone/hexane (Kumar et al., 2012) andaniline/benzene.

5.3.2. Separation of Aromatic/Aliphatic and Aromatic/Alicyclic Mixtures

A major class of organic-organic separations by PV includes the separation of aromatics from other hydrocarbons. While separation is certainly challenging, the existence of the electron clouds on the aromatics gives them a rather hydrophilic nature compared to the hydrophobic hydrocarbons, thus offering a potential key for separation.

Table 3. Literature data of organic/organic separation

	Membrane material	Thickness (μm)	Separation	Feed Composition	T (°C)	Total Flux (kg/m² h)	α	References
Polar/ Non Polar	Polyethylene/ ethylene propylene dyeneterpolymer (50 wt%)	200	n-Hexane/Acetone	50 wt% hexane	28	1.2	1.47	Kumar et al., 2011
	Poly(ether ether ketone) (PEEKWC)	35-40	MeOH/MTBE	14.3 wt% MeOH	30	0.2	14	Zereshki et al., 2011 (a)
	Poly lactic acid	9	EtOH/ETBE	20 wt% EtOH	30	0.05	7	Zereshki et al., 2011 (b)
	Polypyrrole/ hexafluorophosphate	-	MeOH/Toluene	20 wt% MeOH	57.5	0.4	200	Zhou et al., 1996
Aromatic/ Alicyclic	BAPPH+ aromatic diacids	20	Benzene/Cyclohexane	50 wt% benzene	50	1.47	4	Wang et al. (a), 2001
	BAPP+ aromatic diacids	20	Benzene/Cyclohexane	50 wt% benzene	50	0.817	4.9	Wang et al. (a), 2001
	Polyvinyl chloride (PVC) + H-β-zeolite	30±5	Benzene/Cyclohexane	50 wt% benzene	30	0.0232	12,06	Zhang et al., 2008
	Polyvinyl chloride (PVC) + Rh/H-β-zeolite	30±5	Benzene/Cyclohexane	50 wt% benzene	30	0.0388	26.44	Zhang et al., 2008
	PEBAX (poly(ether-block-amide) 4033	150	Benzene/Cyclohexane	10-90 wt% benzene	40	0.8-3.3	3.2-1.5	Yildirim et al., 2008
Aromatic/ Aliphatic	Poly(vinyl acetal)	35	Benzene/Isooctane	42 wt% benzene	35	0.078	25	Mandal and Bhattacharya, 2008
	Polybenzimidazole (PBI) (10 wt%)+ Matrimid® 5218 (hollow fiber configuration)	-	Toluene/Isooctane	50 wt% Toluene	60	1.35	200	Kung et al., 2010
	Polyacrilic acid + α-cyclodextrin (50 wt%)	30	*o*-/*p*-xylene	50 wt% *p*-xylene	25	0.006	1.5	Zusumocahyo et al., 2004
	Palygorskite (2 wt%)/polyacrylamide	20	*m*-/*p*-xylene	10 wt% *m*-xylene	25	0.035	1.39	Yang et al., 2006
Isomers	Polyamide–imide + α- cyclodextrin	10	n-/t- ButOH	50 wt% n-ButOH	60	0.003	1.34	Wang et al., 2009
	Polyamide–imide + β- cyclodextrin	10	n-/t- ButOH	50 wt% n-ButOH	60	0.01	1.30	Wang et al., 2009
	Polyamide–imide + γ- cyclodextrin	10	n-/t- ButOH	50 wt% n-ButOH	60	0.022	1.13	Wang et al., 2009

The mixture of benzene with cyclohexane is a representative of this class of liquid mixtures. The first studies on pervaporative separations involving aromatic-containing mixtures were performed about 50 years ago using modified cellulose ester membranes. In most cases, the aromatic hydrocarbon permeated faster than the aliphatic hydrocarbon, but low selectivities were usually obtained. The separation of aromatic/aliphatic hydrocarbon mixtures by PV has been more intensively studied than the pervaporative separation of aromatic/alicyclic mixtures which started about 35 years ago and to which less research efforts has been devoted meanwhile.

Some of the main studied aromatic/aliphatic and aromatic/alicyclic separations are benzene/cyclohexane (Wang et al. (a), 2001, Zhang et al., 2008, Yildirim et al., 2008), benzene/other paraffins, benzene/olefins, toluene/paraffins and ethyl benzene/ethyl cyclohexane.

5.3.3. Separation of Aliphatic Hydrocarbons

This type of mixture consists of molecules being very similar in chemical nature. For instance, in the case of a hexane/heptane mixture there is no apparent difference in the nature and strength of the interactions between a membrane and the individual permeants. Therefore, in such cases permselectivity is largely governed by the steric effects of the permeants, i.e. diffusional cross-section and molecular length. In the pervaporative separation of binary mixtures of paraffins, usually the component with the smaller molecular weight permeates faster than the other one because of a smaller diffusional cross-section. In binary mixtures of olefins with paraffins, however, there is a relatively large difference in chemical nature of both components. Therefore good separation results may be expected if there is any specifically designed membrane material. PolyAn GmbH, for instance, manufactures olefin-selective target-organophilic PV membranes.

Some of the studied aliphatic hydrocarbon separations are hexane/heptane, hexane /octane, hexane/neohexane, cyclohexane/heptane, cyclohexane/ neohexane, heptane/iso-octane, hexane/iso-octane, 3-methyl-pentane/2,3-dimethyl-pentane, methyl cyclohexane/iso-octane, hexene/hexane, heptene/heptane, cyclohexene/cyclohexane, 1,3-butadiene/iso-butene and 1,3-butadiene/trans-2-butene.

5.3.4. Separation of Aromatic Isomers

This class of mixtures is one of the most difficult to separate. The industrial separation of C_8-aromatics (*o*-xylene, *m*-xylene, *p*-xylene and ethyl benzene) is now performed by super-fractionation, fractional crystallization, adsorption, or a combination of these methods. These processes are very complex and, in general, energy intensive. Therefore the separation of such mixtures by PV has attracted much interest. Since the initial studies almost 30 years ago with cellulose ester based membranes which provided considerably good fluxes but a low selectivity, many more investigations have been performed. Most of the studies have reported a very poor separation, because of the very similar size, shape and chemical nature of the aromatic isomers. Although xylene isomers have almost the same molecular volume, all tested membranes except those polymers containing specific functional groups, showed the following order of selective permeation: *p*-xylene >*m*-xylene >*o*-xylene. This is consistent with the order of decreasing diffusional cross-section area of these isomers. Some of the main studied isomer separations include C8-isomers (mixtures of *o*-, *m*-, and *p*-xylene and ethyl benzene) (Zusumocahyo et al., 2004, Yang et al., 2006), dichlorobenzene isomers, nitrochlorobenzene isomers, nitrotoluene isomers, benzene/toluene, toluene/o-xylene and styrene/ethyl benzene. The separation of aromatic isomers continues to be an active research area since the present separation methods in use are both complex and energy intensive.

CONCLUSION AND OUTLOOK

PV is an attractive separation technology with significant potential to efficiently separate liquid-liquid mixtures, mainly in the chemical, pharmaceutical and food industries. Driven by fluctuating energy prices, more stringent environmental legislation and the request for more selective, mild separation techniques, more and more companies are currently rethinking their processes and considering alternative, membrane based processes. The potential applications of PV range from the well established dehydration of organic solvents to the removal of organics from dilute aqueous bulk streams and challenging organic-organic steams. Besides conventional applications in the downstream purification of chemicals, for instance the dehydration of ethylene glycol, or as an end-of-pipe technology, for instance in waste water treatment, pervaporationis also expected to play a significant role in future biorefineries, for instance in the removal of inhibitory compounds from fermentation broths or the recovery of high added value bio-chemicals from liquid biomass derived streams.

In many cases, PV has to compete with conventional, robust and well-established liquid separation processes like distillation, liquid-liquid extraction and adsorption, the cost of which can be readily calculated. PV is a newer technology that is much less known in industry, and there are often doubts and prejudices about its capabilities and about the investment and operational costs of a PV unit. Despite its clear advantages and demonstrated capabilities in for instance solvent dewatering, in many cases PV as a stand-alone process may not supply products suitable for further processing or waste disposal in accordance with environmental standards, and therefore not make sense from an economical point of view. The main bottlenecks are the relatively large investment costs associated with the PV unit and the membranes, the sometimes relatively low separation efficiencies and doubts about the membrane lifetime. Hybrid processes are regarded as a means of overcoming these limitations. In an attempt to combine the advantages of robust standard thermal separation processes and an alternative, energy-efficient separation technology, hybrid distillation-PV processes have been studied and implemented, as reviewed by Lipnizki and co-workers (Lipnizki et al., 1999). By integrating both separation techniques in a hybrid process, solvent mixtures which cannot be separated by conventional distillation because of the formation of azeotropes and are therefore simply incinerated instead, could be effectively separated, allowing recovery and purification of the solvents to the targeted quality in an energy-efficient, thus competitive way. Therefore, to make PV more applicable in the industrial practice, especially in conservative environments where distillation has acquired a dominant position, more efforts should be devoted to the integration of PV units in existing separation processes. Furthermore, from a membrane point of view, there exists a clear need to develop new membranes demonstrating high fluxes and selectivities. As the number of PV membrane materials is limited and most materials show important drawbacks, great continuous efforts must be made to develop new-film forming materials or to modify available materials to obtain better membranes with both higher fluxes and separation factors. Besides enhanced separation efficiencies, many new studies focus on novel membranes having the requirements of high chemical resistance, respectable sorption capacity and good mechanical strength. In this respect, integration of nano-fillers into traditional or novel polymers offers a particularly attractive route to prepare PV membranes with improved properties.

APPENDIX: NOMENCLATURE

Symbols

- *a* Activity (-)
- C Concentration (wt%)
- D Diffusivity (cm^2/s)
- J Flux (Kg/m^2h)
- *l* Permeate side interface (-)
- L Diffusion coefficient (m^2/sec)
- n_i Mole fraction (-)
- N numbers of moles (-)
- p Pressure (bar)
- P Permeability (m^3/ m h bar)
- R Universal gas constant (J/mol K)
- S Solubility (cm^3 (STP)/cm^3atm)
- T Absolute temperature (K)
- ν Partial molar volume (m^3/mol)

Greek Letters

- α Selectivity (-)
- β Enrichment factor (-)
- γ Activity coefficient (-)
- μ Chemical potential (J/mol)
- θ Membrane thickness (μm)

Subscripts

- *0* Liquid feed phase at point 0
- $0_{(m)}$ Liquid in the membrane at point 0
- *i* component *i*
- *j* component j

Superscripts

- F Feed
- G Gas
- L Liquid
- P Permeate

Acronyms

ABE	Acetone-butanol-ethanol
GA	Glutaraldehyde
MFI	Modernite Framework Inverted
MSE	Molecular Surface Engineered
PEC	Polyelectrolite complex
PV	Pervaporation
TFC	Thin Film Composite
THF	Tetrahydrofuran
VOC	Volatile Organic Compounds

Abbreviations for Polymers

BAPPH	2,2-bis(4-(4-aminophenoxy)phenyl)hexafluoropropane
BAPPP	2,2-bis(4-(4-aminophenoxy)phenyl)propane
CA	Cellulose acetate
CS	Chitosan
HEC	Hydroxycellulose
MATRIMID	(3, 3', 4, 4'- benzophenonetetracarboxylicdianhydride and diaminophenylindane)
PAA	Polyacrylic acid
PAN	Poly (acrylonitrile)
PAni	Polyaniline
PDMS	Poly (dimethylsiloxane)
PEBAX	Polyether-polyamide block copolymers PVA: Poly (vinyl alcohol)
PES	Polyethersulfone
POMS	Poly (octylmethylsiloxane)
PS	Polysulfone
PTMSP	Poly(1-trimethylsilyl-1-propyne)
PVDF	Polyvinylidenefluoride
PVP	Polyvinylpyrrolidone
SA	Sodium alginate

REFERENCES

Adnadjevid D., Jovanovid D., Gajinov S., Effect of different physicochemical properties of hydrophobic zeolites on the pervaporation properties of PDMS-membranes, *J. Membr. Sci.* 136 (1997) 173-179.

Baker R. W., *Membrane Technologu and applications,* 2nd edition (2004) Chapter 9, p.370, John Wiley and Sons, Ltd

Baker, A.W.. Wijmans, J.G.; Huang, Y., Permeability, permeance and selectivity: A preferred way of reporting pervaporation performance data, *J. Membr. Sci.* 348 (2010) 346-352.

Bhat S.D., Aminabhavi T.M., Zeolite K-LTL-loaded sodium alginate mixed matrix membranes for pervaporation dehydration of aqueous–organic misture, *J. Membr. Sci.*306 (2007) 173-185.

Binning R.C., Lee R.J., Jennings J.F., Martin E.C., Separation of liquid mixtures by permeation, *ACS Div.Petrol. Chem.Prepr.* 3 (1958) 131-141.

Binning R.C., James F.E., Permeation: A new way to separate mixtures, Oil Gas J. 56 (1958) 104-105.

Bolto B., Hoang M., Xie Z., A review of membrane selection for the dehydration of aqueous ethanol bypervaporation, *Chem. Eng. Process. Process. Intens.* 50 (2011) 227-235.

Bowen T.C., Noble R.D., Falconer J.L., Fundamentals and applications of pervaporation through zeolite membranes, *J. Membr. Sci.*245 (2004) 1-33.

Burshe M.C., Sawant S.B., Joshi J.B., Pangarkar V.G., Dehydration of ethylene glycol by pervaporation using hydrophilic IPNs of PVA, PAA and PAAM membranes, *Sep. Purif. Technol.* 13 (1998) 47-56.

Brüschke H.E.A., State-of-art of pervaporation processes in the chemical industry, In: Nunes S.P., Peinemann K.V. (Eds.), *Membrane Technology in the Chemical Industry*, Wiley, Weinheim, Germany, 2003, pp. 127-172.

Castricum H.L., Paradis G.G., Mittelmeijer-Hazeleger M.C., Kreiter R., Vente J.F., ten Elshof J.E., Tailoring the separation behavior of hybrid organosilica membranes by adjusting the structure of the organic bridging group, Adv.Funct. Mater. 21 (2011) 2319-2329.

Chapman P.D., Oliveira T., Livingston A.G., Li K., Membranes for the dehydration of solvents bypervaporation, *J. Membr.Sci.* 318 (2008) 5-37 (a).

Chapman P., Loh X.X, Livingston A.G., Li K., Oliveira T.A.C., Polyaniline membranes for the dehydration of tetrahydrofuran by pervaporation, *J. Membr. Sci.* 309 (2008) 102–111 (b).

Claes S., Vandezande P., Mullens S., De Sitter K., Peeters R., Van Bael K., Preparation and benchmarking of thin film supported PTMSP-silica pervaporation membranes, *J. Membr. Sci.* 389 (2012) 265-271.

Dogan H., Hilmioglu N.D., Chitosan coated zeolite filled regenerated cellulose membrane for dehydration of ethylene glycol/water mixtures by pervaporation, *Desalination* 258 (2010) 120-127.

Dong Y.Q., Zhang L., Shen J.N., Song M.Y., Chen H.L., Preparation of poly(vinyl alcohol)-sodium alginate hollow-fiber composite membranes and pervaporation dehydration characterization of aqueous alcohol misture, *Desalination* 193 (2006) 202–210.

Ezeji T.C., Qureshi N., Blaschek H.P., Butanol fermentation research: Upstream and downstream manipulations, *The Chem. Rec.* 4 (2004) 305-314.

Fadeev A.G., Meagher M.M., Kelley S.S., Volkov V.V., Fouling of poly(-1-(trimethylsilyl)-1-propyne) membranes in pervaporative recovery of butanol from aqueous solutions of ABE fermentation broth, *J. Membr. Sci.* 173 (2000) 133-144.

Fadeev A.G., Kelley S.S., McMillan J.D., Selinskaya Y.A., Khotimsky V.S., Volkov V.V., Effect of yeast fermentation by-products on poly(1-(trimethylsilyl)-1-propyne) pervaporative performance, *J. Membr. Sci.* 214 (2003) 229-238.

Farber L., Applications of pervaporation, *Science* 82 (1935) 158.

Fleming H.L., Slater C.S., *Pervaporation, In: Ho W.S.W.,Sirkar K.K. (Eds.), Membrane Handbook,* Van Nostrand Reinhold, New York, USA, 1992, pp. 103-159.

García V., Päkkilä J., Ojamo H., Muurinen E., Keiski R.L., Challenges in biobutanol production: How to improve the efficiency?, *Renew.Sust. Energy Rev.* 15 (2011) 964-980.

Gaykawad S.S., Van der Wielen L.A.M., Straathof A.J.J., Effects of yeast-originating polymeric compounds on ethanol pervaporation, *Biores. Technol.* 116 (2012) 9-14.

Ghazali M., Nawawi M., Huang R.Y.M., Pervaporation dehydration of isopropanol with chitosan membranes, *J. Membr. Sci.* 124 (1997) 53-62.

González-Marcos, J.A., López-Dehesa C., González-Velasco J.R., Effect of operation conditions in the pervaporation of ethanol-water mixtures with poly(1-trimethylsilyl-1-propyne) membranes, *J. Appl. Polym. Sci.* 94 (2004) 1395-1403.

Graham T., On the absorption and dialytic separation of gases by colloid septa. Action of a septum oncaoutchouc, *Philos. Mag.* 32 (1866) 401-420.

Hadler A.B., Ott L.S., Bruno T.J., Study of azeotropic mixtures with the advanced distillation curveapproach, *Fluid Phase Equilib.* 281 (2009) 49-59.

Hao X., Pritzker M., Feng X., Use of pervaporation for the separation of phenol from dilute aqueous solutions, *J. Membr. Sci.* 335 (2009) 96-102.

Heissler E.G., Hunter A.S., Sciliano J., Treadway R.M., Solute and temperature effects in the pervaporation of aqueous alcoholic solutions, *Science* 124 (1956) 77-79.

Hitchens L., Vane L.M., Alvarez F.R., VOC removal from water and surfactant solutions by pervaporation: a pilot study, *Sep. Purif. Technol.* 24 (2001) 67-84.

Hong W., Xianshi L., Mingcheng N., Ben L., Zhingyi J., Integral PVA-PES composite membranes by surface segregation method for pervaporation dehydration of ethanol, *Chinese J. Chem. Eng.* 19 (2011) 855-862.

Huang R.Y.M., Pal R., Moon G.Y., Characteristics of sodium alginate membranes for the pervaporation dehydration of ethanol-water and isopropanol-water mixtures, *J. Membr. Sci.*160 (1999) 101-113.

Ikegami T., Yanagishita H., Kitamoto D., Haraya K., Nakane T., Masuda H., Koura N., Sano T., Highly concentrated aqueous ethanol solutions by pervaporation using silicalite membrane – Improvement of ethanol selectivity by addition of sugars to ethanol solution, *Biotechnol.Lett.* 21 (1999) 1037-1041.

Ikegami T., Kitamoto D., Negishi H., Haraya K., Matsuda H., Nitanai Y., Koura N., Sano T., Yanagishita H., Drastic improvement of bioethanol recovery using a pervaporation separation technique employing a silicone-rubber-coated silicalite membrane, *J. Chem. Technol.Biotechnol.* 78 (2003) 1006-1010.

Ikegami T., Kitamoto D., Negishi H., Iwakabe K., Imura T., Sano T., Haraya V, Yanagishita H., Reliable production of highly concentrated bioethanol by a conjunction of pervaporation using a silicone rubber sheet-covered silicalite membrane with adsorption process, *J. Chem. Technol.Biotechnol.* 79 (2004) 896-901.

Izák P., Schwarz K., Ruth W., Bahl H., Kragl U., Increased productivity of Clostridium acetobutylicum fermentation of acetone, butanol and ethanol by pervaporation through supported ionic liquid membrane, *Appl.Microbiol.Biotechnol.* 78 (2008) 597-602.

Jia M.D., Peinemann K.V., Behling R.D., Preparation and characterization of thin-film zeolite-PDMS composite membranes, *J. Membr. Sci.* 73 (1992) 119-128.

Jiraratananon R., Chanachai A., Huang R.Y.M., Uttapap D., Pervaporation dehydration of ethanol–water mixtures with chitosan/hydroxyethylcellulose (CS/HEC) composite membranes I. Effect of operating conditions, *J. Membr. Sci.* 195 (2002) 143–151.

Jonquières A., Clément R., Lochon P., Néel J., Dresch M., Chrétien B., Industrial state-of-the-art ofpervaporation and vapour permeation in the western countries, *J. Membr. Sci.* 206 (2002) 87-117.

Kalyani S., Smitha B., Sridhar S., Krishnaiah A., Pervaporation separation of ethanol–water mixtures through sodium alginate membranes, *Desalination* 229 (2008) 68–81.

Kang Y.S., Shin E.M., Jung B., Kim J.J. Composite membranes of poly(1-trimethylsilyl-1-propyne) and poly(dimethylsiloxane) and their pervaporation properties for ethanol-water mixture, *J. Appl. Polym. Sci.* 53 (1994) 317-323.

Kanti P., Srigowri K., Madhuri J., Smitha B., Sridhar S., Dehydration of ethanol through blend membranes of chitosan and sodium alginate by pervaporation, *Sep. Purif. Technol.* 40 (2004) 259–266.

Kashigawa T., Okabe K., Okita K., Separation of ethanol from ethanol/water mixtures by plasma-polymerized membranes from silicone compounds, *J. Membr. Sci.* 36 (1988) 353-362.

Kelman S.D., Rowe B.W., Bielawski C.W., Pas S.J., Hill A.J., Paul D.R., Freeman B.D., Crosslinking poly(1-(trimethylsilyl)-1-propyne) and its effect on physical stability, *J. Membr. Sci.*320 (2008) 123-134.

Kickelbick G., Concepts for the incorporation of inorganic building blocks into organic polymers on a nanoscale, *Prog. Polym. Sci.* 28 (2003) 83-114.

Kober P.A., Pervaporation, perstillation and percrystallization, (reprinted from J. Am. Chem. Soc. in 1997) *J. Membr. Sci.* 100 (1995) 61-64.

Kumar P.V., Anilkumar S., Varughese K.T., Thomas S., Separation of n-hexane/acetone mixtures by pervaporation using high density polyethylene/ethylene propylene dieneterpolymer rubber blend membranes, *J. Hazard. Mater.* 199-200 (2012) 336-342.

Kung G., Jiang L.Y., Wang Y., Chung T.S., Asymmetric hollow fibers by polyimide and polybenzimidazole blends for toluene/isooctane separation, *J. Membr. Sci.*360 (2010) 303-314.

Le N.L., Wang Y., Chung T.S., Pebax/POSS mixed matrix membranes for ethanol recovery from aqueous solutions via pervaporation, *J. Membr. Sci.* 379 (2011) 174-183.

Lipnizki F., Field R.F., Ten P.K., Pervaporation-based hybrid processes: A review of process design, applications and economics, *J. Membr. Sci.* 153 (1999) 183-210.

Lipnizki F., Haussmans S., Field R.W., Influence of impermeable components on the permeation of aqueous 1-propanol mixtures in hydrophobicpervaporation, *J. Membr. Sci.* 228 (2004) 129-138.

Liu Y.L., Yu C.H., Lee K.R., Lai J.Y., Chitosan/poly(tetrafluoroethylene) composite membranes using in pervaporation dehydration processes, *J. Membr. Sci.*287 (2007) 230-236.

Mandal M.K., Bhattacharya P.K., Poly(vinyl acetal) membrane for pervaporation of benzene–isooctane solution, *Sep.Purif. Technol.* 61 (2008) 332-340.

Mandal M.K., Sant S.B., Bhattacharya P.K., Dehydration of aqueous acetonitrile solution by pervaporation using PVA-iron oxide nanocomposite membrane, Colloid. *Surf. A Physicochem. Eng. Asp.* 373 (2011) 11-21.

Matsuda H., Yanagishita H., Negishi H., Kitamoto D., Ikegami T., Haraya K., Nakane K, Idemoto Y., Koura N., Sano T., Improvement of ethanol selectivity of silicalite membrane in pervaporation by silicone rubber coating, *J. Membr. Sci.* 210 (2002) 433-437.

Mitchell J.K., On the penetrativeness of fluids, *J. Membr. Sci.* 100 (1995) 11-16.

Mori Y., Inaba T., Ethanol production from starch in a pervaporation membrane bioreactor using Clostridiumacetobutylicum B18, *Biotech. Bioeng.* 36 (1990) 849-853.

Moermans B., De Beuckelaer W., Vankelecom I.F.J., Ravishankar R., Martens J.A., Jacobs P.A. Incorporation of nano-sized zeolites in membranes, *Chem. Commun.* 24 (2000) 2467-2468.

Morigami Y., Kondo M., Abe J., Kita H., Okamoto K., The first large-scale pervaporation plant usingtubular-type module with zeolite NaA membrane, *Sep.Purif. Technol.* 25 (2001) 251-260.

Mulder M., Principles of Membrane Technology, Kluwer Academic Publishers, *Dordrecht,* 1996.

Nagai K., Fundamentals and perspectives for pervaporation, In: Drioli E., Giorno L. (Eds.), *Comprehensive Membrane Science and Engineering*, Volume 2: Membrane Applications in Molecular Separations, Elsevier, Kidlington, 2010, pp. 243-271.

Nagase Y., Sugimoto K., Takamura Y., Matsui K., Chemical modification of poly(substituted acetylene). VI. Introduction of fluoroalkyl group into poly(1-trimethylsilyl-1-propyne) and the improved ethanol permselectivity at pervaporation, *J. Appl.Polym. Sci.* 43 (1991) 1227-1232.

Offeman R.D., Ludvik C.N., Posioning of mixed matrix membranes by fermentation components in pervaporation of ethanol, *J. Membr. Sci.*367 (2011) 288-295.

Oudshoorn A., van der Wielen L.A.M., Straathof A.J.J., Assessment of options for selective 1-butanol recovery from aqueous solution, *Ind. Eng. Chem. Res.* 19 (2009) 7325-7336.

Oudshoorn A., van den Berg C., Roelands C.P.M., Straathof A.J.J., van der Wielen L.A.M., Short-cut calculations for integrated product recovery options in fermentative production of bio-bulk chemicals, *Process Biochem.* 45 (2010) 1605-1615.

Park H.C., Separation of alcohols from organic liquid mixtures by pervaporation, PhD Thesis (1993), Faculty of Science and Technology, University of Twente, Enschede, The Netherlands.

Peng M., Vane L.M., Liu S.X., Recent advances in VOCs removal from water by pervaporation, *J. Hazard.Mater.* 98 (2003) 69-90.

Peng P., Shi B., Lan Y., A review of membrane materials for ethanol recovery by pervaporation, *Sep. Sci. Technol.*46 (2011) 234-246.

Sano T., Yanagishita H., Kiyozumi Y., Mizukami F., Haraya K., Separation of ethanol/water mixture by silicalite membrane on pervaporation *J. Membr. Sci.* 95 (1994) 221-228.

Semenova S.I., Ohya H., Soontarapa K., Hydrophilic membranes for pervaporation: an analytical review, *Desalination* 110 (1997) 251-286.

Schmidt S.L., Myers M.D., Kelley S.S., McMillan J.D., Padukone N., Evaluation of PTMSP membranes in achieving enhanced ethanol removal from fermentations by pervaporation, *Appl. Biochem. Biotechnol.* 63-65 (1997) 469-482.

Shao P., Huang R.Y.M., Polymeric membrane pervaporation, *J. Membr. Sci.*287 (2007) 162-179.

She M., Hwang S.T., Effects of concentration, temperature, and coupling onpervaporation of diluteflavor organics, *J. Membr. Sci.* 27 (2006) 16-28.

Smitha B., Suhanya D., Sridhar S., Ramakrishna M., Separation of organic-organic mixtures by pervaporation-a review, *J. Membr. Sci.*241 (2004) 1-21.

Smitha B., Dhanuja G., Sridhar S., Dehydration of 1,4-dioxane by pervaporation using modified blendmembranes of chitosan and nylon 66, *Carbohydrate Polymers* 66 (2006) 463–472.

Sun H., Lu L., Chen X., Jiang Z., Pervaporation dehydration of aqueous ethanol solution using H-ZSM-5 filled chitosan membranes, *Sep. Purif. Technol.* 58 (2008) 429–436.

Svang-Ariyaskul A., Huang R.Y.M., Douglas P.L., Pal R., Feng X., Chen P., Liu L., Blended chitosan and polyvinyl alcohol membranes for the pervaporation dehydration of isopropanol, *J. Membr. Sci.*280 (2006) 815-823.

Takegami S., Yamada H., Tsuji S., Pervaporation of ethanol/water mixtures using novel hydrophobic membranes containing polydimethylsiloxane, *J. Membr. Sci.* 75 (1992) 93-105.

TeHenneppe H.J.C., Bargeman V, Mulder M.H.V., Smolders C.A., Zeolite filled silicone rubber membranes. Part 1. Membrane preparation and pervaporation results, *J. Membr. Sci.*35 (1987) 39-55.

Thongsukmak A., Sirkar K.K., Pervaporation membranes highly selective for solvents present in fermentation broths, *J. Membr. Sci.* 302 (2007) 45-58.

Van Gemert R.W., Cuperus P.F., Newly developed ceramic membranes for dehydration and separation of organic mixtures by pervaporation, *J. Membr. Sci.*105 (1995) 287-291.

Van Hecke W., Vandezande P., Claes S., Vangeel S., Beckers H., De Wever H., Integrated bioprocess for long-term continuous cultivation of Clostridium acetobutylicum coupled to pervaporation with PDMS composite membranes, *Biores. Technol.* 111 (2012) 368-377.

Van Hoof V., Dotremont C., Buekenhoudt A., Performance of Mitsui NaA type zeolite membranes for the dehydration of organic solvents in comparison with commercial polymeric pervaporation membrane, *Sep. Purif. Technol.* 48 (2006) 304–309.

Vane L.M., Alvarez F.R., Full-scale vibrating pervaporation membrane unit: VOC removal from water and surfactant solutions, *J. Membr. Sci.* 202 (2002) 177-193.

Vane L.M., A review of pervaporation for product recovery from biomass fermentation processes, *J. Chem.Technol.Biotechnol.* 80 (2005) 603-629.

Vane L.M., Separation technologies for the recovery and dehydration of alcohols from fermentation broths, *Biofuels Bioprod. Bioref.* 2 (2008) 553-588.

Vane L.M., Namboodiri V.V., Bowen T.C., Hydrophobic zeolite-silicone rubber mixed matrix membranes for ethanol-water separation: Effect of zeolite and silicone component selection on pervaporation performance, *J. Membr. Sci.* 208 (2008) 230-241.

Varghese J.G., Kittur A.A, Rachipudi P.S., Kariduraganavar M.Y., Synthesis, characterization and pervaporation performance of chitosan-g-polyaniline membranes for the dehydration of isopropanol, *J. Membr. Sci.* 364 (2010) 111-121.

Vauclair C., Tarjus H., Schaetzel P., Permselective properties of PVA-PAA blended membrane used for dehydration of fusel oil by pervaporation, *J. Membr. Sci.*125 (1997) 293-301.

Verkerk A.W., Van Male P., Vorstman M.A.G, Keurentjes J.T.F., Properties of high flux ceramic pervaporation membranes for dehydration of alcohol/water mixtures, *Sep. Purif. Technol.* 22-23 (2001) 689-695.

Villaluenga J.P.G., Tabe-Mohammadi A., A review on the separation of benzene/cyclohexane mixtures by pervaporation processes, *J. Membr. Sci.* 169 (2000) 159-174.

Volkov A.V., Volkov V.V., Khotimskii V.S., Membranes based on poly((1-trimethylsilyl)-1-propyne) for liquid-liquid separation, *Polym. Sci.Serie A* 51 (2009) 1367-1382.

Volkov V.V., Fadeev A.G., Khotimsky V.S, Litvinova E.G., Selinskaya Y.A., McMillan J.D, Kelley S.S., Effects of synthesis conditions on the pervaporation properties of poly(1-(trimethylsilyl)-1-propyne) membranes, *J. Appl. Polym. Sci.* 91 (2004) 2271-2277.

Volkov V., Kujawski W., Fouling in pervaporative membrane bioreactors, International Conference on Pervaporation, Vapor Permeation and Membrane Distillation, September 2011, Torun, Poland.

Von Wroblewski S., Ueber die natur der absorption der gase durch flussigkeiten unten hohendrucken, *Ann .Physik. u Chem.* 8 (1879) 29-52.

Wang Y.C., Li C.L., Huang J., Lin C., Lee K.R., Liaw D.J., Lai J.Y., Pervaporation of benzene/cyclohexane mixtures through aromatic polyamide membranes, *J. Membr.Sci.* 185 (2001) 193-200 (a).

Wang X. P., Modified alginate composite membranes for the dehydration of acetic acid, *J. Membr. Sci.* 170 (2000) 71-79.

Wang X.P., Li N., Wang W.Z., Pervaporation properties of novel alginate composite membranes for dehydration of organic solvents, *J. Membr. Sci.* 193 (2001) 85-95 (b).

Wang Y., Chung T.S., Wang H., Goh S.H., Butanol isomer separation using polyamide-imide/CD mixed matrix membranes via pervaporation, *Chem. Eng. Sci.* 64 (2009) 5198-5209.

Wee S.L., Tye C.T., Bhatia S., Membrane separation process – Pervaporation through zeolite membrane, *Sep.Purif. Technol.* 63 (2008) 500-516.

Wijmans J.G., Baker R.W., The solution-difussion model: a review, *J. Membr. Sci.*107 (1995) 1-21.

Yang Y., Qian J., Xuan L., An A., Zhang L., Gao C., Preparation and pervaporation of a palygorskite/polyacrylamide inorganic-organic membrane for separating m-/p-xylene isomers, *Desalination* 193 (2006) 193-201.

Yang D., Li Jiang Z., Lu L., Chen X., Chitosan/TiO2 nanocompositepervaporation membranes for ethanol dehydration. *Chem. Eng. Sci.* 64 (2009) 3130-3137.

Yildirim A.E., Hilmioglu N.D., Tulbentci S., Separation of benzene/cyclohexane mixtures by pervaporation using PEBA membranes, *Desalination* 219 (2008) 14–25.

Yu C., Zhong C., Liu Y., Gu X., Yang G., Xing W., Xu N., Pervaporation dehydration of ethylene glycol by NaA zeolite membranes. *Chem. Eng. Res. Des.* (2011), doi:10.1016/j.cherd.2011.12.003.

Yuan H.K., Ren J., Ma X.H., Xu Z.L., Dehydration of ethyl acetate aqueous solution by pervaporation using PVA/PAN hollow fiber composite membrane, *Desalination* 280 (2011) 252-258.

Zereshki S., Figoli A., Madaeni S.S., Simone S., Esmailinezhad M., Drioli E., Pervaporation separation of MeOH/MTBE mixtures with modified PEEK membrane: Effect of operating conditions, *J. Membr.Sci.*371 (2011) 1-9 (a).

Zereshki S., Figoli A., Madaeni S.S., Galiano F., Drioli E., Pervaporation separation of ethanol/ETBE mixture using poly(lactic acid)/poly(vinyl pyrrolidone) blend membranes, *J. Membr.Sci.*373 (2011) 29-35 (b).

Zhang X., Qian L., Wang H., Zhong W., Du Q., Pervaporation of benzene/cyclohexane mixtures through rhodium-loaded β-zeolite-filled polyvinyl chloride hybrid membranes, *Sep.Purif. Technol.* 63 (2008) 434-443.

Zhang X.H., Liu Q.L, Xiong Y., Zhu A.M., Chen Y.,. Zhang Q.G, Pervaporation dehydration of ethyl acetate/ethanol/water azeotrope using chitosan/poly (vinyl pirrolidone) blend membranes, *J. Membr. Sci.*327 (2009) 274-280.

Zhao Q., Qian J., An Q., Zhu M., Yin M., Sun Z., Poly(vinyl alcohol)/polyelectrolyte complex blend membrane for pervaporation dehydration of isopropanol, *J. Membr. Sci.* 343 (2009) 53–61.

Zhou M., Persin M., Sarrazin J., Methanol removal from organic mixtures by pervaporation using polypyrrole membranes, *J. Membr. Sci.*117 (1996) 303-309.

Zusumocahyo S.P., Kanamori T., Sumaru K., Iwatsubo T., Shinbo T., Pervaporation of xylene isomer mixture troughcyclodextrins containing polyacrilic acid membranes, *J. Membr. Sci.*231 (2004) 127-132.

In: Membrane Processes for Sustainable Growth
Editors: Angelo Basile and Alfredo Cassano

ISBN: 978-1-62618-446-6

Chapter 4

MEMBRANES FOR SUSTAINABLE FOOD PACKAGING

Alberto Figoli* and Silvia Simone
Institute on Membrane Technology (ITM-CNR), c/o University of Calabria, Rende (Cosenza), Italy

ABSTRACT

Membrane science and technology are currently applied in several fields. It is widely recognized that the knowledge achieved by membranologists in different areas could be exploited at industrial level to reduce the environmental impact and energy requirements of many processes.

In this chapter it will be shown how membranes can be successfully used for food packaging in several cases, especially when novel systems beside traditional barrier packaging are required.

In particular, Gas Separation membranes can be useful for Modified Atmosphere Packaging (MAP), often used in the case of fresh, respiring, products, such as fruits and vegetables; in fact, it requires films having specific permselectivity to gases as oxygen and carbon dioxide. Membranes loaded with specific substances can be used as Active releasing Packaging systems, for controlled delivery of such substances to the packed product; this concept has been already developed in medical field.

Peculiar attention will be devoted to biopolymers and their application for food packaging. Also in this case, the know-how of membranologists could help to improve the properties of films produced from biopolymers (by well developed techniques as use of plasticizers, blending, loading with nanoparticles).

The production of food packaging from biodegradable-non polluting sources, but which also allows better preservation of food quality and safety, is in perfect agreement with the philosophy of sustainability, since it could help reducing the exploitation of natural resources and the environmental impact connected to food/packaging production, delivery and disposal.

Keywords: food packaging; modified active packaging (MAP); selective membranes for food packaging; biodegradable polymers

---

* Corresponding Author; e-mail: a.figoli@itm.cnr.it

1. Introduction

Nowadays, the words "sustainable" and "sustainability" are widely used, often together with the words "development" and "growth". What does it really mean "sustainable development"? According to the Brundtland Report (WCED, 1987) "Sustainable Development" is the growth that meets the needs of present generations without compromising the ability of future generations to satisfy the same requirements. Development and/or sustainable growth means, therefore, to produce without over-exploiting natural resources, reducing waste and environmental impact of farming and industrial production.

According to the definition of the European Council (1994) "packaging shall mean all products made of any materials of any nature to be used for the containment, protection, handling, delivery and presentation of goods, from raw materials to processed goods, from the producer to the used or the consumer". In other words, food packaging can be considered as a very heterogeneous group of papers, envelops, boxes, coatings and so on, which are employed with the aim of increasing foodstuffs shelf life, slowing down/avoiding deterioration and bacterial/fungal contamination, and prolonging maintenance of properties as colour, firmness and flavour.

How the packaging of food products can meet the philosophy of sustainable development? There are at least two different ways in which the production of packaging intended for use on products such as fruits, vegetables, meats and cheeses can become "sustainable".

The first aspect is the production of food packaging from materials that are recyclable, biodegradable, in a process in which both the production and disposal have a reduced environmental impact.

The second one is reduction of the amount of food wasted by consumers. The correct use of packaging can and should help to reduce such waste. Paradoxically, it was estimated that it is more "sustainable" to produce a packaging material that is more pollutant, but that better preserves the contained product, contributing to reduce such waste, rather than a perfectly recyclable or biodegradable package that is less effective in food preservation (Wikström et al., 2010; Williams et al., 2011).

The correct approach to the choice of materials and packaging techniques can contribute not only to minimize the amount of wasted food, but also to reduce the energy consumption (for example, allowing a good preservation of food without refrigeration) and pollution (not only by using biodegradable and/or recyclable materials but also by reducing the environmental impact due to the production of food itself). In other words, less waste means reduced production, which reduces the environmental impact due to crops, livestock and processing of raw products in foods to the consumer.

How, then, scientific research may contribute to the production of sustainable packaging? In our opinion, trying to combine as much as possible these two aspects, in order to produce a packaging, which is not polluting itself, but which also allows a better conservation, reducing the total amount of produced and wasted food.

In this context, membrane science and technology can be helpful to achieve the meeting between these two aspects: *i)* production of packaging materials from natural, biodegradable and low environmental impact materials; *ii)* search for new techniques that allow better preservation and conservation of foods (i.e. the "active" packaging).

In this chapter, we will show how membranes, thanks to their unique properties of permselectivity and ability to sustain controlled delivery, can be successfully applied in food packaging area. The aspects of sustainability and environmental impact reduction of food packaging materials, through the use of non-conventional biodegradable polymers, will be emphasized.

2. Membranes in Food Packaging

Today, market globalization, human habits change and increasing attention towards issues as food quality/safety and human activities impact on the environment bring about new challenges in the field of food packaging.

The development of new materials and techniques for food packaging can be an area of interest for application of innovative membranes.

A *membrane* is, in its more general definition, "a phase or a group of phases that lies between two different phases, which is physically and/or chemically distinctive from both of them and which, due to its properties and the force field applied, is able to control the mass transport between these phases" (Paul and Yampol'skii, 1994). Today, membrane science is present in many aspects of human quotidian life, from water, milk and beverages treatment to biomedical devices. It can be also applied to two crucial and relatively new aspects of food packaging (Figure 1): *i) modified atmosphere packaging* (active and passive MAP) and *ii) active packaging* (AP).

Modified atmosphere packaging is of special interest for fresh and minimally processed fruits and vegetables. It aims at maintaining an optimum gas balance and, especially, selected ratio between O_2 and CO_2, in the package headspace in order to ensure living vegetal tissues respiration and survival but avoiding ageing and deterioration. It is evident that proper materials for MAP should not be impenetrable barriers but, on the contrary, must ensure proper gas and humidity exchange with the environment.

Active packaging is a recent developed strategy in which the product, the package and the environment interact in a positive way in order to extend product shelf life or to achieve some special characteristics.

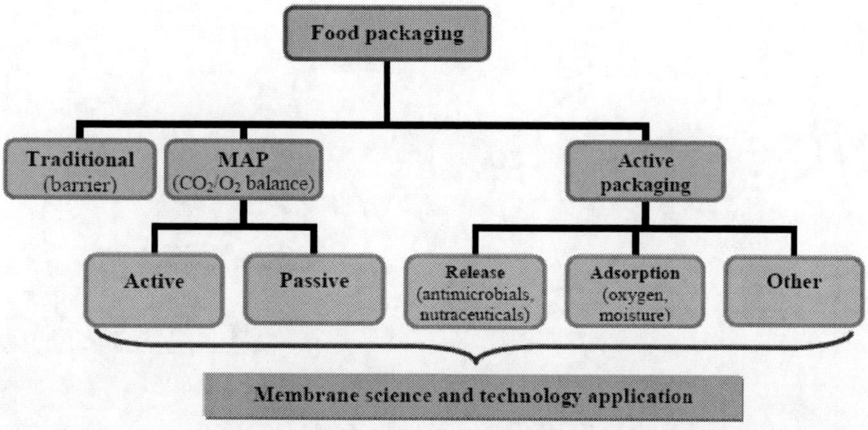

Figure 1. Fields of food packaging relevant to membrane science and technology.

In particular, the package material can be used to deliver some substances to the product (such as antimicrobials, flavours, antioxidants, nutraceuticals) or to adsorb undesirable ones (oxygen, ethylene, moisture) during storage, with optimal rate, in order not only to increase product shelf life but also to improve its nutritional value and/or taste, odour and appearance.

3. Materials and Membranes for Modified Atmosphere Packaging

Dehydrated and processed meat, eggs, cheese and fatty foods are classical examples of products that need to be completely insulated from the environment during storage, since oxygen and light can promote some adverse reactions, i.e. oxidation. In these cases, packaging material should be completely impenetrable to gases and humidity, to avoid oxidation, which may lead to off-flavour production and loss of flavor. Packaging should also avoid both water vapour penetration from the environment, which may lead to sogginess or microbial growth in the food, and humidity escaping from the package, which may cause undesirable textural changes of the product (Lee et al., 2008). Indeed, modified atmosphere and under-vacuum packaging of non-respiring products requires the use of highly impenetrable materials, able to maintain vacuum or the optimal atmosphere initially flushed.

In other cases, modified atmosphere packaging is employed to create and/or maintain a proper atmosphere in the package headspace, through selected gases and humidity exchanges with the environment. In *passive* MAP it is progressively created just thank to the exchanges between the packaging headspace and the environment. In *active* MAP, an initial optimal atmosphere is flushed in and maintained/modified during storage thank to interdiffusion of gases and/or water vapour between the package and the environment.

MAP is often used to control fresh food respiration, i.e. to maintain a proper balance between O_2 and CO_2 in the package headspace, in order to allow respiration but avoid ageing, deterioration and, obviously, contamination of the product. In fact, fresh or minimally processed fruits and vegetables tissues are made up of living cells, which should continue carrying out their metabolic activities (Figure 2).

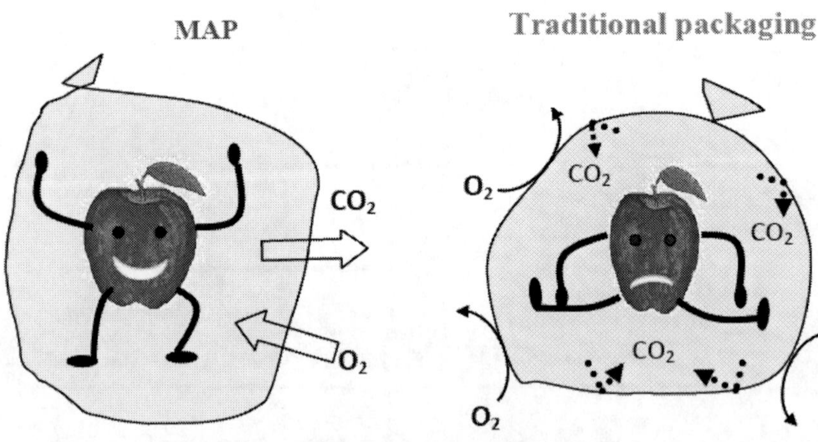

Figure 2. Comparison between MAP and traditional impenetrable barrier packaging in the case of fresh fruits and vegetables.

However, respiratory activity should be maintained as low as possible, in order to reduce oxidation and browning (a consequence of polyphenol oxidase activity (Falguera et al. 2010)).

Low levels of O_2 and increased concentrations of CO_2 are usually referred as positive conditions to improve quality/safety and prolonging shelf life of packaged fresh products. These conditions may slow down food respiration, delay ripening and, hence, preserve food firmness and chlorophyll/pigments retention. Moreover, high levels of CO_2 may inhibit the growth of aerobic organisms, such as bacteria, yeasts and moulds, reducing food spoilage. However, as observed above, the balance between CO_2 and O_2 is often critical for food safety and quality preserving. Absence of O_2 can induce anaerobic metabolism, production of ethylene and, hence, fast product deterioration; moreover, anaerobic conditions are favorable for the growth of dangerous bacteria, as Clostridium.

Therefore, in these cases, the selection of packaging material is of paramount importance, since its properties of selectivity and permeability, in combination with product characteristics, are crucial for the obtaining of the desired optimal atmosphere within the package.

Selectivity and permeability are the two main properties of each membrane used in any separation process, carried out in liquid/vapour and/or gas phase. In particular, gas separation (GS) membrane properties can be exploited in MAP application.

GS is one of the main industrial applications of membrane technology; the advances achieved in this field in the past decades allowed the obtaining of high fluxes and large surface area modules (Murphy et al., 2009). GS requires membranes having a dense, selective layer, and can be carried out using:

- Dense membranes: A homogeneous, often transparent, polymer film usually obtained by solvent casting followed by evaporation and drying (Torchia et al., 2004).
- Asymmetric dense membranes: Anisotropic membranes, having a dense skin and a porous substrate, generated by the same polymeric dope after solvent casting and coagulation in a proper bath (Non solvent Induced Phase Separation, NIPS) (Brunetti et al., 2009).
- Coated membranes: The selective layer is obtained by coating the membranes with a thin layer of a highly permeable polymer (silicone rubber), in order to overcome selectivity losses due to pinholes in the skin layer of asymmetric membranes (Baker et al., 2002; Bernardo et al. 2009)
- Composite membranes: Membranes made up of a porous layer, working as mechanical support, and a dense, selective layer, obtained from different polymeric dopes by NIPS and/or other techniques (Baker et al., 2002; Bernardo et al. 2009).
- Metallic membranes: Palladium membranes, for instance, are the best choice for hydrogen separation due to their high selectivity (Iulianelli and Basile, 2011).

For what concerning food packaging, gas permeation takes place through permeable packages, and it is a chemico-physical phenomenon that can be defined as the movement of gas, vapour or liquid molecules across a homogeneous matrix, driven by a concentration gradient. It can be described by the solution/diffusion model, since the molecules are driven to the polymeric membrane (high concentration, high pressure side), dissolve in the membrane matrix, diffuse down the concentration gradient and desorb on the other membrane side (low concentration, low pressure) (George et al., 2001). The permeability coefficient P depends on the product between the solubility coefficient, S, and the diffusivity, D, which can be derived by the Henry's and Fick's first law respectively (Zolandz and Fleming, 1992):

$$P = S \times D \quad (1)$$

To describe the ability of a membrane to separate two gas species, A and B, it is convenient to calculate the permselectivity ($\alpha_{A/B}$), which is obtained from the ratio between the permeability coefficients of A and B:

$$\alpha_{A/B} = P_A/P_B = (S_A/S_B) \times (D_A/D_B) \quad (2)$$

Obviously, if $P_A = P_B$, $\alpha_{A/B} = 1$, and the membrane is not selective at all. The permselectivity increases as P_A increases with respect to P_B; it means that permeation of the specie A through the membrane is favored with respect to the specie B.

Permselectivity of polymeric materials is often described using the ratio between permeability coefficients of model gas species ($\alpha CO_2/N_2$, and $\alpha O_2/N_2$). Looking at the so-called Robeson plots (Robeson, 2008), it can be seen that there is an inherent relationship between the selectivity and the permeability of a polymeric material. The straight lines referred as upper bounds, above which no data points still exist, are shifting to progressively higher values. However, a typical trade off between selectivity and permeability is observed. Most of membrane research in this field aims at achieving an optimum balance between these two properties.

Concerning modified atmosphere packaging of fresh foods, as fruit and vegetables, the gas species of interest are CO_2 and O_2, since it has been demonstrated that the respiration rate of the product after harvesting depends on the ratio between the partial pressures of these two gases (Lee et al., 1991). For any product, there exists an optimum balance between the partial pressures of oxygen and carbon dioxide, defined as respiratory quotient (RQ) (Cameron et al., 1995), which should be maintained in the package headspace by proper regulation of gas exchanges with the environment, which can be achieved through selective and non selective permeation of both species through the package walls. Therefore, membrane materials can be selected on the basis of their permselectivity between CO_2 and O_2.

For instance, Torchia et al. (2004) prepared dense membranes in modified poly ether ether ketone (PEEK-WC), with and without the additive poly-alpha-pinene, obtaining permselectivities ($\alpha CO_2/O_2$) matching the values required by different fresh fruits and vegetables, as Brussels sprouts, turnip, carrot, apple and green pepper, varying the ratio between polymer and additive concentrations in dope composition.

Moreover, as pointed out by Figoli et al. (2010) and by Mensitieri et al. (2011), while synthetic plastic films (for instance, poly propylene and poly ethylene terephtalate) exhibit excellent barrier properties, most of them does not result in optimal O_2 and CO_2 atmospheres, especially when the product has a high respiration rate. Furthermore, the increase of oil price and the high environmental impact of oil-derived plastic materials push research to move towards "green" packaging materials. The US Environmental Protection Agency (EPA) suggested the reduction of environmentally toxic materials and the use of packaging systems easier to reuse or compost, in order to improve municipal waste management and reduction (EPA, 2009). Also in Italy, plastic shopping bags have been recently banned, while the problem of non-recyclable materials environmental impact is gaining increasing attention and becoming a major concern, also as a consequence of the recent emergency in Naples and other cities. Therefore, the replacement of conventional oil-derived plastics by biopolymers from renewable sources, as biodegradable polyesters, proteins and polysaccharides, seems to be an excellent approach to meet these directives and fits with real sustainable development strategy.

Indeed, as pointed out by Mensitieri et al. (2011), packaging structures based on biopolymers display a wide range of CO_2/O_2 permselectivity values, which makes them suitable for MAP of fresh fruits and vegetables; such range is unattainable with synthetic polymers.

3.1. Biopolymers for MAP

Biopolymers can be defined as biodegradable materials that can be decomposed by microorganisms and enzymatic degradation (ASTM standard D-5488-94d), resulting in water, inorganic compounds, carbon dioxide and/or methane, and biomass.

Among the different materials that can be used for food packaging, biopolymers can be divided in four different categories, on the basis of their origin (Mensitieri et al. 2011; Avérous, 2004), as shown in Figure 3.

In literature (Table 1), there are several studies reporting the use of films based on biodegradable materials for packaging of fresh fruits and vegetables, thanks to their improved gas permeability with respect to oil derived plastics.

On the other hand, however, their performance is sometimes impaired by their high water permeability, which can cause moisture loss and wilting of the product.

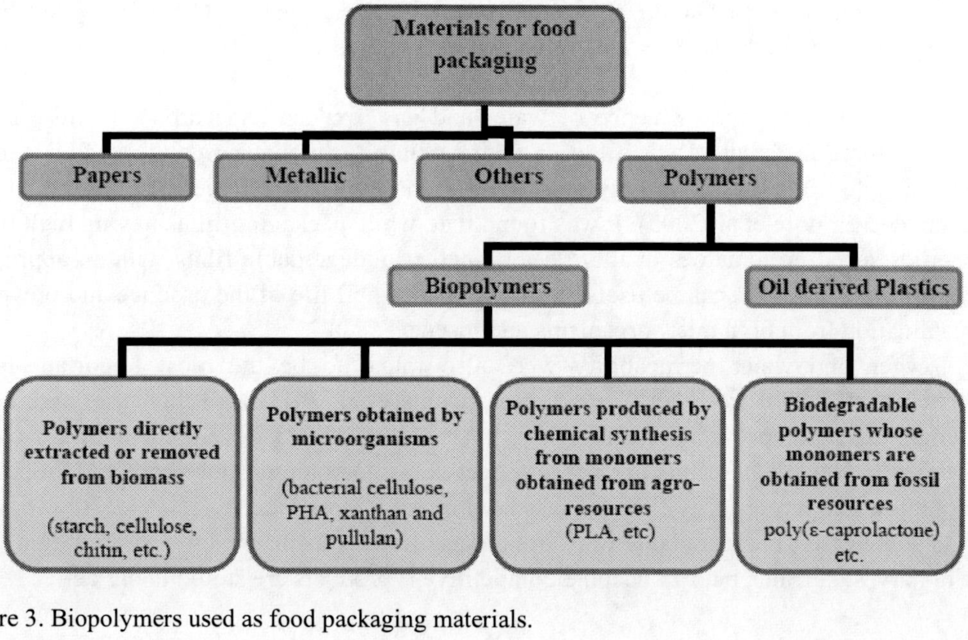

Figure 3. Biopolymers used as food packaging materials.

Conte et al. (2009a) tested four different polymeric materials (two biodegradable films, a commercially available multilayer film (MLF) and polypropylene) for packaging of fresh cut "barattiere" melon slices, untreated and dipped in water solution containing ascorbic acid (1%) and calcium chloride (0.2%). It was found that the MLF was the best material in delaying the detrimental phenomena involved in quality acceptance, showing the higher MAL (Microbiological Acceptability Limit, expressed in days), thanks to its ability of maintaining low O_2 availability and high CO_2 concentration in the package headspace. However, MAL of melon slices pre-treated with $CaCl_2$ and ascorbic acid and packed using a monolayer film based on a blend of biodegradable polyesters was only slightly lower (13 vs. 15 days). Therefore, authors concluded that the biodegradable films, combined with a dipping pre-treatment, could represent a good compromise between performance and environmental impact. On the contrary, oriented polypropylene-based (OPP) package was found to be more effective in prolonging shelf life of fresh-cut zucchini, packed under passive and active MAP, depending on the cultivar type, with respect to a biodegradable film based on polyester (Lucera et al., 2010).

In another study, five types of films, three biodegradable and two based on commercially available traditional polymers, were tested as packaging materials for table grape (Conte et al. 2011a). On the basis of the found oxygen and carbon dioxide concentrations in the package headspace, authors concluded that nylon/polyolefin film showed the best barrier properties; however, among the tested bio-films, the thicker monolayer one, based on polyester, gave interesting results. Moreover, the overall acceptability of packed fruits at the end of the storage time was similar in both cases.

Other works refer that the maintaining of optimal O_2 concentration in the package headspace is necessary for fruit ripening. For instance, availability of O_2 is critical for tomato quality. Kantola and Helèn (2001) compared the performance of four biodegradable packages (corn starch based Mater-Bi® type ZF03U; copolymer poly(HB-co-HV) Biopol®; a polymerization product of lactic acid, polylactide, and regenerated cellulose) with low density polyethylene (LDPE). They monitored weight loss, moisture content, colour, firmness and flavour of tomatoes packed in perforated bags for three weeks. The sensory quality of tomatoes in biodegradable packages remained as good as the quality of tomatoes stored in LDPE bags.

In another study, the influence of water and gas (oxygen and carbon dioxide) permeability of three different plastic films, a commercially available polyolefinic film and two biodegradable films (based on polyester), on the quality decay kinetics of plum tomato was examined (Muratore et al. 2005). It was found that, while packaging films having high barrier properties speed up tomatoes quality decay kinetic, biodegradable films with an appropriate permeability coefficient can be useful to improve the shelf life of the produce and prevent its contamination from both microorganisms and insects.

Oxygen and water permeability were also found to be the most important factors affecting shelf life of fresh-cut Cime di rapa (Conte et al. 2011b). In this case, high barrier materials, as nylon/polyethylene film, are not suitable due to development of off-odours. However, biodegradable films are not as effective as commercial polypropylene because of their high water permeability, which caused wilting of the product.

In some cases, biodegradable films are not as effective as plastic ones in prolonging shelf life of harvested fruits, but can become competitive if products are stored under active MAP.

Conte et al. (2009b) examined the influence of two different packaging systems, made of OPP and of biodegradable co-extruded polyester respectively, on quality loss of ready-to-use cherries.

Table 1. Comparison between performance of traditional and biodegradable packaging for different fresh fruits and vegetables

Fruit or vegetable specimen	Packaging based on oil derived plastic	*VS.*	Packaging based on biodegradable polymer	Conclusions	REF
 *"Barattiere" melon*	1) MLF (Nylon/ polyolefin) 2) OPP film		1) Biodegradable polyesters (Multilayer); 2) Biodegradable polyesters (Monolayer).	MLF was the best. Polyester film performance was improved by product pre-treatment with $CaCl_2$ and ascorbic acid.	Conte et al., 2009 a
 *Zucchini*	OPP film		1) Biodegradable polyesters (Multilayer),	OPP package was more in all cases.	Lucera et al., 2010
 *Table grape*	1) MLF (Nylon/ polyolefin) 2) OPP film		1) Polyester monolayer (100 μm); 2) Polyester monolayer (50 μm); 3) Co-extruded starch and polyesters.	MLF film showed the best barrier properties; 100 μm monolayer gave interesting results.	Conte et al. 2011a
 *Tomatoes*	LDPE		1) Mater-Bi® type ZF03U; 2) Biopol®; 3) Polylactide; 4) Regenerated cellulose.	Tomatoes packed using biodegradable materials were kept as good as ones stored in LDPE bags for three weeks.	Kantola et al., 2001
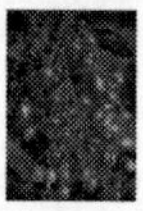 *Pomodoro Datterino ®*	High permeability polyolefinic Film		1) mixture of three biodegradable polyesters; 2) mixture of two biodegradable polyesters.	films with high barrier properties (ternary polyester mixture) speed up the quality decay kinetic. Binary polyester mixture performance was similar to that of reference sample	Muratore et al. 2005
 *Cime di rapa*	1) MLF (Nylon/ polyolefin) 2) OPP		1) Polyester monolayer (100 μm); 2) Polyester monolayer (50 μm).	MLF bags performance impaired by off-odours. Biodegradable materials impaired by water permeability The OPP gave better results.	Conte et al. 2011b
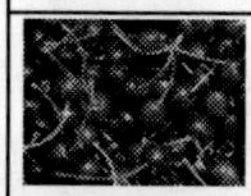 *Cherries*	OPP film		Biodegradable polyesters (Multilayer)	OPP shows the best performances Under MAP, the performance of biodegradable film improved.	Conte et al. 2009b
 *Mango*	Carton boxes covered with LDPE		Carton boxes covered with chitosan film	Chitosan extended shelf life of product without any microbial growth and off flavour.	Srinivasa et al., 2002

While, under ordinary atmosphere conditions, OPP shows the best performances, under MAP, both investigated films exert similar effects on the portioned fruit. Therefore, the combination of a biodegradable film to modified atmosphere packaging could gain widespread acceptance by the industry.

In another interesting work, mango fruits were kept in carton boxes covered with low-density polyethylene or chitosan film (Srinivasa et al., 2002). While fruits covered with LDPE showed off-flavour, due to fermentation, and fungal growth, and were partially spoiled, those stored in chitosan-covered boxes showed an extension of shelf-life of up to 18 days, without any microbial growth and off flavour.

Chitosan film was effective in delaying the ripening process, retarding chlorophyll degradation, due to high CO_2 and low O_2 levels in the headspace, thus allowing extension of fruit shelf-life. These results are also connected to chitosan intrinsic anti-bacterial and anti-fungal properties, which are crucial also for active packaging, and which will be the focus of the following paragraph.

4. Materials and Membranes for Active Packaging

Active packaging is an innovative concept of packaging, which exploits the interaction between the product, the package and the environment in order not only to extend the shelf life of the product itself, but also improve food safety, quality and sensory properties (Yam et al., 2005).

AP systems have been divided in categories by different authors. Yam et al. (2005) divided AP systems into three main categories: adsorbers, releasing systems and others.

Appendini and Hotchkiss (2002), reported the following categories: *i)* Addition of sachets/pads containing volatile agents; *ii)* Incorporation of volatile and non-volatile agents directly into the polymeric matrix; *iii)* Coating or adsorbing active agents on the package surface facing the food atmosphere or the food surface; *iv)* Immobilization of active agents on to the polymer surface by chemical reaction; *v)* Use of polymeric materials showing intrinsic anti-microbial properties; *vi)* Preparation of multilayer films containing an active layer.

Dainelli et al. (2008) classified AP in two main types: *i)* non-migratory active packaging and *ii)* active releasing packaging.

4.1. Non migrating AP

Traditional examples of non migrating AP systems involve O_2 scavengers, CO_2 scavengers and emitters and moisture absorbers (Dainelli et al., 2008; Rodriguez-Aguilera et al., 2009).

O_2 scavenging is effective in avoiding moulds growth and prevents oxidation of certain food constituents as fats; the most common system is based on the oxidation of iron powder.

CO_2 scavengers are sometimes used to adsorb carbon dioxide produced by fermented or roasted products, as coffee. CO_2 adsorption can be either physical (by zeolites or active carbon powder) or chemical (based on properties of Calcium or Magnesium hydroxide or Sodium carbonate).

CO_2 emitters can be used to inhibit microbial growth and reduce fresh products respiration rate (similarly to what happens in MAP). These systems are based on ferrous carbonate, ascorbic acid and sodium bicarbonate.

Moisture scavengers, as silica gel, are also useful to prevent microbial/mould growth, and can be used in impermeable packages to adsorb the humidity released by the product.

Other examples are ethylene scavengers, to slow down the maturation of climacteric fruits, and entrapment of silver ions, with antimicrobial properties, in a silicate network.

4.2. Active Releasing Packaging

Active releasing packaging systems can be used to deliver different substances to the product (Dainelli et al., 2008), such as ethanol, water vapor, sulphur dioxide, antimicrobial agents, essential oils, enzymes and flavors.

In this field, the knowledge gathered by membrane scientists can be useful, in particular thank to the ability of membranes to sustain controlled delivery of active substances.

As observed by Baker (2004), the concept of membranes for controlled delivery has been developed in medical field.

A membrane can act as reservoir system, regulating the release of a substance through its wall, can release a substance directly loaded in its matrix or, in alternative, exploit the osmosis phenomena to release the substance.

Extruded monolayer or multilayer films, obtained by coextrusion or coating, can be used as interesting solutions for active releasing packaging.

For instance, a plastic film containing allyl isothiocyanate (AITC), a strong antimicrobial substance, entrapped in cyclodextrins, for protecting the volatile active agent from thermal degradation during extrusion, is already commercially available in Japan (Lee, 2005). After packaging of the food product, when exposed to high moisture conditions, cyclodextrins change in structure and release the antimicrobial agent in the atmosphere surrounding the food.

Figoli et al. (2004) and Jansen and Figoli (2004) developed a multilayer system in which an asymmetric porous membrane based on modified PEEK-WC is exploited to control the release of an antimicrobial agent (oxalic acid).

In particular, multilayer films were obtained by subsequently casting on the same support (Figure 4a):

1) A dense layer of PEEK-WC with poly-alpha-pinene (ratio from 100/0 to 50/50), obtained by solvent evaporation;
2) An adhesive layer of poly-alpha-pinene, working also as reservoir of oxalic acid;
3) A porous PEEK-WC film, obtained by dry/wet phase inversion.

The properties of this layer can be tailored by acting on the phase inversion conditions, in order to modulate the oxalic acid release. In particular, authors demonstrated that the oxalic acid release increases when increasing the coagulation bath temperature, since more porous structures are such obtained.

Biodegradable membranes and films have been recently reported also as useful devices for active releasing packaging.

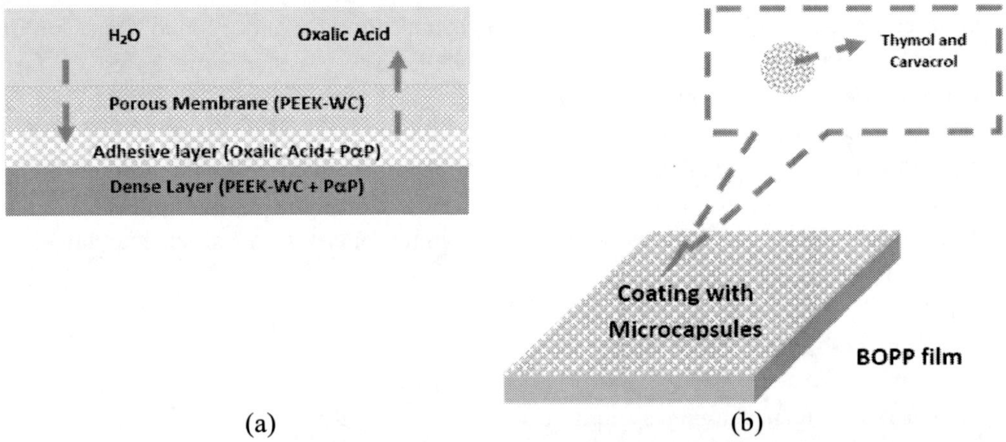

Figure 4. Systems for controlled delivery of antimicrobial substances in Active Packaging Systems. a) Multilayer film (Figoli et al., 2004); b) Film coated with microcapsules (Guarda et al., 2011).

Several examples of biopolymer active releasing systems, based on different materials, such as starch, corn zein and PLA are reported in the review by Mensitieri et al. (2011). These systems can be used to deliver lysozyme, nisin and natural antimicrobial compounds. Other interesting examples of biopolymer based materials for active packaging can be found in Atarés et al. (2011) (antioxidants as ascorbic acid, citric acid and ginger essential oil in hydroxypropylmethylcellulose based edible films), Chana-Thaworn et al. (2011) (hydroxypropyl methylcellulose edible films containing kiam wood (Cotyleobium lanceotatum) extract as natural antimicrobial agent) and Gialamas et al. (2010) (sodium caseinate edible films incorporating Lactobacillus sakei cells).

Chitosan, a natural polycationic polysaccharide, made up of β-(1,4)-linked N-acetyl-D-glucosamine and D-glucosamine units, derived from deacetylation of chitin (found in the crustacean's shells, insect's cuticle and cell wall of fungi), is one of the most promising bio-materials for active packaging. It shows intrinsic antibacterial properties, which have been only partially explained. It can be used alone (Martínez-Camacho et al., 2010) or in combination with other biopolymers as guar gum (Rao et al., 2010) or gelatin (Pereda et al., 2011; Gómez-Estaca et al., 2011), to prepare films that were found to be effective in growth inhibition of different microorganisms. Recently, chitosan/bergamot oil composite films have been prepared and the release of limonene as active agent has been examined (Sánchez-González et al., 2011).

Microencapsulation is another strategy that can be used to control delivery of active agents in food packaging. Guarda et al. (2011) prepared microcapsules containing carvacrol and thymol, two naturally occurring antimicrobial agents, which can be found in oregano and thyme essential oils.

Bi-axially oriented polypropylene (BOPP) films coated with microcapsules were successfully applied to inhibit the growth of 5 different microorganisms (S. *aureus, L. innocua, E. coli, S. cerevisiae and A. niger*) (Figure 4b).

This strategy can be combined also with biopolymers, as demonstrated by Marcuzzo et al. (2010), who encapsulated 10 different aroma compounds and prepared emulsified edible films of carrageenan. These films were able to retain volatile compounds during film processing and to release them gradually.

5. Issues and Challenges for Use of Biopolymers on Industrial Scale

There are still many problems connected with the use of biopolymers for the production of films and membranes for food packaging, mainly due to their poor mechanical properties and hydrophilicity.

Different chemical additives, such as plasticizers, antioxidants, stabilizers, fillers, can be added to tailor the physico-chemical properties and to improve the processability, the mechanical properties and the sealability of biopolymers.

Plasticizers are low molecular weight nonvolatile substances that can reduce chain to chain interactions in the polymeric matrix, therefore improving film flexibility, reducing the glass transition and the melting temperatures, and increasing film permeability (Abdorreza et al., 2011). Plasticizers allow the melt-processing of protein and starches, which, due to their strong intermolecular interactions and high crystallinity, would else require high temperatures that lead to thermal degradation before achieving melt flow (Mensitieri et al. 2011).

For instance, it has been reported that sorbitol can reduce the onset temperatures and improve the heat sealability of sago starch films, while better mechanical properties can be achieved using a mixture of sorbitol and glycerol 3:1 (Abdorreza et al., 2011).

Chemical crosslinking is another method to improve the mechanical properties and the thermal stability of polymers, biopolymers and their blends (Yu et al., 2011). Glutaraldehyde and other aldehydic compounds are often used for chemical crosslinking. However, their use for food packaging is limited due to their high toxicity.

Recently, genipin, a natural compound obtained from jasmine fruits, has been proposed as chemical cross-linker to replace glutaraldehyde. For instance, totally biodegradable films were cast from a solution containing soy protein isolate, genipin and glycerol as plasticizer (González et al., 2011).

Blend with polycaprolactone (PCL) is another strategy that can be used to improve biopolymeric films properties and, in particular, to increase hydrophobicity and, hence, water barrier properties (Mensitieri et al., 2011). Several examples are reported in literature, see for instance John et al. (1998), and Sarazin et al. (2008).

The use of nanofillers has been recently reported as further strategy to tailor and improve biopolymer films properties. Membranes loaded with nanofillers have been already developed for different broad applications (gas separation, waste water treatment etc.) and they are known as Mixed Matrix Membranes (MMM). The strategies employed in their preparation could be successfully applied for development of novel food packaging materials.

The most intensive researches are focused on layered silicates, such as natural Sodium MontMorillonite (MMT) clays. Organomodified clays can be prepared in order to enhance interaction with the polymer chain; several types are commercially available from Southern Clay (Cloisite®), Nanocor (Nanomer®), Elementis Specialities (Bentone®), Sud Chemie (Nanofil®).

For instance, Mascheroni et al. (2010, 2011) reported the preparation of a wheat gluten/sodium MMT system as matrix for controlled delivery of the natural antimicrobial agent carvacrol, with promising application in the field of "green" active packaging.

Progress about the use of nanoclays in combination with polysaccharides has been recently reviewed by Chivrac et al. (2009).

Other fillers that have been used in combination with biopolymers include chitin whiskers (Li et al., 2011), multi-walled carbon nanotubes (Famá et al., 2011), nano-Ag and nano-TiO_2 (Hu et al., 2011), cellulose whiskers and microfibrils (Belbekhouche et al., 2011).

Nanofillers are a promising strategy to improve biopolymer films properties and, therefore, enlarge their use as food packaging materials. However, as pointed out by Falguera et al. (2011), there are many safety concerns about nanomaterials, since, due to their size, nanoparticles can penetrate into living cells and remain in the body. There is a lack of studies about their effects on human health. Therefore, it is necessary to evaluate their toxicity and the effects produced by chronic exposure before these materials could be commercialized for food packaging.

Conclusion and Future Trends

In this chapter, it has been shows that food packaging can meet the philosophy of sustainable development also thanks to membrane science and technology. In fact, membranes can be applied to two novel concepts of food packaging (Modified Atmosphere Packaging and Active Packaging). These two innovative systems can improve food quality/safety, prolonging shelf life and better preserving hygiene and nutritional value. This could help in reducing the amount of food wasted by consumers, thus reducing the exploitation of natural resources.

Moreover, the know-how of membranologists can be exploited in producing food packaging materials based on biopolymers, in order to overcome problems connected with their low mechanical resistance and/or difficult processability.

Therefore, membranes could be successfully applied to produce better packaging, with enhanced performances and reduced environmental impact. In the next years, a growing acceptance of these innovative food packaging systems at industrial level is expected.

Future challenges are replacements of chemical cross-linkers with natural substances as genipin and better investigation of nanofillers effect on human health. If it will be proved that these systems are not dangerous or that they could be embedded into the film avoiding their release, their use will be enlarged as food packaging materials. Nanofillers can be applied to improve properties of biofilms with promising applications in the field of green and sustainable packaging.

Appendix: Nomenclature

Symbols

D	Diffusivity (cm^2/s)
P	Permeability (cm^3 (STP) cm/cm^2 s cmHg)
S	Solubility (cm^3 (STP) / cm^3 cmHg)
$\alpha_{A/B}$	Selectivity (-)
RQ	Respiratory Quotient (-)

Acronyms

AP	Active Packaging
EPA	Environmental Protection Agency
GS	Gas Separation
LDPE	Low Density Polyethylene
MAL	Microbiological Acceptability Limit
MAP	Modified Atmosphere Packaging
MLF	Multilayer Film
MMM	Mixed Matrix Membranes
MMT	Mont-morillonite
WCED	World Commission on Environment and Development

Abbreviations for Polymers and Chemicals

AITC	Allyl isothiocyanate
BOPP	Bi-axially oriented poly propylene
OPP	Oriented poly propylene
PCL	Polycaprolactone
PEEK-WC	Poly ether etherketone
PLA	Poly lactic acid

References

Abdorreza M. N., Cheng L. H., Karim A. A. (2011), 'Effects of plasticizers on thermal properties and heat sealability of sago starch films', *Food Hydrocolloids*, 25, 56-60.

Appendini A., and Hotchkiss J. H. (2002), 'Review in antimicrobial food packaging', *Innov. Food Sci. Emerg. Tech.*, 3, 113-126.

Atarés L., Pérez-Masiá R., Chiralt A. (2011), 'The role of some antioxidants in the HPMC film properties and lipid protection in coated toasted almonds', *Journal of Food Engineering*, 104, 649–656.

Avérous L. (2004). 'Biodegradable Multiphase Systems Based on Plasticized Starch: A Review', *Journal of Macromolecular Science, Part C: Polymer Reviews*, 44, 231-274.

Baker R. W. (2002), 'Future Directions of Membrane Gas Separation Technology', *Ind. Eng. Chem. Res.*, 41, 1393-1411.

Baker R. W. (2004), 'Medical Applications of Membranes' in *Membrane Technology and Applications*, McGraw-Hill, USA, ISBN 0-07-135440-9.

Belbekhouche S., Bras J., Siqueira G., Chappey C., Lebrun L., Khelifi B., Marais S., Dufresne A. (2011), 'Water sorption behavior and gas barrier properties of cellulose whiskers and microfibrils films', *Carbohydrate Polymers*, 83, 1740–1748.

Bernardo P., Drioli E. and Golemme G. (2009), 'Membrane Gas Separation: A Review/State of the Art', *Ind. Eng. Chem. Res.*, 48, 4638–4663.

Brunetti A., Simone S., Scura F., Barbieri G., Figoli A., Drioli E., (2009) 'Hydrogen mixture separation with PEEK-WC asymmetric membranes', *Separation and Purification Technology*, 69, 195-204.

Cameron A. C., Talasila P. C. and Joles D. W. (1995), 'Predicting film permeability needs for modified-atmosphere packaging of lightly processed fruits and vegetables', *Hort. Sci.* , 30, 25.

Chana-Thaworn J., Chanthachum S., Wittaya T. (2011), 'Properties and antimicrobial activity of edible films incorporated with kiam wood (Cotyleobium lanceotatum) extract', *LWT - Food Science and Technology*, 44, 284-292.

Chivrac F., Pollet E., Averous L. (2009), 'Progress in nano-biocomposites based on polysaccharides and nanoclays', *Materials Science and Engineering*, R 67, 1–17.

Conte A., Mastromatteo M., Antonacci D., and Del Nobile M. A. (2011a), 'Influence of cultural practices and packaging materials on table grape quality', *Journal of Food Process Engineering*, 1-7.

Conte A., Scrocco C., Brescia I., Del Nobile M. A. (2009a), 'Different packaging strategies for fresh-cut 'barattiere' melon cultivar (cucumis melo L.)', *International Journal of Food Science and Technology*, 44, 1422–1428.

Conte A., Scrocco C., Brescia I., Mastromatteo M., Del Nobile M. A. (2011b), 'Shelf life of fresh-cut Cime di rapa (Brassica rapa L.) as affected by packaging', *LWT - Food Science and Technology*, 44, 1218-1225.

Conte A., Scrocco C., Lecce L., Mastromatteo M., Del Nobile M. A. (2009b), 'Ready-to-eat sweet cherries: Study on different packaging systems', *Innovative Food Science and Emerging Technologies*, 10, 564–571.

Dainelli D., Gontardb N., Spyropoulos D., Zondervan-van den Beuken E. and Tobback P. (2008), 'Active and intelligent food packaging: legal aspects and safety concerns', *Trends in Food Science and Technology*,19,S103-S112.

EPA, U.S. Environmental Protection Agency (2009). *Sustainable Materials Management: The Road Ahead.* Retrieved from http://www.epa.gov/wastes/inforesources/pubs/vision.htm.

European Council (1994), 'Directive 94/62/EC on packaging and packaging waste', *Official Journal of the European Union*, L365, 10-23.

Falguera V., Gatius F., Pagan J., Ibarz A. (2010), 'Kinetic analysis of melanogenesis by means of Agaricus bisporus tyrosinase', *Food Research International*, 43, 1174-1179.

Falguera V., Quintero J. P., Jimenez A., Aldemar Munoz J. and Ibarz A. (2011), 'Edible films and coatings: Structures, active functions and trends in their use', *Trends in Food Science and Technology*, 22, 292-303.

Famá L. M., Pettarin V., Goyanes S. N., Bernal C. R. (2011), 'Starch/multi-walled carbon nanotubes composites with improved mechanical properties', *Carbohydrate Polymers*, 83, 1226–1231.

Figoli A., Drioli E., Jansen J. J. M. and Wessling M. (2004), 'Film antimicrobico per prolungare la shelf life', ISSN0033-9687, *Rassegna dell'Imballagio*, Nov. 2004, N°16, Year 25th.

Figoli A., Mascheroni E., Limbo S., Drioli E. (2010), 'Membranes for food packaging, Cap. 10', in *Membrane Technology, Volume 3: Membranes for food applications*, edited by Peinemann K. V., Nunez S. P. and Giorno L., Wiley-VCH Verlag GmbH and Co. KGaA, Weinheim, 223-240.

George S. C., Thomas S. (2001), 'Transport phenomena through polymeric systems', *Progr. Polym. Sci.*, 26, 985-1017.

Gialamas H., Zinoviadou K. G., Biliaderis C. G., Koutsoumanis K. P. (2010), 'Development of a novel bioactive packaging based on the incorporation of Lactobacillus sakei into sodium-caseinate films for controlling Listeria monocytogenes in foods', *Food Research International*, 43, 2402–2408.

Gómez-Estaca J., Gómez-Guillén M. C., Fernández-Martín F., Montero P. (2011), 'Effects of gelatin origin, bovine-hide and tuna-skin, on the properties of compound gelatine chitosan films', *Food Hydrocolloids*, 25, 1461-1469.

González A., Strumia M. C., Alvarez Igarzabal C. I. (2011), 'Cross-linked soy protein as material for biodegradable films: Synthesis, characterization and biodegradation', *Journal of Food Engineering*, 106, 331–338.

Guarda A., Rubilar J. F., Miltz J., Galotto M. J. (2011), 'The antimicrobial activity of microencapsulated thymol and carvacrol', *International Journal of Food Microbiology*, 146, 144–150.

Hu Q., Fang Y., Yang Y., Ma N., Zhao L. (2011), 'Effect of nanocomposite-based packaging on postharvest quality of ethylene-treated kiwifruit (Actinidia deliciosa) during cold storage', *Food Research International*, 44, 1589–1596.

Iulianelli A. and Basile A. (2011), 'Hydrogen production from ethanol via inorganic membrane reactors technology: a review', *Catal. Sci. Technol.*, 1, 366–379.

Jansen J. J. M. and Figoli. A. (2004), *Preparation and characterization of a new antimicrobial polymer film for food packaging application, Master thesis*. Faculty of Science and Technology, University of Twente, Enschede (NL) and Institute on Membrane Technology (ITM-CNR), Italy.

John J., Tang J., Bhattacharya M. (1998), 'Processing of biodegradable blends of wheat gluten and modified polycaprolactone', *Polymer*, 39, 2883-2895.

Kantola M. and Helen H. (2001), 'Quality changes in organic tomatoes packaged in biodegradable plastic films', *Journal of Food Quality*, 24, 167-176.

Lee D. S. (2005), 'Packaging containing natural antimicrobial or antioxidative agents', in J. H. Han (Ed.), *Innovations in food packaging*, Elsevier Academic Press.

Lee D. S., Haggar P. E., Lee J., and Yam K. L. (1991), 'Model for fresh produce respiration in modified atmospheres based on principles of enzyme kinetics', *J. Food. Sci.* , 56, 1681.

Lee D. S., Yam K., Piergiovanni L. (2008), 'Overview of Food Packaging System, Ch. 1', In *Food Pack. Sci. Tech.*, CRC Press, 5-7.

Li X., Li X., Ke B., Shi X., Du Y. (2011) 'Cooperative performance of chitin whisker and rectorite fillers on chitosan films', *Carbohydrate Polymers*, 85, 747–752.

Lucera A., Costa C., Mastromatteo M., Conte A., Del Nobile M. A. (2010), 'Influence of different packaging systems on fresh-cut zucchini (Cucurbita pepo)', *Innovative Food Science and Emerging Technologies*,11,361–368.

Marcuzzo E., Sensidoni A., Debeaufort F., Voilley A. (2010), 'Encapsulation of aroma compounds in biopolymeric emulsion based edible films to control flavour release', *Carbohydrate Polymers*, 80, 984–988.

Martínez-Camacho A. P., Cortez-Rocha M. O., Ezquerra-Brauer J. M., Graciano-Verdugo A. Z., Rodriguez-Félix F., Castillo-Ortega M. M., Yépiz-Gómez M. S., Plascencia-Jatomea

M. (2010), 'Chitosan composite films: Thermal, structural, mechanical and antifungal properties', *Carbohydrate Polymers*, 82, 305–315.

Mascheroni E., Chalier P., Gontard N., Gastaldi E. (2010), 'Designing of a wheat gluten/montmorillonite based system as carvacrol carrier: Rheological and structural properties', *Food Hydrocolloids*,24, 406–413.

Mascheroni E., Chalier P., Gontard N., Gastaldi E. (2011), 'Anti-microbial effectiveness of relative humidity-controlled carvacrol release from wheat gluten/montmorillonite coated papers', *Food Control*, 22, 1582-1591.

Mensitieri G., Di Maio E., Buonocore G. G., Nedi I., Oliviero M., Sansone L. and Iannace S. (2011), 'Processing and shelf life issues of food packaging materials and structures from renewable resources', *Trends in Food Science and Technology*, 22, 72-80.

Muratore G., Del Nobile M. A., Buonocore G. G., Lanza C. M., Nicolosi Asmundo C. (2005), 'The influence of using biodegradable packaging films on the quality decay kinetic of plum tomato (PomodorinoDatterino®)', *Journal of Food Engineering*, 67, 393–399.

Murphy T. M., Offord G. T., Paul D. R., 'Fundamentals of Membrane Gas Separation, Ch. 4', in *Membrane Operations* (eds. Drioli E. and Giorno L.), Germany, Wiley-VCH, 63-82.

Paul D. R., Yampol'skii Y. P. (1994), 'Introduction and perspective, Ch. 1', in *Polymeric Gas-separation membrane*, eds Paul D. R. and Yampol'skii Y. P., CRC Press Inc, 2-15.

Pereda M., Ponce A. G., Marcovich N. E., Ruseckaite R. A., Martucci J. F. (2011), 'Chitosan-gelatin composites and bi-layer films with potential antimicrobial activity', *Food Hydrocolloids*, 25, 1372-1381.

Rao M. S., Kanatt S. R., Chawla S. P., Sharma A. (2010), 'Chitosan and guar gum composite films: Preparation, physical, mechanical and antimicrobial properties', *Carbohydrate Polymers*, 82, 1243–1247.

Robeson L. M. (2008), 'The upper bound revisited', *Journal of Membrane Science*, 320, 390.

Rodriguez-Aguilera R., Oliveira J. C. (2009), 'Review of Design Engineering Methods and Applications of Active and Modified Atmosphere Packaging Systems', *Food Eng. Rev.*, 1, 66–83.

Sánchez-González L., Cháfer M., González-Martínez C., Chiralt A., Desobry S. (2011), 'Study of the release of limonene present in chitosan films enriched with bergamot oil in food simulants', *Journal of Food Engineering*, 105, 138–143.

Sarazin P., Li G., Orts W. J., Favis B. D. (2008), 'Binary and ternary blends of polylactide, polycaprolactone and thermoplastic starch', *Polymer*, 49, 599-609.

Srinivasa P. C., Baskaran R., Ramesh M. N., Harish Prashanth K. V. (2002), Tharanathan R. N., 'Storage studies of mango packed using biodegradable chitosan film', *Eur. Food Res. Technol.* , 215, 504–508.

Torchia A. M., Clarizia G., Figoli A., Drioli E. (2004), 'Potentiality of PEEKWC as new material in food packaging', *Ital. J. Food Sci.*, XV, 173-184.

WCED, World Commission on Environment and Development (1987), *Our Common Future*, Oxford, Oxford University Press.

Wikström F., Williams H. (2010), 'Potential environmental gains from reducing food losses through development of new packaging - a life-cycle model', *Packaging Technology and Science*, 23, 403–411.

Williams H., Wikström F. (2011), 'Environmental impact of packaging and food losses in a life cycle perspective: a comparative analysis of five food items', *Journal of Cleaner Production*, 19, 43–48.

Yam K. L., Takhistov P. T., Miltz J. (2005), 'Intelligent packaging: concepts and applications', *J. Food Sci.* , 70, R1–R10.

Yu Q., Song Y., Shi X., Xu C., Bin Y. (2011), 'Preparation and properties of chitosan derivative/poly(vinyl alcohol) blend film crosslinked with glutaraldehyde', *Carbohydrate Polymers*, 84, 465–470.

Zolandz R. R., Fleming G. K. (1992), 'Gas Permeation theory', in *Membrane Handbook* (eds Ho W. S. W. and Sirkar K. K.), Van Noostrand Reinhold, ISBN 0-442-23747-2.

In: Membrane Processes for Sustainable Growth
Editors: Angelo Basile and Alfredo Cassano
ISBN: 978-1-62618-446-6

Chapter 5

MEMBRANE AND MEMBRANE REACTORS FOR SUSTAINABLE GROWTH

Angelo Basile*, Sudip Chakraborty and Adolfo Iulianelli
Institute on Membrane Technology (ITM-CNR),
University of Calabria, Rende (Cosenza), Italy

ABSTRACT

During the last years, the growing interest towards inorganic membranes for hydrogen production/purification as well as CO_2 sequestration is justified by their different benefits such as high resistance to chemical degradation, well-defined stable pore structure (in the case of porous inorganic membranes), high mechanical stability and elevated resistance at high operating temperatures. This chapter describes how the inorganic membrane technology can be applied fruitfully both to generate and separate hydrogen and concentrate CO_2 to be further stored. In the meanwhile, by using polymeric membranes, CO_2 can be removed at relatively low temperature with good selectivity and in a cost-effective approach. Furthermore, this chapter also pays attention to the membrane bioreactor technology, pointing out the new findings in this area and describing the potentiality of its application in true life.

Keywords: membrane technology; membrane bioreactors; sustainable growth

LIST OF SYMBOLS AND ACRONYMS

A	Membrane surface area
BSMBR	Biomass separation membrane bioreactor
CD	Cryogenic distillation
CFC	Clorefluoro carbon
D_{eff}	Effective diffusivity

* Corresponding Author: e-mail: a-basile@itm.cnr.it

EDC	Endocrine Disruption Chemicals
EMBR	Extractive membrane bioreactor
FBR	Fixed bed reactor
IEMBR	Ion exchange membrane bioreactor
J_{H2}	hydrogen flux through membrane
K_n	Knudsen Number
K_p/K_r	Restricted diffusion roughly
MABR	Membrane aeration bioreactor
MEBR	Membrane enzyme bioreactor
MPBR	Membrane pervaporation bioreactor
MR	Membrane reactor
MS	Membrane separation
MSR	Methane steam reforming
NF	Nanofiltration
Pe_{H2}	hydrogen permeability
PEMFC	Proton exchange membrane fuel cells
$p_{H2,perm}$	hydrogen partial pressures in permeate side
$p_{H2,ret}$	hydrogen partial pressures in retentate side
P_i	permeability of gas
POP	Persistent Organic Pollutants
PSA	Pressure swing adsorption
PSS	porous stainless steel
Q	Molar Flow Rate
RO	Reverse Osmosis
TMP	Transmembrane pressure
WGS	Water gas shift
ε_p	Porosity (or area of support/ area of pores);
τ	Tortuosity
Δp	Pressure difference

1. INTRODUCTION

In the view point of sustainable growth, it should be taken into account that, since the economic boom of the 1990s, the increased worldwide energy demand, coupled with the growing attention towards environmental issues, has led to renewed emphasis on the so-called "hydrogen economy" (Bockris, 1971). In the meanwhile, fuel cell technology is strictly linked to hydrogen economy because proton exchange membrane fuel cells (PEMFCs) are able to convert the chemical energy of hydrogen as a feeding fuel directly into electric energy at moderate temperatures as well as useful also as batteries for storing energy (Steele and Heinzel, 2001). Furthermore, PEMFCs offer several benefits such as high efficiency compared to conventional internal combustion systems, zero pollutant emissions, high energy density, quiet operation and easy control. In this contest, the main industrial process for hydrogen production is methane steam reforming (MSR), by exploiting off-gas and natural gas. Nevertheless, hydrogen needs to be purified before being supplied to PEMFCs. Indeed, the final CO concentration present in the reformed hydrogen-rich stream is well above the allowable levels (CO concentration <10 ppm for on-board vehicle applications – 2007 DOE Guidelines).

Therefore, hydrogen needs to be purified and, conventionally, various processes are available such as pressure swing adsorption (PSA), cryogenic distillation (CD) or membrane separation (MS). In particular, the conventional processes accomplish carbon dioxide removal by absorption followed by methanation or oxidizing remaining carbon oxides (in addition to water gas shift reaction). Since the 1980s, most of the existing hydrogen plants utilize PSA technology to recover and purify hydrogen at levels above 99.9% (Stocker, 1998). Commonly, a commercial PSA unit typically consists of a multi-column system formed by 4-12 adsorbent beds, even if more beds are useful to provide higher hydrogen recovery and/or higher capacity (Miller, 1999).

In the last decades, a growing interest has been paid towards membrane separation process. Indeed, nowadays membrane-based process is considered one of the most promising technologies for producing high purity hydrogen. The MS process is based on the selective permeation of hydrogen through such membrane. Table 1 summarizes the comparative characteristics and performance of three technologies for purifying hydrogen: PSA, CD and MS. It appears quite evident that MS seems to represent the most interesting and cost effective way to recover high purity hydrogen compared to the conventional systems. Therefore, this chapter emphasizes the role of the membrane technology in the field of hydrogen production and its purification, combined to the PEMFCs technology as a possible and sustainable scenario for green energy production in the future. In the meanwhile, another important issue related to the CO_2 emissions caused by the industrial processes and many other variables affecting the climate change is considered and discussed. Also in this case, membrane technology applied to CO_2 sequestration may play an important role, particularly if linked to pre-combustion and post-combustion industrial processes (Iulianelli et al., 2011; Basile et al. 2010).

Table 1. Comparison between different hydrogen purification processes

Index	PSA	Cryogenic Distillation	Membranes [a]
H_2 recovery / vol.%	75-90	90-98	< 90
H_2 purity / vol. %	$\geq$ 99.9	95-99	< 95
H_2 product pressure	Feed Pressure	Variable	<<Feed pressure
By products available	No	Yes	No
Feed Pressure / bar	10-50	15-35	15 -125
Energy Consumption	Intensive	Very Intensive	Low
Flexibility to expansion	Very Good	Good	Limited
Reliability	High	Moderate	High
Capital Cost	High	Very High	Low

[a] Values typically used in refineries (Patel et al., 2005).

In the last part of the chapter, the bio-membrane technology concludes the investigation on the possible applications of membrane technology in the view of sustainable growth, paying special attention to the bio-membrane systems.

2. ZERO EMISSIONS FUEL CELLS AND HYDROGEN PRODUCTION/PURIFICATION VIA MEMBRANE TECHNOLOGY

In the last years, the scientific community has done many efforts in the issue involving the decrease of oil reserves and the increase of the environmental pollution caused by the emissions of CO_2 and other greenhouse gases. Various strategies have been proposed to make possible a sustainable growth, which indicated that it is strongly necessary to develop new technologies and to exploit alternative/renewable sources with respect to the fossil fuels. On this route, the combination of hydrogen and electricity represents an important solution to realize sustainable energy. Hydrogen fuelled fuel cells represent the most efficient devices for converting hydrogen into electricity, Figure 1. In particular, PEMFCs may produce electricity directly from hydrogen and oxygen, without combustion and with zero pollutants emission. Several benefits may be noticed for PEMFCs such as: 1) low operative temperatures (80 – 100 °C), 2) high current density, 3) compactness, 4) fast start-up, 5) suitability for discontinuous operation. These features make PEMFCs the most promising and attractive alternative technologies for a wide variety of power applications ranging from portable/micropower and transport to large-scale stationary power systems for buildings and distributed generation, Figure 2 (Costamagna et al., 2001).

At industrial level, hydrogen is produced as a hydrogen-rich stream and, as a particular case, the "syngas" is characterized by high hydrogen content (around 50%) besides other by-products. Generally, as a first step the production of syngas takes place in a gasifier. Subsequently, syngas is cooled down to remove its contaminant, especially H_2S and, then, it is reheated to be further reacted with steam in a WGS reactor to maximize its hydrogen content. WGS reaction is exothermic and its equilibrium conversion decreases with temperature.

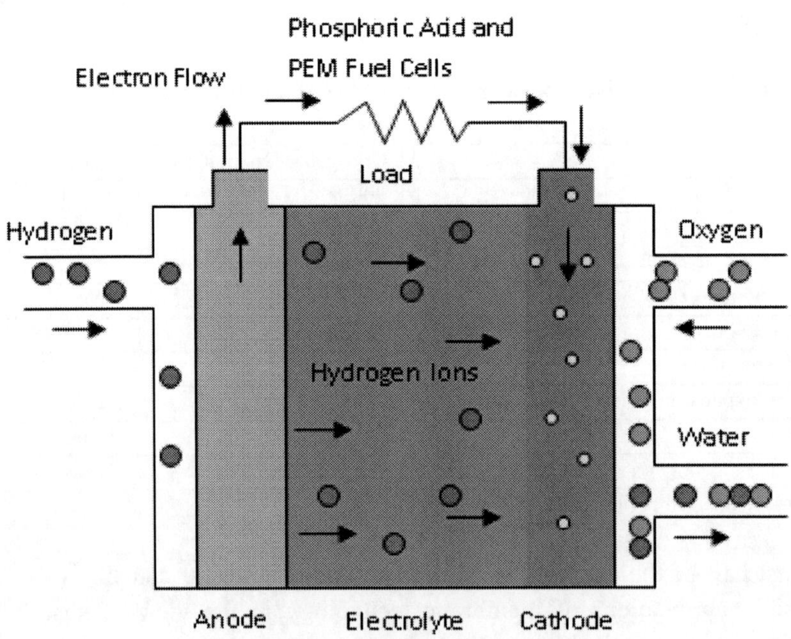

Figure 1. Scheme of a generic PEMFC.

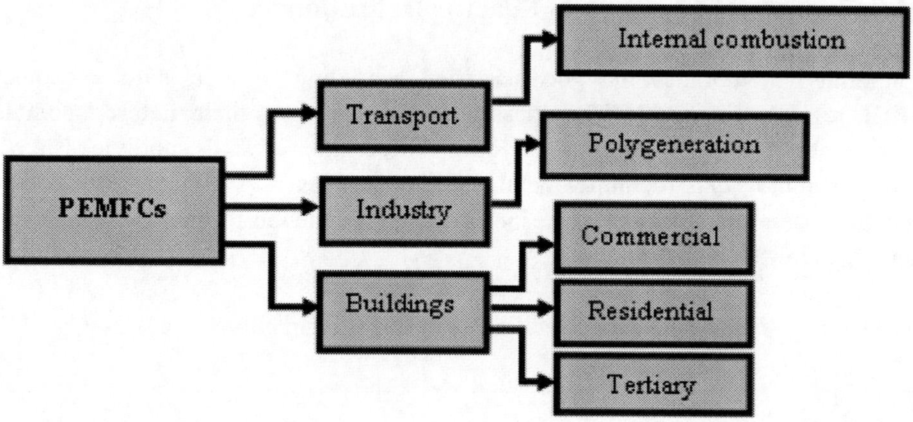

Figure 2. Scheme of possible PEMFC applications.

Therefore, commonly two reactors, the first one at high temperature and the second one at lower temperature, are useful to increase CO conversion, which represents the main contaminants of hydrogen when supplied to PEMFCs. The stream coming out from WGS stage, although it contains a CO concentration smaller than 0.5%, must be treated in further separation units to produce hydrogen with a CO content tolerable by PEMFCs, typically ≤ 10 ppm of CO for avoiding to poison the anodic Pt-based catalyst, (Basile and Iulianelli, 2007).

Otherwise, it would be more convenient to convert CO into hydrogen via WGS reaction using inorganic membrane reactor technology. Indeed, at scientific level, palladium-based membrane reactors (MRs) represent a sustainable technology able to perform simultaneously both WGS reaction and hydrogen separation/purification in the same stage.

In the last decades, the inorganic membranes have attracted the scientific interest owing to their high hydrogen permeation selectivity (Grashoff et al., 1983).In particular, dense self-supported palladium membranes are full selective to hydrogen permeation through the membrane with respect to all other gases. Therefore, they are usually assembled in a membrane reactor with the intent to recover a pure hydrogen stream.

In the meanwhile, palladium-based MRs allow to concentrate in CO_2 the retentate stream (the stream not permeated through the membrane), which could be further separated via polymeric membrane technology. In the following paragraph, a short overview about membranes and membrane reactors is given, pointing out the main characteristics of the inorganic membranes used for hydrogen separation, their benefits and drawbacks.

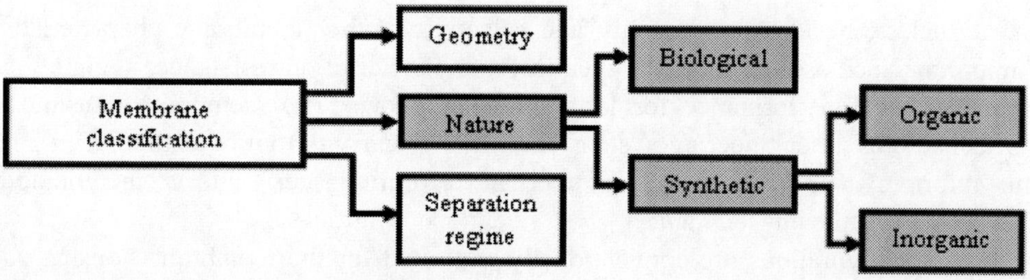

Figure 3. Scheme of membranes classification.

2.1. Membrane and Membrane Reactor Definitions

Commonly, membranes may be classified according to their nature, geometry and separation regime, Figure 3. The classification based on their nature subdivides the membranes into *biological* and *synthetic* ones. These latter are still subdivided into *organic* (polymeric) and *inorganic* (ceramic, metal, etc.) membranes.

The key parameters, which correspond to membrane performances, are in particular summarized below:

Permeability (P), the flux of a specific gas through the membrane (1):

$$\frac{P}{\delta} = \frac{Q}{A\Delta p} \tag{1}$$

where δ is the effective thickness of the membrane, Q the molar flow rate of the gas permeating through the membrane, A the membrane surface and Δp is the pressure difference across the membrane; *ideal gas selectivity (α)*, the ratio between the gas permeating flux, J, of pure gas "i" over that of pure gas "j"(2):

$$\alpha_{ij} = \frac{J_i}{J_j} \tag{2}$$

For many membranes, there is a trade-off between selectivity and permeability. A highly permeable membrane tends to have low selectivity and vice versa. In the field of polymeric membranes, as schematically shown in Figure 4, Robeson (2008) suggested that this trade-off may be illustrated as the "log" of selectivity versus "log" of permeability of the more permeable gas for a binary mixture. In this figure, the upper-bound sets the limit to reach both high selectivity and high permeability. The upper-bound relationship is expressed by the equation reported below (3):

$$P_i = k\alpha_{ij}^n \tag{3}$$

where P_i is the permeability of the more permeable gas and "n" is the slope of the log–log limit.

Nevertheless, besides permeability and selectivity, other membrane characteristics are very important such as their thermal, chemical and plasticization resistance, aging effects for ensuring continual performance for long time operations. The membrane housing into a reactor constitutes an engineering system, the MR, for carrying out contemporary a reaction (steam reforming, dry reforming, autothermal reforming, etc.) and a membrane-based separation in the same physical unit.

Various configurations are conventionally exploited for the membrane housing such as plate-and-frame or tubular systems. For example, in Figure 5, a tubular MR configuration is illustrated.

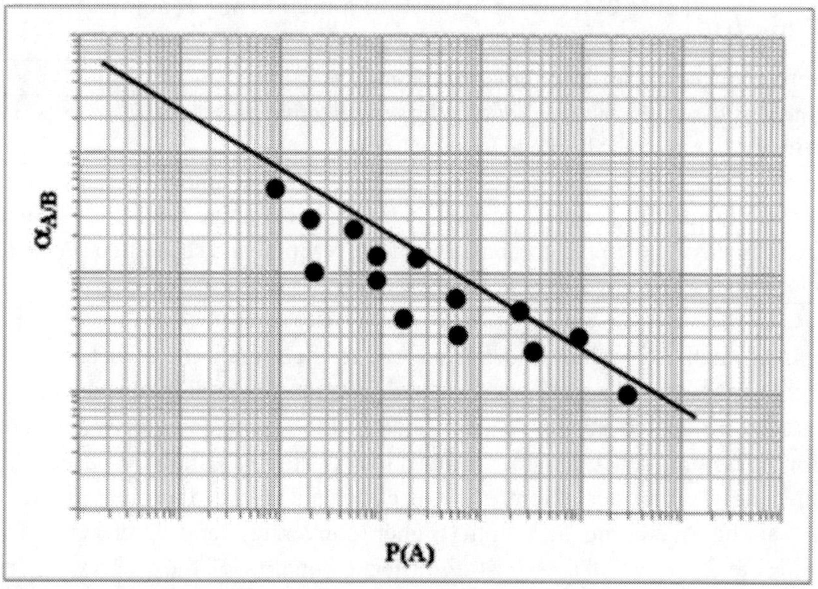

Figure 4. Upper bound correlation for general binary mixture separation.

Commonly, a MR is a device able to remove selectively a product from the reaction system or introduces a reactant, reaching better performances in terms of conversion and/or yield than a conventional fixed bed reactor (FBR) operating at the same MR conditions.

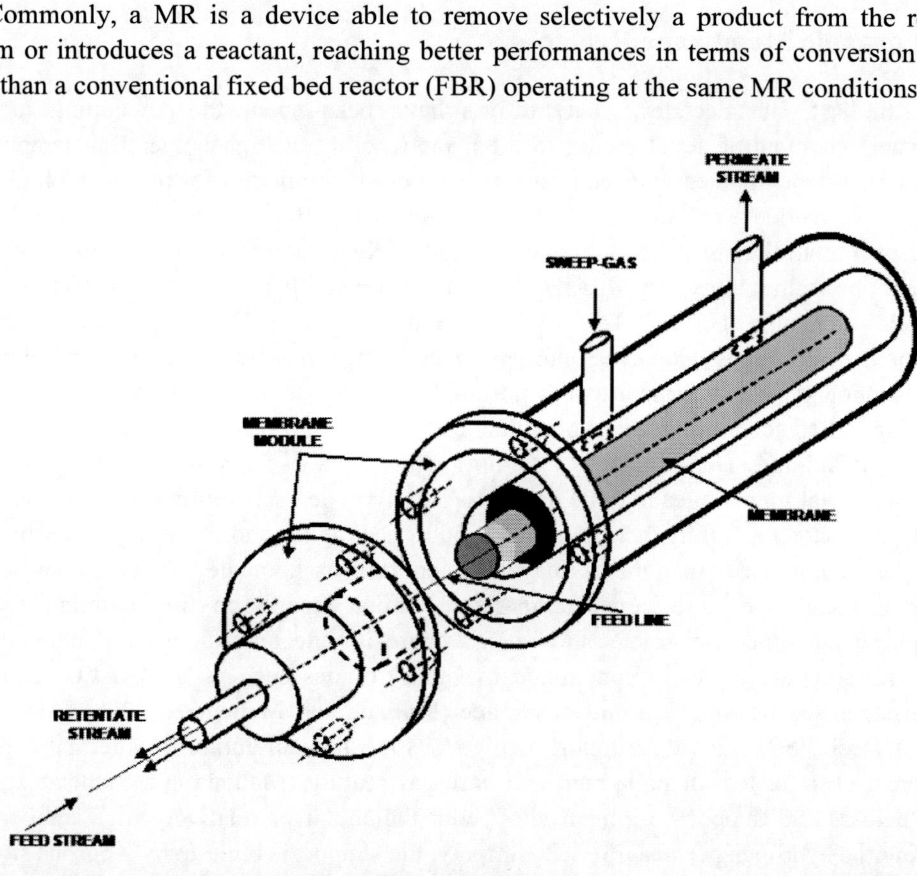

Figure 5. Generic scheme of a tubular membrane reactor.

Table 2. Generic benefits of a membrane reactor

Conversion enhancement of the equilibrium limited reactions
Enhancement of the hydrogen yield and hydrogen selectivity (in the field of MRs used for hydrogen production)
Achievement of the same performances attained in the FBR at milder operative conditions or, instead, achievement of better performances at the same operative conditions as in the FBR
Reduced capital costs due to the combination of the reaction and separation process in only one system

This enhancement of performances is related to the membrane effect, which favours the shifting of conversion of an equilibrium limited reaction towards the products, by selectively removing a such product from the reaction side to the permeate side. When inorganic MRs are used for reforming reactions to produce hydrogen, by using Pd-based membranes to remove selectively hydrogen from the reaction side, the thermodynamic equilibrium restrictions can be overcome and both higher conversion and hydrogen yields may be reached. In Table 2, some of the most significant benefits of the MRs with respect to the FBRs are summarized.

2.2. Inorganic Membrane Reactors

In the last four decades, many efforts have been spent for developing inorganic membranes at scientific level owing to their applicability at higher operating temperatures than polymeric membranes. In particular, the most used membranes for producing high purity hydrogen are the dense palladium-based membranes due to the full hydrogen perm-selectivity with respect to all the other gases (Gao et al., 2004). Nevertheless, the recent literature on the argument shows that alloys such as Zr- V- and Nb-based, Ni-Nb-Zr (Yamaura et al., 2003), V-Al alloys (Zhang et al., 2002) or V-Ni (Amano et al., 1991) may represent cheaper solutions than Pd-alloys. However, inorganic membranes may be subdivided on the basis of their morphology and/or structure into porous and dense. In particular, porous membranes may be classified according to their pore diameter in: microporous (d_p< 2 nm), mesoporous (2 nm <d_p< 50 nm) and macroporous (d_p> 50 nm), while the dense membranes have d_p< 0.5 nm.

Dense metal membranes may be useful for the hydrogen separation task and, commonly, they operate at temperature between 300 - 600 °C (Paglieriand Way, 2002).Furthermore, dense palladium-based membranes may be supported or laminated. Then, in the case of supported membranes, the upper temperature limits depend on the material, chemical atmosphere and support characteristics such as porosity and pore diameter (Uemiya, 1999). Porous supports are typically constituted by binary oxides such as SiO_2, Al_2O_3 and B_2O_3, having excellent anchor effect and adherence (Mallada and Menendez, 2008). Also porous stainless steel (PSS) is a valid support owing to its mechanical durability, thermal expansion coefficient close to that of palladium and ease gas sealing (Mallada and Menendez, 2008). Nevertheless, PSS supports form an alloy with palladium at relatively high temperatures, depleting the hydrogen permeability. Commonly, the support is built up by a porous structure, where the main part has relatively large pores (commonly 2–20 μm) and one or more thin layers, with decreasing pore size. In the case of laminated membranes, thin palladium (or palladium alloys) layer avoids the formation of oxides on the metallic surfaces, resulting in a hydrogen adsorption activation energy reduction and in an increase of the hydrogen permeation flux. Hydrogen permeation through these membranes may take place by means of a solution-diffusion mechanism, constituted by a complex process with several stages (Adhikari and Fernando, 2006; Dittmeyer et al., 2001):

- dissociation of molecular hydrogen at the gas/metal interface;
- adsorption of the atomic hydrogen on membrane surface;
- dissolution of atomic hydrogen into the palladium matrix;
- diffusion of atomic hydrogen toward the opposite side;
- re-combination of atomic hydrogen to form hydrogen molecules at the gas/metal interface;
- desorption of hydrogen molecules.

Therefore, the permeation of hydrogen through a metal membrane is a function of the solubility and diffusion of hydrogen in the metal, besides the membrane thickness and temperature influence. At a fixed temperature, the hydrogen permeation flux through a dense palladium membrane is generically described by the following equation (4):

$$J_{H2} = Pe_{H2}\,(p^{n}_{H2,ret} - p^{n}_{H2,perm})/\delta \qquad (4)$$

where: J_{H2} is the hydrogen flux through the membrane; Pe_{H2} the hydrogen permeability; δ the membrane thickness; $p_{H2,ret}$ and $p_{H2,perm}$ the hydrogen partial pressures in the retentate (the reaction side) and permeate sides (the volume in which the hydrogen permeating through the membrane is collected), respectively, and n (in the range 0.5–1) is the dependence factor of the hydrogen flux on the hydrogen partial pressure.

This equation even points out the inverse proportionality to the membrane thickness. Indeed, the role of the membrane thickness is very important: on one hand, a thinner membrane offers a higher permeability; on the other hand, thicker membranes are necessary in order to ensure the mechanical resistance and strength.

In the case of relatively low pressure, the diffusion is assumed to be the rate-limiting step and the factor n is equal to 0.5. In this case, the equation (4) becomes the Sieverts-Fick law (5):

$$J_{H2,Sieverts} = Pe_{H2}\,(p^{0.5}_{H2,ret} - p^{0.5}_{H2,perm})/\delta \qquad (5)$$

Otherwise, at high pressures the hydrogen-hydrogen interactions in the palladium bulk are not negligible. In this case, n becomes equal to 1(6):

$$J_{H2} = Pe_{H2}\,(p_{H2,ret} - p_{H2,perm})/\delta \qquad (6)$$

One of the main issues concerning the utilization of pure palladium membranes is the "hydrogen embrittlement" phenomenon. At temperature lower than 300 °C and pressure below 2.0 MPa, theβ-hydride phase may nucleate from theα-phase, resulting in severe lattice strains. Therefore, pure palladium becomes brittle after a few cycles of α=β transitions (Grashoff et al., 1983).

These transitions do not involve in a change of the lattice structure, but in a lattice dilatation. β-hydride phase formation is represented by a clustering of hydrogen atoms, whose energy of attraction, being associated with the lattice, strains around the dissolved hydrogen atom (Brodowsky, 1972). In order to avoid this phenomenon, a possible solution is represented by the use of a Pd-alloy containing other metals, such as silver. Indeed, the role of silver is explained by its electron donating behaviour, being largely similar to the one of the hydrogen atom in palladium. Silver and hydrogen atoms would compete for the filling of electron holes (Shu et al., 1991).

As reported in Table 3, another critical issue is represented by the palladium surface contamination of some compounds such as Hg vapour, hydrogen sulfide, SO_2, thiophene, arsenic, unsaturated hydrocarbons, chlorine carbon from organic materials. These chemical compounds can poison the membrane surface reducing the effective hydrogen fluxes by 20 to 100%.

Porous membranes (ceramics, carbon, etc.) may be also useful to separate hydrogen from other gases such as CO_2, CO, CH_4 by allowing the smaller hydrogen molecules (diameter = 2.83 Å) to diffuse through the membrane pores and, thus, by leaving behind larger molecules (Phair and Badwal, 2006). Indeed, the membrane acts as a molecular sieve, whose efficiency is determined by the pore size and distribution relative to the molecule size. As a consequence, the hydrogen purity is not so high. (Phair and Badwal, 2006).

Table 3. Palladium surface contamination of some compounds

Effect of contaminant	
Poisoning of sulphur compounds	It is irreversible process and causes the conversion of palladium in palladium sulphide, whose lattice constant is twice than pure Pd and, thus, the structural stress leads to the formation of cracks. For example, Pd-Cu and Pd-Au increase the resistance to H_2S.
Inhibition due to CO	It could cause a decrease in the hydrogen permeation flux, because the CO is adsorbed on the palladium surface blocking available dissociation sites for hydrogen. Moreover, this reduction becomes more significant at low temperature (below 150 °C) or high CO concentration.
Inhibition due to H_2O	The adsorbed water molecules dissociate on the surface of the Pd film: $H_2O_{ads} \rightarrow OH_{ads} + H_{ads}$ and/or $H_2O_{ads} \rightarrow O_{ads} + 2H_{ads}$ where the Hads may permeate into the bulk of the Pd film. H2O is recombined through these reactions: $2OH_{ads} \rightarrow H_2Ogas + O_{ads}$ and/or $O_{ads} + 2H_{ads} \rightarrow H_2Ogas$ Therefore, the process of H_2O dissociation/recombinative desorption contaminates the palladium surface with adsorbed O.
Inhibition due to coke	Hydrogen selectivity of a thin palladium membrane decrease after it is brought in contact with coke at elevated temperature. This phenomenon can be addressed to the fact that carbon atoms penetrate into the palladium lattice and cause the expansion of the palladium lattice.

These membranes can be classified in:

- **Carbon membranes**
- *Adsorption selective carbon membranes* separate non-(or weakly) adsorbable from adsorbable gases (such as H_2S, NH_3 and CFCs) on the basis of an adsorption-surface migration mechanism. Carbon membranes may be exploited in non-oxidising environments with temperatures in the range of 500-900 °C. Nevertheless, the performance of these membranes deteriorate severely if feed streams contain organic traces, if they are brittle and, therefore, difficult to package in the case of membrane surfaces become larger. Furthermore, the price of carbon membranes is still high (Kluiters, 2004).
- **Ceramic membranes**
- They possess a two-layer structure: the separation layer is usually made by alumina, zirconia, titania or silica. They can operate in the range of 300-800 °C, in some cases also up to 1000 °C. The main transport mechanisms for porous inorganic membranes are resumed in Table 4. In some cases, the gas molecules can move through the membrane by means of more than one mechanism. The separation factor for these mechanisms depends strongly on pore size distribution, temperature, pressure and interactions between gases being separated and the membrane surfaces.

Table 4. Description of the transport mechanisms for porous inorganic membranes

Knudsen diffusion	The Knudsen number Kn is defined as the ratio of the average free path of fluid molecules (average distance between collisions) and a representative physical length (e.g., the pore radius). The smaller the Knudsen number, the larger the pores become (relative to the average free path of fluid molecules), the lower selectivity becomes. For Knudsen numbers < 1 the dominant transport mechanism is viscous flow, which is non-selective.
Surface diffusion	Gas molecules are adsorbed on the pore walls of the membrane and migrate along the surface. Surface diffusion increases the permeability of the components adsorbing more strongly to the membrane pores. At the same time, the effective pore diameter is reduced. Consequently, transport of non adsorbing components is reduced and selectivity is increased. This positive contribution of surface diffusion only works for certain temperature ranges and pore diameters. Moreover, it can occur in parallel with Knudsen diffusion.
Capillary condensation	It occurs if a condensed phase (partially) fills the membrane pores. If the pores are completely filled with condensed phase, only the species soluble in the condensed phase can permeate through the membrane. Fluxes and selectivities are generally high for capillary condensation. The appearance of capillary condensation, however, strongly depends on gas composition, pore size, and uniformity of pore sizes.
Molecular sieving	If pore sizes become sufficiently small (3.0-5.2 Å), molecular sieving can be used to separate molecules that differ in kinetic diameter: the pore size becomes so small, that only the smaller gas molecules can permeate through the membrane.

Furthermore, it should be considered that inorganic membranes, although more expensive than the polymeric ones, are chosen in the hydrogen purification field owing to their benefits such as high resistance towards solvents, high mechanical stability and elevated resistance at high operating temperatures, as shortly resumed in Table 5.

Table 5. Benefits and drawbacks of inorganic membranes with respect to the polymeric ones

Benefits	Drawbacks
Long-term stability at high temperatures	High capital cost
Resistance to harsh environments (chemical degradation, pH, etc.)	Embrittlement phenomenon (in the case of dense Pd membranes)
Resistance to high pressure drops	Low membrane surface per module volume
Inertness to microbiological degradation	Difficulty of achieving high selectivities in large scale micro-porous membranes
Easy cleanability after fouling	Generally low permeability of the highly H2 selective (dense) membranes at medium temperatures
Easy catalytic activation	Difficult membrane-to-module sealing at high temperature

3. Polymeric Membrane Technology for CO_2 Capture

As another critical issue is represented by the overabundance of CO_2 in the atmosphere, which constitutes one of the main contributors to global climate change. Indeed, it is responsible for about 64% of the greenhouse effect and global warming (Zhou et al., 2009). This has led to scientific interest in development of new methods for reducing the amount of CO_2 emitted into the atmosphere from industries or power plants. Conventionally, there are three families of processes for CO_2 capture: sorption processes (absorption in solvents or adsorption on solid sorbents), cryogenic processes and membrane processes. Among them, polymeric membrane processes show the potentiality of relatively higher capture temperature by eliminating the intermediate cooling and reheating steps, necessary, for example, in the amine-based absorption process. Furthermore, membranes could offer several benefits such as low capital investment, compactness, ease of installation and operation, simplicity, flexibility and ability to perform separations at low energy penalties (Abertz et al., 2006; Basu et al., 2004).

In the field of CO_2 separation, polymeric membranes may be used, for example, in the treatment of CO_2-rich streams, going out from inorganic MRs, previously utilized for hydrogen production/separation by means WGS reaction, as schematized in Figure 6.

Polymeric membranes may be categorized into glassy (prepared at temperatures below the glass transition temperature) and rubbery (prepared at temperatures above the glass transition temperature) membranes.

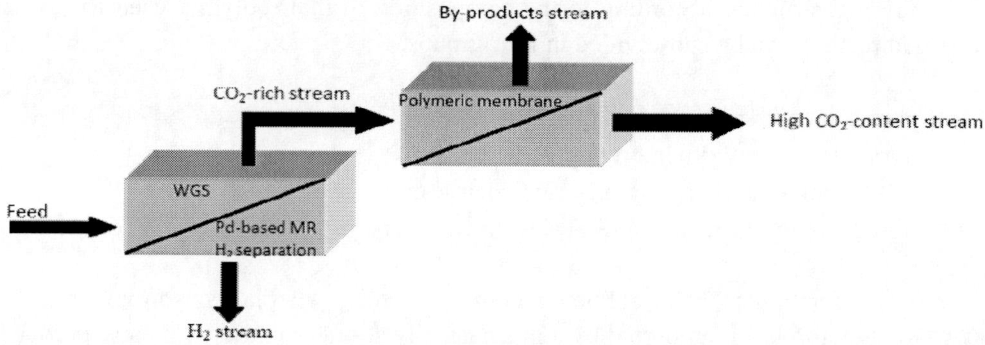

Figure 6. Scheme for theCO_2-rich stream treatment after Pd-based MR stage of WGS reaction.

Glassy membranes show relatively high selectivity and low flux, while rubbery membranes high flux and low selectivity. Operating temperatures are limited up to around 180°C. Among the polymeric membranes with reasonable CO_2 permeability and selectivity, much attention should be paid on the following typologies:

- *Polyamides* are commonly synthesized by the reaction of a diamine with a diahydride (PMDA, 6FDA and BTDA) in an aprotic solvent to form a polyamic acid. Polyimides combine excellent thermal and chemical stability with a wide range of CO_2permeabilities.
- *Polyarylates* are synthesized by reaction of a diol, such as bisphenol A, with an acid chloride. The majority of polyarylates possesses a CO_2 permeability lower than 25 barrer. Nevertheless, a small number shows permeabilities as high as 85 (Pixton and Paul, 2003), such as the sterically bulky polyarylate, TBHFBPA/tBIA, which exhibits the highest CO_2 permeability with a CO_2/N_2 selectivity equal to 19.
- *Polycarbonates* are commonly synthesized by reaction between a diol and phosgene under a variety of conditions. Most of the polycarbonates have a CO_2 permeability under 40 barrer and selectivities range from 15 to over 25. One notable exception to this is the polycarbonate TMHFPC, which possesses a CO_2 permeability of 111 barrer and a CO_2/N_2 selectivity of 15.
- *Polypyrrolones* are synthesized by reaction of di-anhydride and a tetra-amine functionalized monomer (Yang et al., 2008). They are characterized by high thermal and chemical resistance. Nevertheless, owing to their high rigidity, the polypyrrolones may act as a molecular sieve.
- *Polysulfones* are generally synthesized by condensation reaction between bisphenol and dihalogenated diphenylsulfone. The most extensively studied polysulfone is PSF formed using bisphenol A and other polysulfones are structurally related to this polymer.

4. Membranes and Bioreactors

Apart of the general classification previously introduced, another classification of membranes is based on the subdivision of synthetic membranes, which show a large variety in their physical structure according to the composition of their polymer used for. Based on their structure, they can be sub-divided in four groups:

- porous membranes;
- homogeneous solidmembranes;
- solid membranes carrying electrical charges;
- liquid or solid films containing selective carriers.

Isotropic membranes have a homogeneous structure, whereas anisotropic membranes consist of a thin top layer supported by a mechanically stronger underlying more porous layer membrane material, which can be either hydrophobic or hydrophilic. In Table 6, the most common membranes used for proteins treatment and their iso-electric points are reported. The electrostatic nature of both membrane and macromolecules determines attraction and repulsion phenomena. If the membrane and the biomolecule have the same electrostatic charge, repulsion naturally occurs. If the membrane surface has an opposite charge of the protein molecule, the surface will attract a monolayer of charged protein molecules, to form a self-repulsive layer.

Table 6. Membrane materials with their iso-electric points

Membrane material	Iso-electric pH	References
Aluminum oxide	4.4	Bowen and Hughes, 1990
Polysulfone	3.6	Nystrom et al., 1989
Polycarbonate	2.5	Keesom et al.,1988
Sulfonated polysulfone	<2.5	Nystrom et al., 1989
Cellulose acetate/nitrate	2.2	Bowen et al.,2002
Polyethersulfone	3.1	Pontié et al.,1997
Polysulfone	0.5	Pontié et al.,1997
SPG	1.5	Ryosuke et al.,1996

Therefore, if the convective flows overcome the electrostatic repulsion, the protein molecules can pass through; otherwise they are retained (Liu et al., 2005).

4.1. Enzymes

Different enzymes have been discovered and already applied in many different bio-catalytic applications within porous support. It is well understood that the enzyme has some specific advantages over inorganic, metal-derived or chemical catalysts. The interest in biocatalysts oriented research is increasing over the past decades. This is primarily due to their ability to utilize a wide spectrum of substrates, high stability towards extremes of temperature, pH and organic solvents, and chemo-, regio- and enantio-selectivity. In many cases, the biocatalyst catalyzed reactions resemble closely the pathways designed by nature for the metabolism of live beings. Thus, the reaction mechanisms and processes of immobilization are viewed as more environmental friendly in separation processes. Also, products with high purity and high added value are observed by the aid substrate specificity and stereo-specificity properties of biocatalysts. Membrane bioreactor (MBR) is a process unit to be used for different purposes. One of the main features of MBR technology is the ability of the membrane to remove pathogenic organisms, providing disinfection of the effluent. This is particularly important when considering reuse options. The membrane offers a physical barrier to the organisms that is unaffected by the influent quality. Reductions in bacteria and viruses of 4 - 8 log have been reported in Table 7.

Table 7. Reduction in microorganisms using different membrane systems

Membrane used in MBR	Pore Size [μm]	Average log Reduction	Bacteria / Virus	References
PS	0.1	4 to 6	Total Coliforms	Chiemchaisri et al., 1992
PE	0.5	5	Coliphage QB	Gander et al., 2000
Memtec	0.2	Not found	Total Coliforms	Kolega et al., 1991
Memcor	0.2	3.8	Faecal Coliforms	Till, 1998
Renovexx	0.5-1.5	3.3	Faecal Coliforms	Till, 1998
Stork	0.05-0.2	2.5	Faecal Coliforms	Till,1998
Starcosa	0.2	Not found	Total Coliforms	Till, 1998
Dow	0.2	<7	Total Coliforms	Till, 1998

The enzyme processes are potentially energy saving and save investing in special equipment resistant to heat, pressure or corrosion. Due to their efficiency, specific action, the mild conditions in which they work and their high biodegradability, enzymes are very well suited for a wide range of industrial applications for example, within the detergent, textile and starch industries. The use of recombinant gene technology has further improved manufacturing processes and enabled the commercialization of enzymes that could previously be not produced. Furthermore, the latest developments within modern biotechnology, introducing protein engineering and directed evolution, have further revolutionized the development of industrial enzymes (Figure 7).

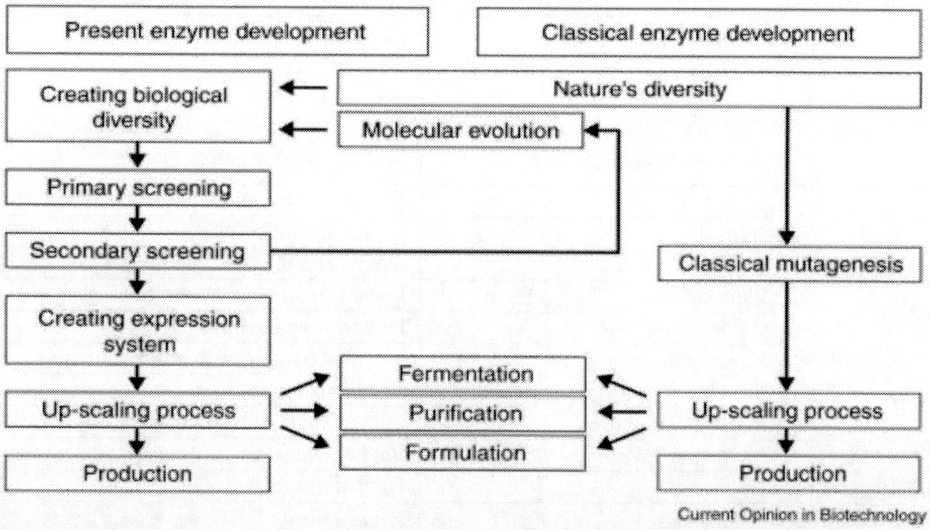

Figure 7. Some industrial applications of enzymes.

Due to considerably unstable nature of enzymes, it has been necessary to develop a different and efficient immobilization technique, which can protect enzymes from different physical as well as chemical stresses such as pH, temperature, salts, solvents, inhibitors and poisons. Immobilization of enzymes in membrane's micro and macro environment has further advantages such as elimination of long and expensive procedures for enzymes separation, purification and their application by enabling easy separation and purification of products from reaction mixtures to catalyst recovery (Stolarzewicz et al., 2011). It has then applied in many industrial applications as shown in Table 8, where most of the application protocol has been developed with the polymeric membrane support, which works simultaneously as a micro-reactor as well as a separator.

Table 8. Industrial applications of biocatalysts and their classes

Industry	Enzyme class	Application
Detergent (laundry and dish wash)	Protease	Protein stain removal
	Amylase	Starch stain removal
	Lipase	Lipid stain removal
	Cellulase	Cleaning, color clarification, anti-redeposition (cotton)
	Mannanase	Mannanan stain removal
Starch and fuel	Amylase	Starch liquefaction and saccharification
	Amylo glucosidase	Saccharification
	Pullulanase	Saccharification
	Glucose isomerase	Glucose to fructose conversion
	Cyclodextrin-glycosyltransferase	Cyclodextrin production
	Xylanase	Viscosity reduction (fuel and starch)
	Protease	Protease (yeast nutrition – fuel)
Food (including dairy)	Protease	Milk clotting, infant formulas (low allergenic), flavor
	Lipase	Cheese flavor
	Lactase	Lactose removal (milk)
	Pectin methylesterase	Firming fruit-based products
	Pectinase	Fruit-based products
Industry	Enzyme class	Application
	Trans glutaminase	Modify visco-elasticproperties
Baking	Amylase	Bread softness and volume, flour adjustment
	Xylanase	Dough conditioning
	Lipase emulsifier	Dough stability and conditioning
	Phospholipase	Dough stability and conditioning
	Glucoseoxidase	Dough strengthening
	Lipoxygenase	Dough strengthening, bread whitening
	Protease	Biscuits, cookies
	Trans glutaminase	Laminated dough strengths
Animalfeed	Phytase	Phytate digestibility – phosphorus release
	Xylanase	Digestibility
	β-Glucanase	Digestibility

Industry	Enzyme class	Application
Beverage	Pectinase	De-pectinization, mashing
	Amylase	Juice treatment, low calorie beer
	β-Glucanase	Mashing
	Acetolactatede carboxylase	Maturation (beer)
	Laccase	Clarification (juice), flavor (beer), cork stopper treatment
Textile	Cellulase	Denim finishing, cotton softening
	Amylase	De-sizing
	Pectatelyase	Scouring
	Catalase	Bleachtermination
	Laccase	Bleaching
	Peroxidase	Excess dye removal
Pulp and paper	Lipase	Pitch control, contaminant control
	Protease	Bio film removal
	Amylase	Starch-coating, de-inking, drainage
	Xylanase	Bleach boosting
	Cellulase	De-inking, drainage improvement, fiber modification
Fats and oils	Lipase	Trans esterification
	Phospholipase	De-gumming, lyso-lecithin production
Organic synthesis	Lipase	Resolution of chiral alcohols and amides
	Acylase	Synthesis of semisynthetic penicillin
	Nitrilase	Synthesis of enantio-pure carboxylic acids
Leather	Protease	Unhairing, bating
	Lipase	De-pickling
Personal care	Amylo glucosidase	Antimicrobial (combined with glucose oxidase)
	Glucose oxidase	Bleaching, antimicrobial
	Peroxidase	Antimicrobial

4.2. Introduction to Membrane Bio-Reactors (MBRs)

Biological membrane reactors or so called membrane bioreactors are commonly seen as the combination of membrane separation and biological treatment unit, where the membrane primarily acts as a reactor as well as separator in the system. They are compact-built purification systems combining the biological degradation with the membrane separation. The selective mass transport with reaction and rejection of the specific product from the reaction environment is the most desirable within the bio-catalytic membrane reactor technology. Among the biocatalysts, the enzyme has been attracted nowadays for the bio-catalytic membrane reactor. Indeed, enzymes can speed up chemical processes that, otherwise, would run very slowly. The most interesting property of a bio-catalytic membrane reactor is that the biocatalyst (*i.e.* enzymes) can work at atmospheric pressure in mild conditions with respect to temperature and acidity (pH). Most of the enzymes operate optimally at a temperature of 30 – 70 °C and at neutral pH values.

The main configuration both for chemical and biological reactor is where the membrane is a barrier that does not participate to the reaction but allows a selective separation or when the membrane contains the catalyst and is catalytically active. The name of bioreactors can be based on the type of solvent and/or separation process used as well as on the type of membrane (e.g. flat-sheet membrane bioreactor, hollow fiber MBR unit, etc.) (Strathmann et al., 2006). The concept of bio-catalytic membrane reactors consists of two different configurations in which the membrane operates as the separation process (Giorno and Drioli, 2000; Chakraborty et al., 2010). In a MBR, membrane filtration occurs either within the bioreactor (submerged configuration, or externally through recirculation subject to a pressure drop across the membrane) either the hydraulic head or a pump. The ultrafiltration (UF) or microfiltration (MF) membranes utilized by MBRs have pore sizes such that water and most solute species pass through the membrane whilst other larger species, such as solids and microorganisms, are retained.

In the MBR side-stream, the biocatalyst is continuously flowed along the membrane or it is held within the membrane unit but, in case of submerged, the MR is an immersed unit. In this kind of reactor, in addition to separation action, the membrane is also a catalytic unit. The different kind of bio-catalytic membrane reactor configuration depends on which way the complex biocatalyst–loaded membrane is realized (Prazeres and Cabral, 1994).

4.2.1. Basic of Different MBRs (Side-Stream and Submerged Configurations)

The membrane step can be implemented into an MBR system in two ways: side-stream and submerged or immersed configuration. In side-stream configuration, the membranes are placed externally from the bioreactor, as schematically reported in Figure 8a. The first generation MBRs was designed according to the side-stream configuration.

Fouling was traditionally prevented by applying a high liquid cross flow through the membrane tubes. Applying a high cross flow velocity, the membranes are less sensitive to fouling and, thus, allow using a relatively high trans-membrane pressure or flux.

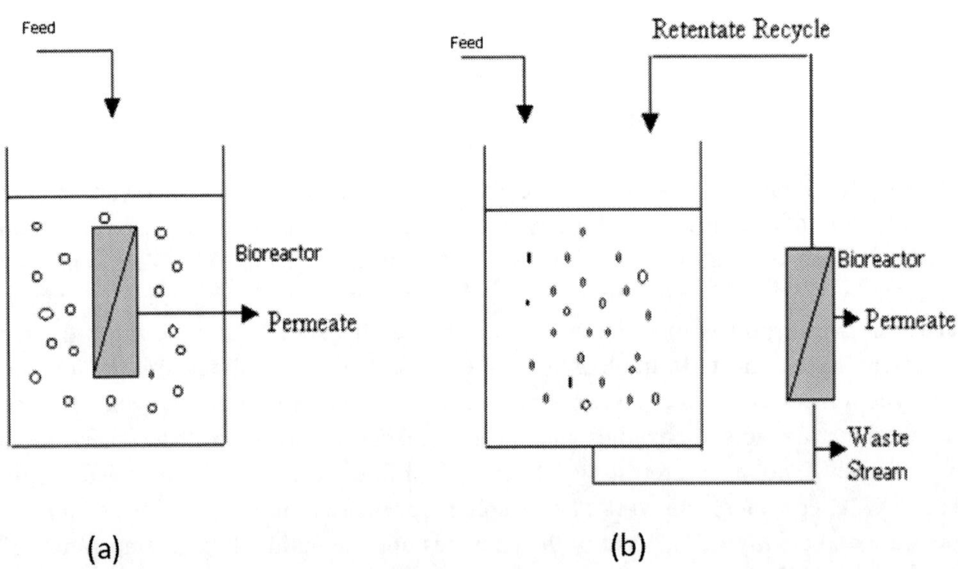

Figure 8. Configuration of different MBR system. (a) Submerged system; (b) Side stream system.

The main disadvantage of the traditional side-stream configuration is the high energy consumption required for the cross flow. In the immersed or internal configuration, the membrane modules are submerged directly into the activated sludge, either in a separate membrane tank or directly in the bioreactor. The immersed MBR configuration is schematically presented in Figure 8b. Nowadays, the membranes are more commonly placed in a separate membrane tank. Indeed, this allows better control over the filtration process and, furthermore, the wastewater undergoes superior biological treatment before reaching the membrane. The disadvantage of a separate membrane tank is the higher energy consumption for recirculation and aeration. Within immersed MBR systems, two different membrane configurations are applied: hollow fiber and flat sheet membranes.

The choice between operating options whether it is side-stream MBR or submerged MBR is dependent on the application, by considering that both side stream and submerged systems show advantages and disadvantages (Table 9). The configuration helps to maintain the flux through the membrane, by reducing the build-up of material at the membrane surface and to increase the operational cycle of the system.

Membrane bioreactors have several intrinsic advantages that make them as a possible alternative when compared with other more conventional bioreactors. One of the main benefits is the possibility to use the biocatalyst in a continuous and, thus, intensive way, which contributes to an increase of productivity and possibly to the economic viability of the process (Prazeres and Cabral, 1994; Chang and Furusaki, 1991). Another important advantage is the continuous selective removal of the products from the reaction media. Nevertheless, actually the key difference between the two configurations is the means of creating shear force along the membrane surface: liquid cross flow versus two phased air-liquid configuration, which is very important to control the different operating parameters in MBR processes. Some specific characteristics of the traditional side-stream and the immersed MBR configurations are summarized in Table 10 (Evenblij, 2006; Judd and McAdam, 2006; Mulder, 1996).

Table 9. Difference between submerged and side stream MBRs

Submerged MBR	Side Stream MBR
Low flux	High flux
Less frequency of cleaning	More frequency of cleaning
Lower operation cost	Higher operating cost
Higher capital cost	Lower capital cost
Low energy consumption	High energy consumption
High aeration cost	Low aeration cost

Table 10. Characteristics of side-stream and submerged MBR

Characteristic	Side Stream	Submerged
Membrane Placement	Outside reactor	Inside the porous matrix
TMP (bar)	2.0-6.0	0.2-0.5
Filtration type	Inside -outside	Outside-inside
Flux ($L/h.m^2$)	5-100	20-40
Energy Consumption (kWh/m^3)	2.0-5.0	0.2-0.5
Cross flow Speed (m/s)	1.0-6.0	0.5

MBR types are generally subdivided into four categories based on the purpose of membrane usage (Stephenson et al., 2002; Gu and He, 2002): (1) biomass separation membrane bioreactor (BSMBR), (2) membrane aeration bioreactor (MABR), (3) extractive membrane bioreactor (EMBR) and (4) another newly developed ion exchange membrane bioreactor (IEMBR) (Oehmen et al., 2006). It can be observed that the majority of these studies focused on the application of BSMBRs, where the membranes were used for solid–liquid separation.

The current research in most of the places is mainly involved in wastewater treatment and reclamation by employing BSMBR with its success in commercial applications. However, MABR and EMBR, due to their expense and immaturity, still need a long period to achieve commercial success and, thus, the research on these two types lagged behind. These two types still have their own potential to attract researchers' attention for specific applications. Toxic organic wastewaters (Xiao and Xing, 2001; Xiao et al., 2001 and 2006) and surface/drinking water (Cao et al., 2006) are two basic areas for the application of EMBRs. Two other innovative processes, membrane enzyme bioreactor (MEBR) for hydrolysis of special organic substances such as olive oil and whey protein (Deng et al., 2005; Cheison et al., 2007) and membrane pervaporation bioreactor (MPBR) for volatile organic wastewater treatment (Wu et al., 2004), were also investigated. It should be pointed out that, although the study of all other types of MBRs was carried out, the BSMBRs have successively achieved commercial applications and the rest types of MBR are still at the level of lab-scale research.

4.2.2. Biocatalysts and Their Immobilization Technique

Immobilized biocatalysts have widespread applications in areas like organic synthesis, pollution control and for diagnostic purposes (Butterfield, 1996; Amounas et al., 2000, 2002). The selection of the membrane to be used in enzymatic membrane reactors should take into account the size of the bio-catalyst, substrate and product as well as the chemical species of the species in solution and of the membrane itself. An important parameter to be used in this selection is the solute rejection coefficient, which should be zero for the product to facilitate permeation and should be one for the enzyme to insure a complete retention of the catalyst inside the reaction system. The selectivity is normally associated with a discrimination based on size exclusion, but a steric exclusion process may be present for molecule with size close to the pore size. Immobilization eliminates the need to separate an enzyme from the product solution and allows these expensive compounds to be reused.

Different enzymes have been already applied in many different biocatalytic applications. It is well understood that the enzyme has some specific advantages over inorganic, metal-derived or chemical catalysts. The interest in lipase oriented research is increasing over the past decades. This is primarily due to their ability to utilize a wide spectrum of substrates, high stability towards extremes of temperature, pH and organic solvents and chemo-, regio- and enantio-selectivity. Products with high purity and high added value can be obtained. Moreover, the use of enzyme decreases the side reactions and simplifies post-reaction separation problems (Pandey et al., 1999). For example lipases (EC.3.1.1.3, triacylglycerol acylhydrolyse) have been used for many years. Therefore, customization of lipases by chemical and physical modifications has been attempted to improve their catalytic properties in hydrolysis and synthesis involving aqueous, non-aqueous and organic solvents. In addition, the cost of enzyme can be reduced by the application of molecular biological tools, allowing the production of lipases in large quantities and with genetically enhanced properties

(Villeneuve et al., 2000). These properties allow them to catalyze reactions with reduced side products, lower waste treatment costs and under mild temperature and pressure conditions. However, there are five major types of immobilization techniques, which are used to bind the enzyme to the membranes described later on.

4.2.2.1. Entrapment

The entrapment method of immobilization (Figure 9), is based on the localization of an enzyme within the membrane. It is done in such a way as to retain protein while allowing penetration of substrate. Asymmetric hollow fiber can provide an interesting support for enzyme immobilization. The enzyme can be entrapped in the sponge layer by cross-flow filtration in the case in which the pore size in the dense layer must retain the enzyme, permitting the passage of the substrate. The amount of biocatalyst loaded, its distribution and activity through the support and its lifetime are very important parameters to properly orientate the development of such systems. The enzyme, mediators and additives can be simultaneously deposited in the same sensing layer. There is no modification of the biological element so that the activity of the enzyme is preserved during the immobilization process.

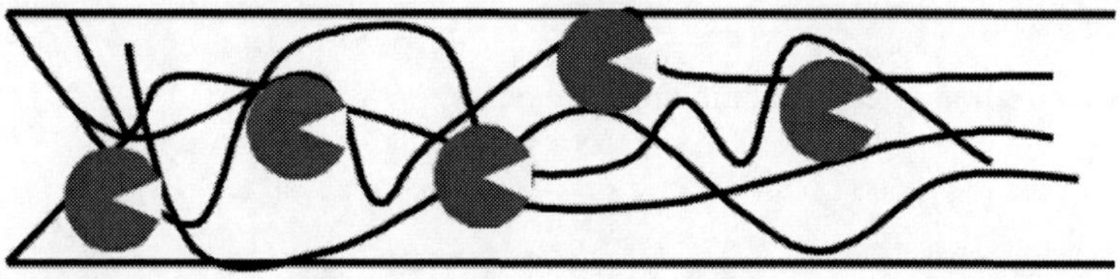

Figure 9. Schematization of entrapment method.

4.2.2.2. Covalent Binding

The covalent binding method (Figure 10), is based on the binding of enzymes and membrane by covalent bonds. The functional groups that take part in this binding of enzyme to membrane can be amino, carboxyl, sulfhydryl, hydroxyl, imidazole or phenolic groups, which are not essential for the catalytic activity. In order to protect the active site, immobilization can be carried out in the presence of its substrate or a competitive inhibitor.

Activity of the covalent bonded enzyme depends on the size and shape of carrier material, nature of the coupling method, composition of the carrier material and specific conditions during coupling. The main advantage of the covalent attachment is that such immobilization is very stable. Unlike physical adsorption, the binding force between enzyme and carrier is so strong that no leakage of the enzymes occurs, even in the presence of substrate or solution of high ionic strength. Moreover, the obtained immobilized enzymes are usually very stable and resistant to extreme conditions.

4.2.2.3. Cross Link

Immobilization of enzymes can be also achieved by intermolecular cross-linking of the protein (Figure 11) either to other protein molecules or to functional groups on an insoluble support matrix. Cross-linking an enzyme is both expensive and insufficient, as some of the protein material will inevitably be acting mainly as a support, resulting in relatively low enzymatic activity.

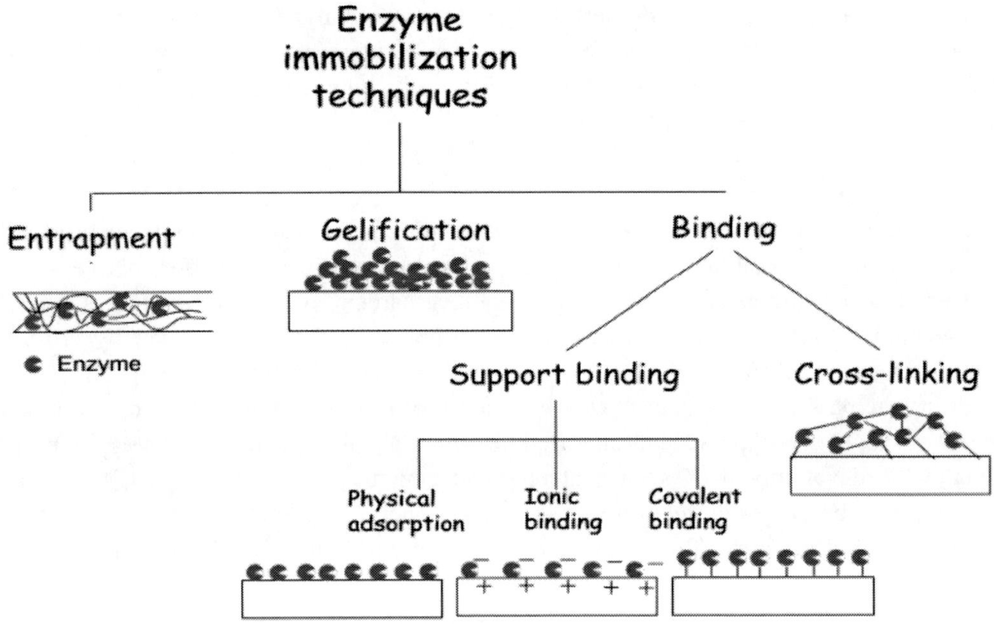

Figure 10. Schematization of covalent binding method.

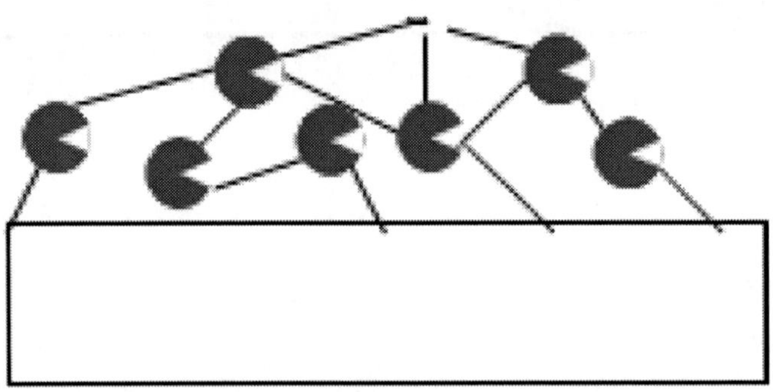

Figure 11. Schematization of cross linking method.

Since the enzyme is covalently linked to the support matrix, very little desorption is likely obtained by using this method. The latter is attractive due to its simplicity and strong chemical binding achieved between biomolecules. The main drawback is the possibility of activity losses due to the distortion of the active enzyme conformation and the chemical alterations of the active site during crosslinking.

4.2.2.4. Affinity

It is a method of enzyme immobilization by siting specific group of bio molecules. This method allows controlling the biomolecule orientation in order to avoid enzyme deactivation and/or active site blocking. Several affinity methods (Figure 12) have been described for the immobilization of enzymes. Improvement can be achieved by introducing a spacer molecule (Nouaimi et al., 2001). In this type, the aim is to create (bio) affinity bonds between an activated support and a specific group of the protein sequence. An enzyme can contain affinity tags in its sequence (e.g. a sugar moiety) but, in some cases, the affinity tag (e.g. biotin) needs to be attached to the protein sequence by genetic engineering methods such as site-directed mutagenesis, protein fusion technology and post-transcriptional modification (Andreescu and Marty, 2006).

4.2.2.5. Adsorption

Enzyme adsorption (Figure 13) onto the solid porous supports represents the easiest method of physical immobilization (Choi, 2004). Enzyme is dissolved in the solution and the solid support is placed in contact with the enzyme solution for a fixed period of time. The unadsorbed enzyme is then removed by washing with buffer.

The adsorption mechanisms are based on weak bonds such as Van der Waal's forces and electrostatic and hydrophobic interactions between the enzyme and support. This technique does not involve any functionalization of the support and is generally non-destructive for enzyme activity.

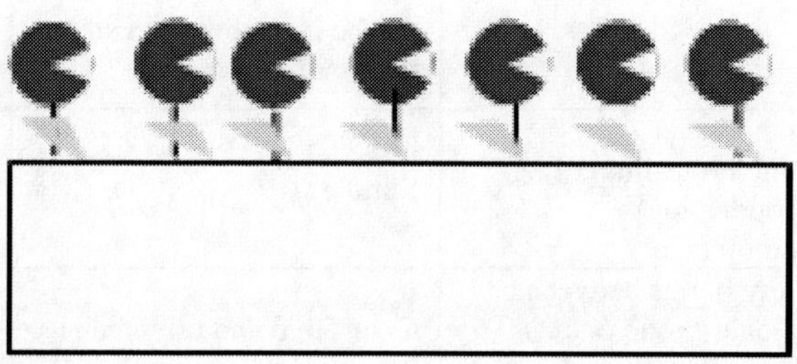

Figure 12. Schematization of affinity method.

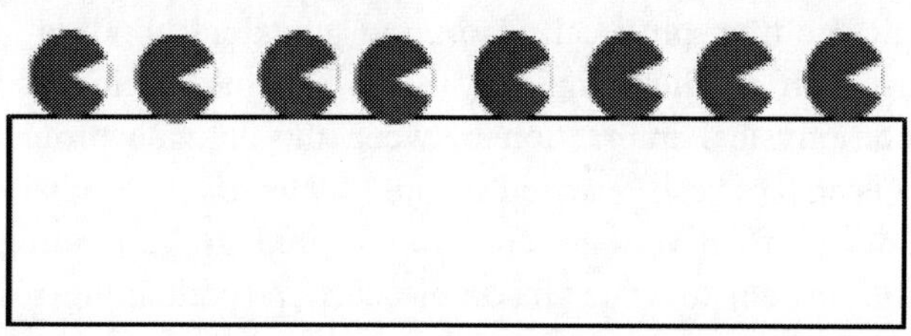

Figure 13. Schematization of adsorbed method.

The entrapment involves localization of the enzyme within the membrane. The aim here is to retain the protein while allowing penetration of substrate. The enzyme is trapped in the sponge layer by cross-flow filtration. There are two kinds of entrapment, Lattice type and Microcapsule type.

One possibility is the gelification of the biocatalyst on the membrane; it is the outcome of concentration polarization. A disadvantage is represented by the possibility of reduced catalytic efficiency due to mass transport limitations and to preferential pathway in the enzyme gel layer. In Table 11, the advantages and disadvantages of different immobilization methods are summarized.

Table 11. Immobilization method and their advantages and disadvantages

Binding Methods	Binding nature	Advantages	Disadvantages
Adsorption	Weak bonds	Simple and easy	Desorption
		Limited loss of enzyme activity	Non-specific adsorption
Covalent Binding	Chemical binding between functional groups of the enzyme and those on the support	No diffusion barrier	Matrix not regenerable
		Stable and Short response time	Coupling with toxic product
		High enzyme activity loss	
Entrapment	Incorporation of the enzyme within a gel or a polymer	No chemical reaction between the monomer and the enzyme that could affect the activity of enzyme	Enzyme leakage and Diffusion barrier
		Several types of enzymes can be immobilized within the same polymer	High concentrations of monomer and enzyme for electro polymerization
Cross-linking	Bond between enzyme/cross-linker (e.g. glutaraldehyde)/ inert molecule	Simple	High enzyme activity and activity loss
Affinity	Affinity bonds between a functional group on a support and affinity tag on a protein sequence	Controlled and oriented immobilization	Need of the presence of specific groups on enzyme

In contrast to this, the catalyst can be directly bound to the membrane. This bound is of many kinds viz. ionic binding, physical adsorption, covalent bonding, and cross-linking. In physical adsorption-which is the simplest way of preparing immobilized enzymes, the reliance is on the non-physical interaction between the enzyme protein and surface of the matrix. This method is cheap, easily carried out and tends to be less disruptive to the enzyme; also a minimum number of activation steps are required. Ionic binding, relies on the ionic bound of the enzyme protein to water-insoluble carriers containing ion-exchange residues. Polysaccharides and synthetic polymers, having ion-exchange centers are usually used for carriers. The binding of enzyme to the carrier is easily carried out and the conditions are much milder than those used in covalent binding method. Hence this method causes little changes in the conformation and the active site of the enzyme and so yields immobilized enzymes with high activity in most of cases.

Covalent binding method is based on the binding f enzymes and membrane by covalent bonds. The functional groups that take part in this binding of enzyme to membrane can be amino sulfydryl, hydroxyl, inidazole or phenolic groups which are not essential for the catalytic activity.

A competitive inhibitor may be used during the process of immobilization to protect the active site. A main advantage of this kind of binding is that it is very stable; there is no leakage of enzyme as compared to physical adsorption. However experimental procedures to carry out covalent binding are much more difficult.

4.3. Enzyme Immobilized into the Porous Matrix of Membranes

To calculate the observed substrate conversion through an enzyme-loaded-support immobilized into the internal surface, it is necessary take into account the concentration profile within the diffusion layer. Beside substrate diffusivity in the bulk phase, the diffusion rate (D_{eff}) through a porous support is influenced by several factors. The effective diffusion coefficient is described by Eq.7:

$$D_{eff} = D_{S_0} \frac{\varepsilon_p}{\tau} \frac{K_p}{K_r} \tag{7}$$

ε_p is the porosity (or area of support/ area of pores); τ is the tortuosity (if the geometry of the pore is not tubular, diffusion occurs changing direction continuously). The tortuosity factor can assume a value in a range of 1.4 to 7; K_p/K_r is the restricted diffusion roughly estimated as $[1 - (r_{substrate}/r_{pore})]4$, which takes into account the dimension of the pore in relation with the dimension of the substrate that can have similar dimension and lead to a situation of restricted diffusion.

The influence of diffusion within porous catalysts upon reaction kinetic was studied in 1930 (Bailey and Ollis, 1986); this effect was studied on a planar membrane with an immobilized enzyme uniformly distributed. Eq. 8 is given as a combination of the steady-state diffusion equation with the applicable kinetics rate expression.

$$D_{eff} \frac{d^2[S]}{dx^2} - \frac{V_{max}[S]}{K_m + [S]} = 0 \tag{8}$$

D_{eff} is the effective diffusivity, this means that at the steady state the substrate diffusion rate, through a porous matrix, is equal to the rate conversion. Also in an immobilized system is possible to evaluate if the reaction is limited by kinetics or mass transport by the Thiele modulus ø, given by Eq. 9, which has the meaning of a reaction rate.

$$\phi = L\left(\frac{V_{max}}{D_{eff} K_m}\right)^{1/2} \tag{9}$$

4.3.1. Applications

The application of biological MR has been summarized previously, where most of the applications of conventional MBRs has been mentioned. Regarding the submerged polymeric reactor, most of the applications is concentrated on the wastewater treatment (Stephenson et al., 2000) such as removal of contaminant, nitrate (Haugen et al., 2002; Kimura et al., 2002), heavy metals, persistent organic pollutants (POPs), endocrine disruption chemicals (EDCs) or other hospital waste (Xianghua et al., 2004), trace mono-valent inorganic pollutants (Velizarov et al., 2003), etc. Due to higher cost and energy consumption the biological treatment is combined with the membrane technology and several membrane bioreactor configurations have been successfully applied. Although the science behind each configuration has been justified, issues related to secondary contamination of the product water remain and require further investigation (Judd and McAdam, 2006).

SMBRs not only have the advantages over common MBR systems, such as complete solid removal from effluent, effluent disinfection, high loading rate capability, low/zero sludge production, rapid start-up, etc., but also have a more compact size and lower energy consumption (Stephenson et al., 2000). The SMBR has been proved to be a promising wastewater treatment technology and has been used for treating a variety of wastewaters such as domestic wastewater (Ueda et al., 1996), tannery wastewater (Yamamoto et al., 1991) and oily wastewater (Knoblock et al., 1994).

Conclusion and Future Trends

In the viewpoint of an enlargement of the hydrogen economy, hydrogen production and its purification could take place with a growing interest in the future. Today, hydrogen upgrading in refinery applications can be achieved by using PSA, membrane, or cryogenic separation processes. Membranes are excellent candidates for hydrogen purification, especially when incorporated into a membrane reactor, combining the reaction/separation process in a single unit. Despite difficulties in the formation of a uniform metallic thin film, Pd-based membranes show very high hydrogen selectivity. In this chapter, particular attention has been devoted to Pd-based membrane technology, pointing out the main benefit of these membranes in shifting the reaction towards the products (in applications in membrane reactors), yielding this way conversions above those based on thermodynamic equilibrium determined from feed composition.

Future research studies should be devoted to the improvement of the mechanical resistance characteristics of the Pd-membranes during the reaction process, at relatively high reaction temperatures and pressures. Furthermore, more experimental studies on the lifetime of the membranes should be realized in order to validate them as good candidates to substitute conventional systems at larger scales.

MBR process becomes more and more attractive for municipal/domestic wastewater treatment when a compact technology is required due to a lack of space or the high cost of additional land in urban areas, or when high effluent quality is required for water reuse or pretreatment for reverse osmosis (RO) or nano-filtration (NF) processes (Yang et al., 2006). The increasingly stringent discharge standards and the great need of water reclamation and reuse will further push forward the application of MBRs for ever-larger municipal wastewater treatment plants. Original research on MBR technology for various industrial wastewater treatment is becoming a hot topic, and the application areas include food-processing wastewater, petrochemical wastewater, hospital wastewater, printing and dyeing wastewater, slaughterhouse wastewater, etc. (Huang et al., 2005; Gu et al., 2002). It is worth noting that the application in landfill leachate treatment is also promising. MBR systems combined with pretreatment anaerobic biological process and/or with post-treatment steps like nanofiltration (NF) were commonly used in landfill leachate treatment in order to achieve good effluent quality (Wang et al.,2003). Another area for MBR application is surface/drinking water treatment for nitrate removal (Cao et al., 2006; Zhu et al., 2006) and organic substances removal. It is worth pointing out that studies on the application of MBR technology for surface/drinking water treatment obviously lagged behind the research on municipal/domestic and industrial wastewater treatment; however in most of the country the surface water pollution is very severe and the watersheds have been contaminated. Therefore, in the future, the research and application of MBR process in surface/drinking water treatment will be boosted profoundly in world. The skilled management and optimized operation of large-scale MBRs are further needed, through which MBR technology will become more cost-effective.

REFERENCES

Abertz V., Brinkmann T., Dijkstra M., (2006) Developments in membrane research: from material via process design to industrial application, *Adv. Eng. Mater.*, 8, 328-358.

Adhikari S., Fernando S., (2006) Hydrogen membrane separation techniques, *Ind. Eng. Chem. Res.,* 45, 875-881.

Amano M., Komaki M., Nishimura C., (1991) Hydrogen permeation characteristics of palladium-plated V-Ni alloy membranes, *J. Less-Common. Met.,* 172, 727-731.

Amounas M., Innocent C., Cosnier S., Seta P., (2000) A membrane based reactor with an enzyme immobilized by an avidin–biotin molecular recognition in a polymer matrix, *J. Membrane. Sci.,*176, 169-176.

Amounas M., Magne V., Innocent C., Dejean E., Seta P., (2002) Elaboration and chemical reactivity of enzyme modified ion exchanging textiles, *Enzyme and Microb. Tech.,* 31, 171-178.

Andreescu S, Marty JL., (2006) Twenty years research in cholinesterase biosensors: from basic research to practical applications, *Biomol. Eng.,* 23, 1-15.

Bailey J.E., Ollis D.F., (1986) Biochemical engineering fundamentals, 2nd ed. New York: *McGraw Hill,* 334–410.

Basile A., Iulianelli A., (2007) Alternative sulfonated polymers to Nafion for PEM fuel cell, pp. 135-160, Chapter 3, in *Fuel Cell Research Trends*, L.O. Vasquez (Ed.), Nova Science Pub. Inc., New York, ISBN: 1-60021-669-2.

Basile A., Iulianelli A., Gallucci F., Morrone P., (2010) Developments and innovation in carbon capture and storage (CCS) technology, Ch. 7: Development of advanced CO_2-separation membranes and technology, *Woodhead Pub.,* UK, 203-243, ISBN 978-1-84569-533-0.

Basu A., Akhtar J., Rahman M., Islam M., (2004) A review of separation of gases using membrane systems", *Petrol. Sci. Techn.*, 22, 1343-1368.

Bockris, J.O.M., (1971) Hydrogen economy, *Environment,* 1, 13 – 51.

Bowen W.R., Doneva T.A., Yin H.B., (2002) Atomic force microscopy studies of membrane—solute interactions (fouling), *Desalination,* 146, 97-102.

Bowen W.R., Hughes D.T., (1990) Properties of microfiltration membranes. Adsorbition of bovine serum albumin at aluminium oxide membranes, *J. Membrane Sci.,* 51, 189-200.

Brodowsky H., (1972) On the non-ideal solution behavior of hydrogen in metals, *Ber. Bunsenges Physik. Chem.,* 76, 740-749.

Butterfield D.A., (1996) Biofunctional membranes, *Plenum Press* NY, 117-129.

Cao J.H., Zheng X.L, Pan M.X., Li P., Hou J.H., Song X.Q., (2006) Nitrate removal in groundwater using extractive membrane biological reactors, *J. Xi'an Univ. Architect. Technol.,* 38, 574–579.

Chakraborty S., (2010) Development of submerged biocatalytic membrane reactors for innovative production systems, PhD Thesis, University of Calabria, Italy.

Chang H.N., Furusaki S., (1991)Membrane bioreactors: Present and prospects, *Adv. Biochem. Engin./Biotechnol,* 44, 27-64.

Cheison S.C., Wang Z., Xu S.Y., (2007) Use of response surface methodology to optimize the hydrolysis of whey protein isolate in a tangential flow filter membrane reactor, *J. Food Eng.,* 80, 1134–1145.

Chiemchaisri C., Womg Y.K., Urase T., Amamoto K., (1992)Organic stabilization and nitrogen removal in membrane separation bioreactor for domestic wastewater treatment, *Water Sci. Technol.,* 25 (10) 231-240.

Choi M.F.M., (2004) Progress in enzyme-based biosensors using optical transducers, *Microchim. Acta,* 148, 107–132.

Costamagna P., Srinivasan S., (2001) Quantum jumps in the PEMFC science and technology from the 1960s to the year 2000 Part II. Engineering, technology, development and application aspects, *J. Power Sou.,* 102, 253-269.

Deng H.T., Xu Z.K., Dai Z.W., Wu J., Seta P., (2005) Immobilization of candida rugosa lipase on polypropylene microfiltration membrane modified by glycol polymer: hydrolysis of olive oil in biphasic bioreactor, *Enzyme Microb. Technol.,* 36, 996–1002.

Dittmeyer R., Höllein V., Daub K., (2001) Membrane reactors for hydrogenation and dehydrogenation processes based on supported palladium, *J. Mol. Catal. A: Chem.,* 173, 135-184.

Evenblij H., (2006) *Filtration Characteristics in MBR*, PhD Thesis, Delft University of Technology, The Netherlands.

Gander M., Jefferson B., Judd S., (2000) Aerobic MBRs for domestic wastewater treatment: a review with cost considerations, *Sep. Purif. Techn.,* 18, 119-130.

Gao H., Lin Y.S., Li Y., Zhang B., (2004) Chemical stability and its improvement of palladium-based metallic membranes, *Ind. Eng. Chem. Res.,* 43, 6920-6930.

Giorno L., Drioli E., (2000) Bio catalytic membrane reactors: applications and perspectives, *Trend Biotechnol.,* 18, 339-349.

Grashoff G.J., Pilkington C.E., Corti C.W., (1983). The purification of hydrogen – A review of the technology emphasizing, the current status of palladium membrane diffusion, *Plat. Met. Rev.,* 27, 157-168.

Gu G.W., He Y.L., (2002) *Membrane Bioreactors: Application and Research in Wastewater Treatment Chemical Industry Press,* Beijing, China.

Haugen K.S., Semmens M.J., Novak P.J., (2002) A novel in situ technology for the treatment of nitrate contaminated groundwater, *Water. Res.,* 36, 3497–3506.

Huang X., (2005) Overview on MBR's research and application in China, in: *Proceedings of the International Symposium on Membrane Technologies for Water and Wastewater Treatment,* Beijing, China, 41–50.

Iulianelli A., Basile A., Li H., Van Den Brink R.W., (2011) Inorganic membranes for pre-combustion carbon dioxide (CO_2) capture, in Advanced membrane science and technology for sustainable energy and environmental applications, ed. A. Basile and S.P.

Nunes, *Woodhead Publishing Series in Energy – Cornwall (UK),* Ch. 7, pp. 184-213. ISBN: 978-1-84569-969-7, ISSN: 2044-9364.

Judd S.J., McAdam E.J.,(2006) A review of membrane bioreactor potential for nitrate removal from drinking water, *Desalination*, 196(1–3), 135–148.

Keesom W.H., Zelenka R.L., Radke C.J., (1988) A zeta-potential model for surfactant adsorbtion on an ionogenic hydrophobic surface, *J. Colloid Interface Sci.,* 125, 575-585.

Kimura K., Nakamura M., Watanabe Y., (2002) Nitrate removal by a combination of elemental sulphur based denitrification and membrane filtration, *Water Res.,* 36, 1758–1766.

Kluiters, S.C.A., (2004). Status review on membrane systems for hydrogen separation; *Intermediate Report for EU Project MIGREYD NNE5*-2001- 670.

Knoblock M.D., Sutton P.M., Mishra P.N., Gupta K., Janson A., (1994) Membrane biological reactor system for treatment of oily wastewater, *Water Env. Res.,* 66 (2), 133–139.

Kolega M., Grohman G.S., Chiew R.F., Day A.W., (1991)Disinfection and clarification of treated sewage by advanced microfiltration, *Water Sci. Technol.,* 23 (7-9), 1609-1618.

Liu Z. M., Dubremez J., Richard V., Yang Q., Xu Z.K., Seta P.,(2005)Useful method for the spatial localization determination of enzyme (peroxidase) distribution on microfiltration membrane, *J. Membrane Sci.,* 267, 2-7.

Mallada R., Menéndez M., (2008) Inorganic membranes: synthesis, characterization and applications, *Technology and Engineering,* 460.

Miller J.Q., Stocker J., Selection of a hydrogen separation process; *UOP Report:* January 1999.

Mulder M., (1996)Basic Principles of Membrane Technology, *Kluwer Academic Publishers, Dordrecht,* The Netherlands, 149-158.

Nouaimi M., Möschel K., Bisswanger H., (2001) Immobilization of trypsin on polyester fleece via different spacers', *Enzyme Microb. Tech.,* 29, 567–574.

Nystrom M., Lindstrom M., Matthiasson E., (1989) Streaming potential as a tool in the characterization of ultrafiltration membranes, *Colloid. Surface,* 36, 297-312.

Oehmen A., Viegas R., Velizarov S., Reis M.A.M., Crespo J.G., (2006) Removal of heavy metals from drinking water supplies through the ion exchange membrane bioreactor, *Desalination,* 199, 405–407.

Paglieri S.N., Way J.D., (2002) Innovations in palladium membrane research, *Sep. Purif. Reviews,* 31, 1-169.

Pandey A., Benjamin S., Soccol C.R., Nigam P., Krieger N., Soccol V.T., (1999) The realm of microbial lipases in biotechnology, *Biotechnol. Appl.Bioc.,* 29, 119-131.

Patel N., Ludwig K., Morris P., (2005) Insert flexibility into your hydrogen network - Part 1. *Hydrocarbon Process. Int.* Ed., 84, 9.

Phair J.W., Badwal S.P.S., (2006) Materials for separation membranes in hydrogen and oxygen production and future power generation, *Sci. Tech. Adv. Mat.,* 7, 792-805.

Pixton M.R., Paul D.R. (2003), Gas transport properties of polyarylates. Part I. Connector and pendant group effects, *J. Polym. Sci. Part B: Polym. Phys.,* 33, 1135-1149.

Pontié M., Chasseray X., Lemordant D., Lainé J.M., (1997) The streaming potential method for the characterization of ultrafiltration organic membranes and the control of cleaning treatments, *J. Membrane Sci.,* 129, 125-133.

Prazeres D.M.F., Cabral J.M.S., (1994). Enzymatic membrane bioreactors and their applications, *Enzyme Microb. Tech.,* 16 (9), 738-750.

Robeson L.M., (2008) The upper bound revisited, *J. Membrane Sci.,* 320, 390-400.

Ryosuke T., Masayuki N., (1996). Membrane charge of microporous glass membrane determined by the membrane potential method and its pore size dependency, *J. Membrane Sci.,* 111, 19-26.

Shu J., Grandjean B.P.A., (1991) Van Neste A., Kaliaguine S., Catalytic palladium-based membrane reactors: A review, *Can. J. Chem. Eng.,* 69, 1036-1060.

Steele B.C.H., Heinzel A., (2001) Materials for fuel-cell technologies, *Nature,* 414, 345-352.

Stephenson T., Judd S.J., Jefferson B., Brindle K., (2000) Membrane Bioreactors for Wastewater Treatment, *IWA Publishing,* London, UK, ISBN 1-900222-07-8.

Stocker J.,Whysall M., Miller G.Q., 30 Years of PSA Technology for Hydrogen Purification; *UOP Report:* July, 1998.

Stolarzewicz I., Bialecka-Florjañczyk E., Majewska E., Krzyczkowska J., (2011) Immobilization of Yeast on Polymeric Supports, *Chem. Biochem. Eng. Q.,* 25, 135–144.

Strathmann H., Giorno L., Drioli E., (2006) *An introduction to Membrane Science and Technology,* Rome, ISBN 88-8080-063-9.

Till, S., (1998) Cross flow microfiltration of sewage effluents. *PhD Thesis*, Cranfield University, UK.

Ueda T., Hata K., Kikuoka Y., (1996) Treatment of domestic sewage from rural settlements by a membrane bioreactor, *Water Sci. Technol.,* 34, 189–196.

Uemiya S., (1999) State-of-art-the-art of supported metal membranes for gas separation, *Sep. Pur. Methods,* 28, 51-85.

Velizarov S., Reis M.A., Crespo J.G., (2003) Removal of trace mono-valent inorganic pollutants in an ion exchange membrane bioreactor: analysis of transport rate in a denitrification process, *J. Membrane. Sci.,* 217, 269–284.

Villeneuve P., Muderhwa J.M., Graille J., Haas M.J., (2000) Customizing lipases for biocatalysis: a survey of chemical, physical and molecular biological approaches, *J. Mol. Catal. B-Enzym.,* 9(4-6), 113-148.

Wang J., Ji M., Li Z., (2003) Municipal landfill leachate treatment by using combined UASB and MBR system and its problems, *Urban Environ. Urban Ecol.,* 16, 215–217.

Wu Y., Xiao Z.Y., Huang W.X., Zhong Y.H., (2004) Enhancements of fermentation and pervaporation in a coupling silicone rubber membrane bioreactor, *J. Chem. Eng. Chin. Univ.,* 18, 241–245.

Xianghua W., Hangjiu D., Xia H., Ruopeng L., (2004) Treatment of hospital wastewater using a submerged membrane bioreactor, *Process Biochem,* 39(11), 1427–1431.

Xiao M., Zhou J.T., Tan Y., Zhang A.L., Xia Y.H., Ji L., (2006) Treatment of highly-concentrated phenol wastewater with an extractive membrane reactor using silicone rubber, *Desalination,* 195, 281–293.

Xiao Z.Y., Guan Q., Huang W.X., Xing X.H., (2001) Growth kinetics of the bacterium used in a silicone rubber membrane bioreactor to biodegrade the toluene as an organic pollutant in wastewater, *J. Chem. Eng. Chin. Univ.,* 15, 557–562.

Xiao Z.Y., Xing X.H., (2001) Mass transfer kinetics of a new silicone rubber membrane bioreactor for organic wastewater treatment, *J. Chem. Eng. Chin. Univ.,* 15, 71–77.

Yamamoto K, Win K.M, (1991) Tannery wastewater treatment using a sequencing batch membrane reactor, *Water Sci. Technol.,* 23 (7–9), 1639–1648.

Yamaura S.I., Shimpo Y., Okouchi H., Nishida M., Kajita O., Kimura H., Inoue A., (2003) Hydrogen permeation characteristics of melt-spun Ni-Nb-Zr amorphous alloy membranes, *Mater. Trans.,* The Japan Institute of Metals (JIM), 44, 1885-1890.

Yang H., Xu Z., Fan M., Gupta R., Slimane R.B., Bland A.E., Wright I. (2008), Progress in carbon dioxide separation and capture: *A review, J. Env. Sci.,* 20, 14-27.

Yang W., Cicek N., Ilg J., (2006) State-of-the-art of membrane bioreactors: worldwide research and commercial applications in North America, *J. Membrane. Sci.,* 270, 201–211.

Zhang Y., Ozaki T., Komaki M., Nishimura C., (2002) Hydrogen permeation characteristics of vanadium–aluminum alloys, *Scripta Mater.,* 47, 601-606.

Zhou Q., Chan C.W., Tontiwachiwuthikul P., (2009) A monitoring and diagnostic expert system for carbon dioxide capture, *Exp. Systems Appl.,* 36, 1621-1631.

Zhu X.S., Choo K.H., Park J.M., (2006) Nitrate removal from contaminated water using polyelectrolyte-enhanced ultrafiltration, *Desalination,* 193, 350–360.

In: Membrane Processes for Sustainable Growth
Editors: Angelo Basile and Alfredo Cassano
ISBN: 978-1-62618-446-6

Chapter 6

PRESSURE-DRIVEN MEMBRANE APPLICATIONS: MICROFILTRATION AND ULTRAFILTRATION

Alfredo Cassano* and Carmela Conidi
Institute on Membrane Technology (ITM-CNR),
University of Calabria, Rende (Cosenza), Italy

ABSTRACT

An overview of microfiltration (MF) and ultrafiltration (UF) processes is provided. Basic principles, membrane materials and membrane module configurations involved in these operations are described, giving emphasis to advantages and shortcomings. Membrane fouling and concentration polarization phenomena, as well as methods to control and reduce their effects on membrane performance, are also discussed. Finally, well-established applications within food and beverages industries, including milk, fruit juices and wine processing, are reviewed highlighting their significant advantages over traditional methodologies and those which can be potentially achieved through the integration with other membrane operations.

Keywords: microfitration (MF); ultrafiltration (UF); membrane fouling; concentration polarization; membrane processes

NOMENCLATURE

A	membrane surface area (m^2)
a	constant [-]
b	constant [-]
c	constant [-]
C	concentration [mol m^{-3}, mol L^{-1}]
d	constant [-]
d_h	hydraulic diameter [m]

* Corresponding Author: e-mail: a.cassano@itm.cnr.it

D	diffusion coefficient [m^2 s^{-1}]
G	(=k_0) parameter defined in text (s m^{-2})
J	permeate flux [kg m^{-2} s^{-1}]
J_0	initial permeate flux [kg m^{-2} s^{-1}]
$J_{\lim}$	limit value of the permeate flux attained in steady-state conditions [kg m^{-2} s^{-1}]
J_s	solute flux [mol m^{-2} s^{-1}]
J_v	volume flux [m s^{-1}]
k_n	phenomenological coefficient (units depend on mechanism)
k	mass transfer coefficient [m s^{-1}]
K'_s	Hérmia's parameter for standard blocking filtration
$(K'_s/2)A^{0.5}$	(=$k_{1.5}$) parameter defined in text
l	membrane thickness [m]
L	length [m]
L_v	hydrodynamic permeability [m^2 Pa^{-1} s^{-1}]
n	general index (units depend on mechanism)
p	pressure [Pa]
R	rejection [-]
R_m	membrane resistance [Pa m^2 s kg^{-1}]
R'_m	intrinsic membrane resistance [Pa m^2 s kg^{-1}]
R_f	fouling resistance [Pa m^2 s kg^{-1}]
R_g	concentration polarization resistance [Pa m^2 s kg^{-1}]
Re	Reynolds number [-]
Sc	Schmidt number [-]
Sh	Sherwood number [-]
t	time [s]
V	volume [m^3]
VRF	volume reduction factor [-]
x	directional co-ordinate [m]
u	velocity [m s^{-1}]

Greek Letters

δ	boundary layer thickness [m]
Δ	difference or recovery rate [-]
ϕ	gel permeability [m^2 s kg^{-1}]
ν	viscosity [m^2 s^{-1}]
π	osmotic pressure [Pa]

Subscripts

b	bulk
f	feed
g	membrane wall (gel)

p	permeate
r	retentate
s	solute
v	volume
w	water

Abbreviations

CIP	clean-in-place
DE	diatomaceous earth
EDTA	ethylenediaminetetraacetic acid
LCR	low concentrated retentate
MBR	membrane bioreactor
MCR	medium concentrated retentate
MEUF	micellar-enhanced ultrafiltration
MF	microfiltration
MP	milk protein
MPC	milk protein concentrate
MWCO	molecular weight cut-off
PA	polyamide
PE	polyethilene
PES	polyethersulphone
PP	polypropylene
PS	polysulphone
PTFE	polytetrafluoroethylene
PV	pervaporation
PVDF	poly(vinylidene fluoride)
RO	reverse osmosis
UTP	uniform transmembrane pressure
UF	ultrafiltration
WPC	whey protein concentrate

1. INTRODUCTION

Membranes science has gained an important place in chemical technology, offering interesting opportunities in the design, rationalization and optimization of innovative productions, particularly appropriate for a sustainable industrial growth.

The basic properties of membrane operations make them ideal for industrial production: they are generally athermal and do not involve phase changes or chemical additives; they are simple in concept and operation, modular and easy to scale-up; furthermore, they are characterized by a low energy consumption permitting a rational utilization of raw materials and recovery and reuse of by-products. Membrane technologies respond efficiently to the requirements of the so-called "*process intensification*", since they permit drastic improvements in manufacturing and processing, substantially decreasing the equipment-size /production-capacity ratio, the energy consumption, and/or the waste production and resulting in cheaper sustainable technical solutions (Drioli and Romano, 2001). The membranes used in various applications differ widely in their structure, in their function and in the way they are operated. The demand for different membranes is growing year by year due to the development of new applications and new membrane technologies.

Pressure-driven membrane processes such as microfiltration (MF) and ultrafiltration (MF) have been widely applied in various chemical and biochemical processes with many advantages over traditional filtration methods. The possibility of redesigning innovative MF and UF membrane processes with low environmental impact and high quality of the final products has been studied, and in some cases realized, in various industrial sectors. This chapter will focus on general properties of both processes and on selected applications that highlight the wide potential of these technologies in different fields. Case studies related to some applications of interest in food processing have also been included.

The utilization of MF and UF processes on industrial scale is relatively recent, spanning only 40 years. The development of asymmetric reverse osmosis membranes in cellulose acetate in the early 60s (Loeb and Sourirajan, 1962) is considered a milestone in membrane science and technology. Soon, other synthetic polymers were used as basic materials for synthetic membranes. UF was an offshoot of reverse osmosis (RO) and really came into use only in the 1960s when MF on a smaller scale was already well established. The first commercially successful industrial UF system was installed in 1969 by Abcor to recover electrocoat paints from automobile paint shop rinsewater (Goldsmith et al., 1971).

Today the market of UF and MF membranes and related equipment is well developed in a wide range of industrial applications with annual sales higher than one billion US $.

2. Definitions and General Considerations

MF and UF are well-known membrane processes based on the use of a permselective barrier through which fluids and solutes are selectively transported when a hydrostatic pressure is applied across it. As a result the feed solution is converted in a permeate containing all components which have permeated the membrane and a retentate containing all compounds rejected by the membrane (Figure 1).

The term MF is used when particles with diameters of 0.1-10 μm are separated from a solvent or other low molecular weight compounds. The separation mechanism is based on a sieving effect and particles are separated according to their dimensions although the separation is influenced by interactions between the membrane itself and the particles being filtered. Membranes used for MF have a symmetric structure and pores of 0.1 to 10 μm in diameter. The macroporous structure of the membrane with a narrow pore size distribution ensures a quantitative retention of particles of a given size and type (Ripperger and Almann, 2002). These particles are generally larger than those separated by UF and RO. Consequently, the osmotic pressure for MF is negligible and hydrostatic pressure differences used in MF are relatively small (in the range of 0.05-0.2 MPa).

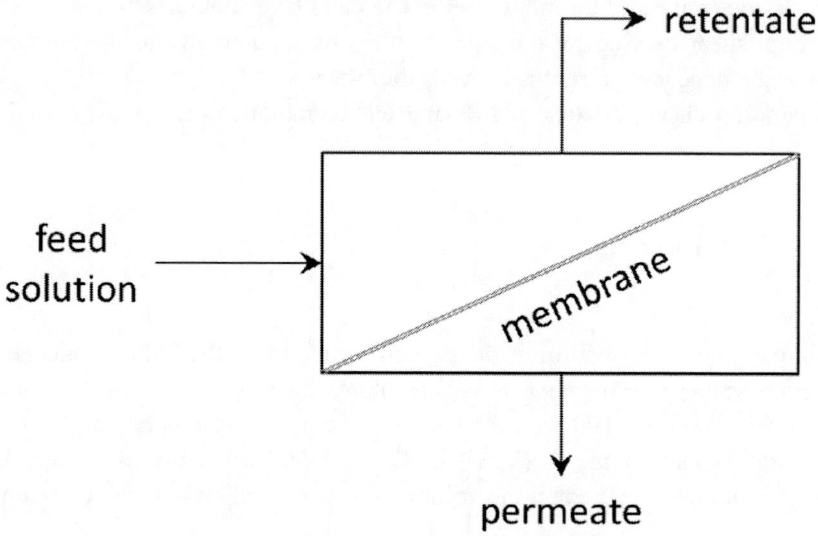

Figure 1. Process scheme of microfiltration and ultrafiltration.

Another important characteristic is the high porosity of MF membranes which leads to high filtration rates at small pressure differences.

UF is a membrane process similar to MF in the mode of separation, based on the use of asymmetric membranes with pore sizes in the skin layer of 2-10 nm. Typically dissolved molecules or small particles not larger than 0.1 μm in diameter are retained. An UF membrane is typically characterized by its molecular weight cut-off (MWCO), defined as the equivalent molecular weight of the smallest species that exhibit 90% rejection. The MWCO of UF membranes is between 10^3 and 10^6 Dalton. Hydrostatic pressures of 0.1-0.5 MPa are typically used. The primary basis for separation is the molecular size although secondary factors, such as molecule shape and charge, can play an important role.

In both MF and UF processes the osmotic pressure due to the presence of macromolecules is usually so low that is negligible. Therefore, the filtration rate can be expressed by the following equation:

$$J_v = L_v \frac{\Delta p}{l} \quad (1)$$

where J_v is the volume flux across the membrane, Δp the pressure difference between the feed and permeate solution, l the membrane thickness and L_v is the hydrodynamic permeability of the membrane.

MF and UF processes can be operated either in dead-end or in cross-flow configurations (Figure 2). In the dead-end mode, the feed flow is perpendicular to the membrane surface. It is forced through the membrane, which causes the retained particles to accumulate and form a type of cake layer at the membrane surface. The thickness of the cake layer increases with the filtration time. Therefore, the permeation rate decreases by increasing the cake layer thickness. In the cross-flow mode, the fluid to be filtered flows tangentially to the membrane surface and permeates through the membrane due to the imposed transmembrane pressure difference. Unlike dead-end filtration, rejected particles form a cake layer on the membrane surface which does not build up indefinitely. The cake layer thickness remains relatively thin due to the high shear exerted by the solution flowing tangentially to the membrane surface keeping away the deposited particles towards the filter exit.

The separation characteristics of MF and UF membranes can be expressed in terms of membrane rejection o retention (R):

$$R = (1 - \frac{C_p}{C_f}) \cdot 100 \tag{2}$$

where C_p is the solute concentration in the permeate and C_f the solute concentration in the feed. Rejection values are between 0% (for solutes having the highest probability to pass through the membrane) and 100% (when solutes are completely retained by the membrane).

The volume reduction factor (VRF) in UF and MF processes is defined as the ratio between the initial feed volume and the volume of the resulting retentate given by:

$$VRF = \frac{V_f}{V_r} = 1 + \frac{V_p}{V_r} \tag{3}$$

where V_f, V_p and V_r are the volume of feed, permeate and retentate, respectively. Sometimes data are also presented in terms of recovery rate (Δ) which is given by the ratio of the permeate volume (V_p) to the feed volume (V_f):

$$\Delta = \frac{V_p}{V_f} \tag{4}$$

Recovery rate values are in the range 0-1. In practical applications the solvent of the feed solution cannot be completely recovered as permeate due to some factors, such as osmotic pressure and viscosity, limiting the retentate concentration (Strathmann et al., 2006).

3. Concentration Polarization

Since MF and UF membranes are partially or completely retentive towards one or more components in the feed solutions, these components tend to accumulate on the membrane surface during filtration causing a decline of the transmembrane flux. This phenomenon is referred to as concentration polarization. Equally devastating for the performance of MF and UF processes is the membrane fouling which can be considered as an irreversible alteration of the membrane caused by specific physical and/or chemical interactions between the membrane and specific components of the feed solution. Membrane fouling manifests itself as a permeate flux decay and an alteration of membrane selectivity (Cheryan, 1998; Strathmann et al., 2006).

While the concentration polarization can be minimized by hydrodynamics means, such as the feed flow velocity and the membrane module design, the control of membrane fouling is more difficult.

When a solution is pressurised through a MF or UF membrane, solutes are convectively transported towards the membrane surface by the solvent which flows through the membrane. If solutes are partially or completely retained by the membrane the initial solute flux towards the membrane will be greater than that through the membrane causing a build-up of retained solutes at the upstream surface of the membrane. The schematic representation of the concentration polarization phenomenon is given in Figure 3. The flux decay due to concentration polarization can be explained assuming two different mechanisms. First, the increased solute concentration on the membrane surface determines a significantly higher osmotic pressure, causing a decrease in the driving force (Δp-$\Delta\pi$, where Δp is the applied transmembrane pressure difference and $\Delta\pi$ the osmotic pressure difference between the feed and permeate) and flux. This effect is generally more pronounced with small solutes having high osmotic pressures (e.g. retained salts in RO). However, some macromolecules, e.g. proteins, can also produce high osmotic pressures when they accumulate at high concentration near the membrane surface due to protein-protein interactions.

According to another view, the accumulated solute can provide a hydrodynamic resistance to the solvent flow in addition to the membrane resistance, reducing the trans-membrane flux. Generally, small particles tend to form a dense particle cake layer, while large macromolecules provide a "gel" layer.

The convective transport of solute to the membrane causes a solute build-up at the membrane/solution interface. The established concentration gradient between the solution at the membrane surface and the bulk causes a diffusive back-transport of the solute into the bulk. A steady state is reached when the two phenomena balance each other.

At high Δp values the solute concentration at the membrane surface reaches a maximum value (C_g) which depends on the physicochemical characteristics of the solution.

The gel layer provides an additional resistance to the filtrate flow affecting the membrane separation characteristics (i.e. by changing the rejection of lower molecular weight components).

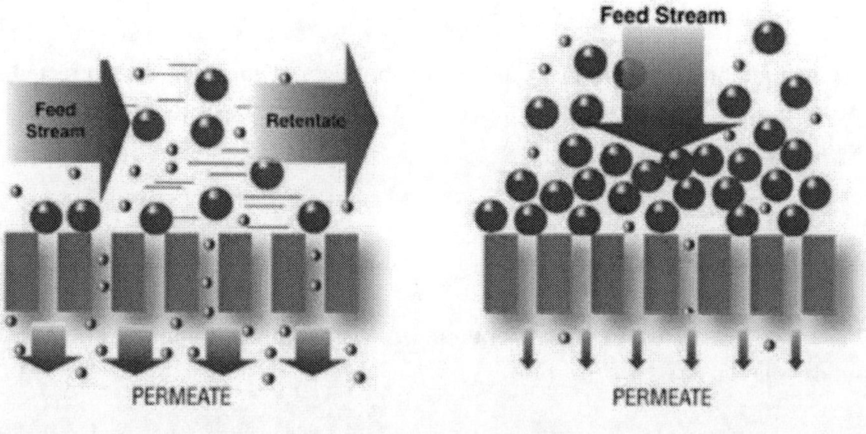

Figure 2. Comparison between cross-flow and dead-end configurations.

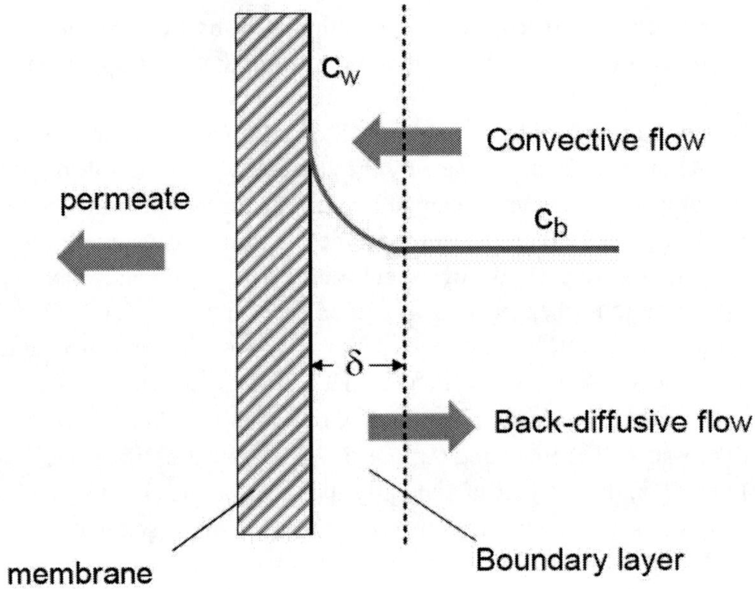

Figure 3. Schematic representation of concentration polarization phenomenon (c_w = gel concentration; c_b = bulk concentration).

At low pressures shear and lift forces are sufficient to minimize particle deposition and the flux increases linearly with pressure. When retained solutes start to accumulate on the membrane the rate of increase in flux decreases. A further increase in pressure causes an increasing of the particle layer without a corresponding increase in flux (Cheryan, 1998).

Unlike fouling the gel-layer is assumed to be dynamic: an increasing of feed velocity and temperature or a decreasing of operating pressure and solute concentration should revert the system back to the pressure-controlled region (Figure 4).

A mathematical model generally accepted to describe the concentration polarization in MF and UF processes is the well-known *film theory*. It assumes that the solute is brought to the membrane surface by convective transport at a rate J_s defined as:

$$J_s = J_v C_b \tag{5}$$

where J_v is the membrane volume flux and C_b the bulk concentration of the rejected solute.

The resulting concentration gradient causes a diffusive flux of the retained material back into the bulk solution at a rate given by:

$$J_s = -D\frac{dC}{dx} \tag{6}$$

where D is the solute diffusion coefficient and dC/dx the concentration gradient of the solute in the boundary layer.

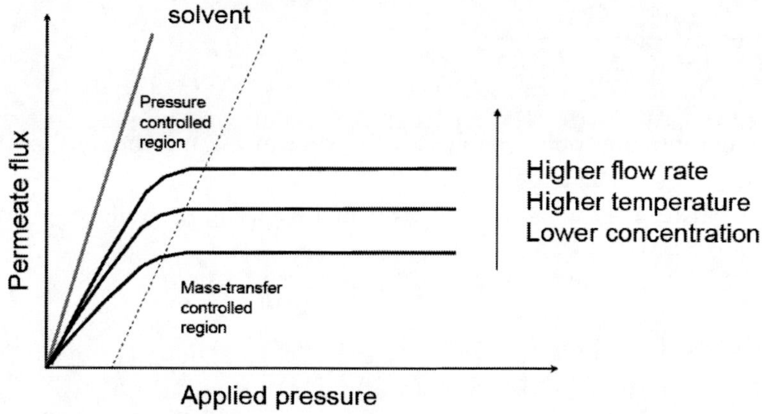

Figure 4. Schematic flux versus applied pressure plot comparing typical UF behaviour with pure solvent flux.

At steady state, the convective transport is counterbalanced by the diffusive flux and Equations (5) and (6) can be equated and integrated over the boundary layer leading to:

$$J_v = \frac{D}{\delta} \ln \frac{(C_g - C_p)}{(C_b - C_p)} \tag{7}$$

where C_g is the solute concentration at the membrane surface (gel concentration), C_P is the permeate solute concentration, δ is the boundary layer thickness.

Since the membrane rejection R is given by Equation (2), the combination of Equations (2) and (7) leads to:

$$\frac{C_g}{C_b} = \frac{\exp(J_v \delta / D)}{R + (1 - R)\exp(J_v \delta / D)} \tag{8}$$

According to Equation (8), the concentration polarization is expressed as the ratio of the solute concentration at the membrane surface and the solute concentration in the bulk. It is a function of the membrane flux J_v, the boundary layer thickness δ, the solute diffusion coefficient in the boundary layer D and the membrane solute rejection R.

The concentration polarization can be described in various feed flow conditions introducing a solute mass transfer coefficient k (having the same units as the flux J_v) given by:

$$k = \frac{D}{\delta} \tag{9}$$

Introducing k into Equation (8) leads to:

$$\frac{C_g}{C_b} = \frac{\exp(J_v/k)}{R + (1-R)\exp(J_v/k)} \tag{10}$$

in which the concentration polarization is a function of the membrane flux, the membrane rejection and the mass transfer coefficient.

The mass transfer coefficient can be estimated by using the following general equation (Sherwood et al., 1975):

$$Sh = a\,\mathrm{Re}^b\, Sc^c (\frac{d_h}{L})d \tag{11}$$

where L is the length of the feed flow channel (or tube); d_h is the hydraulic diameter of the feed channel; a, b, c and d are characteristic constants for different geometries to be experimentally determined.

The Sherwood number is given by:

$$Sc = \frac{\nu}{D} \tag{12}$$

where ν is the viscosity.

The Reynolds number in a channel (or in a tube) is given by:

$$\mathrm{Re} = \frac{d_h u}{\nu} \tag{13}$$

where u is the fluid velocity.

In general, Re values less than 1800 are considered to be laminar flow and Re values greater than 4000 are turbulent flow. The film model can be used to describe the concentration polarization both in turbulent and laminar flow regimes. In the turbulent flow regime entrance effects can be neglected: in this case the exponent d in Equation (11) is zero. In the laminar flow regime entrance effects cannot be neglected and the Sherwood number is expressed as (Bird et al., 1965):

$$Sh = (aSc\,\mathrm{Re}\frac{d_h}{L})^{0.33} \tag{14}$$

and the concentration polarization can be obtained as a function of the flow velocity, the channel height and the channel length by combining Equations (10) and (14).

The film model is not always applicable in MF and UF because the feed solutions contain macromolecular compounds or suspended particles with low diffusion coefficients; consequently, the diffusive mass transport of retained compounds from the membrane surface back into the bulk solution is slower than the convective mass transport towards the membrane. When the solute concentration at membrane surface exceeds the solubility of the feed·constituents it can precipitate forming a solid layer on the membrane surface. This layer results in an additional hydrodynamic resistance to the membrane flux. In these conditions, if Δp is increased, more solutes are transported towards the membrane increasing the layer thickness at the membrane surface without a corresponding increase in flux (Figure 4).

A more simple approach to describe the membrane flux in case of a gel or a cake layer formation is based on the use of the "resistance-in-series" concept.

The intrinsic membrane resistance R_m can be determined using pure water as a feed according to the following equation:

$$J_w = \frac{\Delta p}{R_m} \tag{15}$$

in which Δp is the transmembrane pressure difference and J_w the water flux.

In the filtration of a real feed the membrane resistance is only a small part of the total resistance. If membrane fouling occurs, an additional resistance, named R_f, has to be included:

$$J = \frac{\Delta p}{(R_m + R_f)} \tag{16}$$

R_f is a consequence of physicochemical interactions between particles and membrane structure; therefore it is unaffected by operating parameters and is lumped with the intrinsic membrane resistance as R'_m where $R'_m = R_m + R_f$.

An additional resistance related to the concentration polarization layer indicated as R_g can be included in the flux equation:

$$J = \frac{\Delta p}{(R_m + R_f + R_g)} \tag{17}$$

R_g is affected by operating parameters and physical properties. It is a function of the permeability of the gel (ϕ) and its thickness, which is a function of the applied pressure:

$$R_g = \phi \Delta p \tag{18}$$

The resistance model can be expressed as:

$$J = \frac{\Delta p}{(R'_m + \phi \Delta p)} \tag{19}$$

According to this equation at low pressures the term R_g will be small if compared with R'_m and flux is a function of pressure. At higher pressures R_g becomes relatively large and flux is less dependent on pressure approaching the limiting value $1/\phi$.

Methods to reduce the effect of concentration polarization in MF and UF include: increasing of the cross flow velocity, insertion of turbulence promoters, chemical modification of membrane surface (polymerization of hydroxypropyl acrylate onto PVDF-base membrane in situ, deposition of hydroxypropyl cellulose onto polysulphone membranes, etc.), use of special designs (rotating cylinders or discs, vibratory shear enhanced units, annular fluid flow with vortices), application of electrical fields influencing charge on macromolecules, use of ultrasonics inducing cavitation at surface and inert gas inducing turbulence.

4. Membrane Fouling

Membrane fouling in crossflow MF and UF is a key factor affecting the economic and technological viability of the processes, which essentially depends on the permeate fluxes obtained and their stability with time (Cheryan and Alvarez, 1995). Consequently, the identification and quantification of the prevalent fouling mechanism and efforts to minimize its effects during a continuous filtration process are extremely important.

The term *membrane fouling* is used to describe a long term flux decline caused by the deposition of retained particles (colloids, macromolecules, dissolved organics, etc.) onto the membrane surface or in the membrane pores. It may occur due to a concentration polarization layer development over the membrane surface, the formation of a cake layer and/or a blockage of the membrane pores. The pore blocking can be further characterized by complete, intermediate and standard pore blocking.

The fouling behavior depends on physical and chemical parameters such as concentration, temperature, pH, ionic strength and specific interactions. Consequently, it varies according to the type of foulant agent. The degree of membrane fouling determines the frequency of cleaning, the lifetime of the membrane and the membrane area needed and consequently it has a significant effect on the cost, design and operation of membrane plants. Rejection and yields may also be affected by fouling. If the buildup of solids on the membrane is significant enough, it may act as a secondary membrane and change the effective sieving and transport properties of the system.

Fouling substances can be divided into different categories:

- colloidal or particulate matter;
- inorganic compounds such as $CaSO_4$, $Fe(OH)_x$, and other metal hydroxides;
- organic materials such as proteins, humic acids and other macromolecular materials;
- biological components such as microorganisms and products of their metabolism;
- chemical reactants.

Membrane properties strongly affect the membrane fouling phenomenon. Hydrophobic materials can absorb hydrophobic or amphoteric compounds; consequently, for aqueous feedstreams membrane fouling can be reduced selecting hydrophilic membranes.

The introduction of hydrophilic functional groups, such as hydroxyl groups (-OH), on the surface of a specific membrane can be exploited to reduce membrane fouling by aqueous feedstream components. Methods include surface coating by adsorption, microwave or plasma treatment, grafting of hydrophilic monomers, coating with oriented monolayers.

Most membranes exhibit a net negative charge under usual process conditions; consequently, electrostatic forces can be involved if solutions containing charged particles are processed. The surface topography is another factor influencing membrane fouling: for example, the presence of protuberances on the surface of polyamide-based membranes seems to be responsible of membrane fouling due to a suspended matter capture. On the contrary, cellulose acetate membranes exhibit smoother surfaces and are less prone to fouling especially with regard to biofoulants.

The membrane pore size is another important factor influencing membrane fouling. A physical blockage of pores due to particles with sizes of the same order of magnitude as the range of membrane pore size is typically observed in MF processes. On the contrary, particles much larger than membrane pore sizes can be removed from the surface under the shear forces generated by the flow.

Membrane-solute interactions involved in membrane fouling are strongly dependent on solute properties. The presence of different functional groups and the varying degree of hydrophobicity allow proteins to interact with different feed components and membrane itself. Mineral salts can precipitate on the membrane due to a poor solubility or bind to the membrane through charge interactions. Some fat substances are characterised by functional groups similar to those of poly (vinylidene fluoride) (PVDF) and polysulphone (PS) membranes resulting in considerable fouling phenomena. Polysaccharides, polyhydroxy aromatics, humic substances, antifoams have also been reported as fouling agents.

Finally, process parameters such as temperature, flow rate, pressure and feed concentration have a great influence on membrane fouling. Generally, an increase in temperature results in an increase in flux since the viscosity will be lower and the diffusion higher. However, for some feeds, such as cheese whey, a decreasing in flux has been observed when the temperature is raised up to 30°C due to a decrease in solubility of calcium phosphate. Further, high temperatures determine both denaturation of proteins and microbial growth which will result in a lowering of the flux.

A higher flow rate tends to remove deposited materials and, consequently, reduces the hydraulic resistance of the fouling layer. However, in some cases, the increase in cross flow velocity determines a removal of greater particles from the membrane surface and a stratification of smaller particles on the membrane surface with a consequent pore plugging. Further, the effect of feed flow rate is strictly correlated to the Δp: at high pressures the rate of fouling increases by increasing the cross flow velocity.

For small pressures, the permeate flux increases linearly with the Δp up to a limit value which depends on physical properties of the suspension. As the pressure is increased, the concentration polarization layer reaches a limiting concentration and the flux becomes independent of pressure. In these conditions a limiting flux is reached and any further increase in pressure determines no significant increase of the permeate flux. When the pressure is increased above a critical point a compaction of the gel layer occurs resulting in a lower flux. According to the concept introduced by Field et al. (1995) below the *critical flux* there is no flux decline with the time and the flux depends linearly on the Δp. When the critical flux is reached the flux increases slowly and approaches a constant value named as *limiting flux*. Therefore, it is important to operate below the critical flux.

Although fouling and concentration polarization are characterised by similar aspects there are substantial differences between them. Concentration polarization is a reversible process which takes place over a few seconds. It can be easily controlled by the proper process design and described adequately by simple mathematical models. When the steady-state is reached a further decrease in flux is not observed and the flux becomes constant as a function of time. Fouling is generally irreversible and the flux decline takes place over many minutes, hours or days. It is more difficult to describe and to control experimentally. A continuous flux decline can often be observed.

There are several methods which can be used to minimise fouling to some degree. They can be summarised as: feed pretreatment, selection of appropriate module configuration and process conditions, change of membrane properties, cleaning. Pretreatment methods aim at removing or to stabilizing foulants by changing the properties of the feed solution. They include pH adjustment, heat treatment, addition of complexing agents, chlorination, adsorption onto active carbon.

The selection of appropriate module configurations and operating conditions aims at increasing the convective transport of solute back to the bulk solution. The use of turbulent promoters, ultrasonic vibrations, rotating modules, pulsation of feed flow (backflush and backshock techniques) are all methods commonly used for this purpose (Bertran et al., 1993).

Membrane surface treatments aim at creating a surface less prone to adsorb foulants. Since fouling is more severe in porous membranes, such as those used in MF and UF operations, it can be minimised by using a narrow pore size distribution. The selection of hydrophilic membranes rather than hydrophobics and the use of negatively charged membranes can also be useful to reduce membrane fouling.

When concentration polarization is minimised (i.e. by increasing flow velocity or by using turbulence promoters) also fouling is reduced. Basically, cleaning procedures are based on hydraulic, mechanical, electrical and chemical methods. The method used depends upon the module configuration, the chemical resistance of the membrane and the type of foulant.

The hydraulic cleaning is typically achieved by backflushing consisting in a periodic reversing of the transmembrane flow to remove the fouling layer from both inside and surface membrane. The process is carried out, at regular intervals, by reversing the direction of flow of the permeate usually for a few minutes. The backflushing medium can be the permeate, another liquid or gas. The procedure is carried out with tubular and hollow fibre membranes which can withstand the reversal in pressure difference and it is efficient for many types of foulants. The back-pulsing method, as an alternative method to backflushing, permits to reduce the permeate loss and optimize the operation time during the filtration process. It consists in applying short bursts of back pressure reversing the permeate flow with the permeate exit closed.

Mechanical cleanings include the use of foam balls forced down tubular membrane systems at high velocities which create turbulence at the membrane surface and dislodge fouling materials. The foam balls, slightly smaller than the bore of the tube, are injected into one end of the membrane tube and collected from the other end.

Electrical methods are based on the use of electrical pulsing producing a movement of charged species away from the membrane surface. Special module designs are required to introduce the charge to the membrane surface which is generally of metallic type.

When the membrane flux decreases to unacceptable values, chemical cleaning is required to restore partially or completely the flux. The choice of chemicals depends on the type of membrane and fouling agent and the severity of fouling. Several companies produce chemical cleaning compounds specifically for membranes. Basically chemical agents can be summarised in the following categories: acids, alkalis, detergents, enzymes, complexing agentssuch as ethylenediaminetetraacetic acid (EDTA) and disinfectants.

Chemical cleaning is generally realised by clean-in-place (CIP) methodologies by filling the retentate channels of the membrane module with the cleaning solution contained in a separate tank. The composition of the cleaning solution is strongly dependent on the nature of fouling agents and the properties of the membrane.

Chemical concentration, recirculation time, temperature and pH are all important factors affecting the efficiency of a cleaning protocol and should be optimised case-by-case. Cleaning solutions should be pumped through the system at high velocity and low operating pressure in order to avoid the penetration of foulants in membrane pores. Generally, most chemical cleaners complete their action in 30-60 min. Enzyme cleaners may require longer times.

The temperature of cleaning solutions should be as high as possible in agreement with the temperature limitation of the membrane module. The pH of the cleaning solution depends on the type of fouling agent and membrane. Alkaline solutions such as NaOH and KOH up to pH 12 are useful in solubilising organics and proteins. Acid cleaning solutions at pH 1.5-2.8 are used to remove inorganic salt fouling.

Different mathematical models have been proposed in order to describe the flux decline in membrane fouling based on equations applied to dead-end filtration mechanisms at constant pressure (Hérmia, 1982) and opportunely modified in order to describe the fouling mechanism involved in cross-flow filtration (Field et al., 1995 and 1996). The Field model describes the permeate flux decline with time by the following differential equation:

$$-\frac{dJ}{dt}(J^{n-2}) = k_n(J - J_{\lim}) \tag{20}$$

in which $J_{\lim}$ represents the limit value of the permeate flux obtained in steady-state conditions; k_n and n are phenomenological coefficient and general index, respectively, depending on fouling mechanism.

The model expressed by Eq. (20) permits, on the basis of experimental data, to point out the fouling mechanism involved in the filtration process, according to the estimated value for n as follows:

Complete pore blocking (n= 2)

This situation corresponds to a condition of complete pore obstruction by means of sealing (blocking) which occurs when the particles are larger than pore size. The resulting equation representing this situation is:

$$J = J_{\lim} + (J_0 - J_{\lim})\exp(-k_2 t) \tag{21}$$

where J is the permeate flux, J_0 the initial permeate flux at t=0.

Partial pore blocking (n= 1)

In these conditions a dynamic situation of blocking/unblocking occurs in which particles may bridge a pore by obstructing the entrance but not completely blocking it.

The resulting equation in this case is:

$$\sigma t = \frac{1}{J_{\lim}\left[\ln((J_0 - J_{\lim})/J_0)(J - J_{\lim}))\right]} \tag{22}$$

where $\sigma = k_1$.

Cake filtration (n = 0)

When solid particles or macromolecules do not enter the pores they form a cake on the membrane surface. In this case the overall resistance to the mass transport is composed by a cake resistance and a membrane resistance which is assumed to remain unchanged.

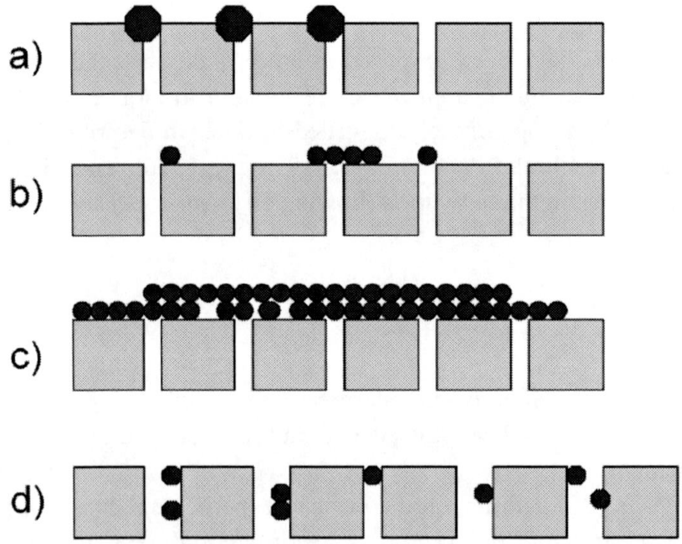

Figure 5. Schematic representation of fouling mechanism: complete pore blocking (a); partial pore blocking (b); cake filtration (c); internal pore blocking (d) (adapted from Field et al., 1996).

The resulting equation is:

$$Gt = \frac{1}{J_{\lim}^2}\left[\ln\left(\frac{J}{J_0}\frac{J_0 - J_{\lim}}{J - J_{\lim}}\right) - J_{\lim}\left(\frac{1}{J} - \frac{1}{J_0}\right)\right] \tag{23}$$

where $G = k_0$.

Internal pore blocking (n= 1.5; J_{lim}= 0)

Particles enter the pores reducing pore volume. In this case membrane resistance increases as a consequence of pore size reduction. In these conditions fouling phenomenon becomes independent of cross flow velocity and a limit value of the permeate flux is not reached, i.e. $J_{\lim} = 0$.

The resulting equation is:

$$J^{\frac{1}{0.5}} = \frac{1}{J_0^{0.5}} + (\frac{K_s^{'}}{2})A^{0.5}t \qquad (24)$$

where $(K_s^{'}/2)A^{0.5} = k_{1.5}$ and A is the membrane area.

A schematic representation of different situations above mentioned is reported in Figure 5.

5. Membrane Materials and Modules

All kinds of different synthetic materials can be used to prepare membranes. Thus the material can be either inorganic, such as ceramic, glass or metal, or organic including all types of polymers (Mulder, 1991).The basic principle involved is to modify the material in such a way by means of an appropriate technique in order to obtain a membrane structure with a morphology suitable for a specific separation.

For UF and MF membranes the most used are organic hydrophobic polymers such as polytetrafluoroethylene (PTFE), polypropylene (PP), polyethersulphone (PES) and polyethylene (PE). PVDF is another hydrophobic material widely used for UF and MF membranes due to its excellent chemical resistant, good thermal and mechanical properties. However, the high hydrophobicity of PVDF leads to rather low permeability and high fouling. Many studies have been carried out in order to enhance the performance of hydrophobic membranes, by introducing hydrophilic functional groups, like $-SO_3H$ or $-COOH$, to the surface by chemical modification, photo-modification or by adding water-soluble polymers such as poly (vinylpyrrolidone) or poly(vinyl methyl ether) to the membrane casting solutions (Baker, 2000; Dave and Reddy, 2011; Reddy et al., 2003 and 2005).

PS is also widely used in the production of UF membranes with good mechanical properties and good resistance in terms of heat and pH; however, it exhibits a poor resistance to solvents.

Regenerated cellulose membranes are more hydrophilic, reducing both protein adsorption and fouling. However, these polymeric membranes are susceptible to chemical degradation by strong caustic and hypochlorite cleaning solutions leading to a significant reduction in membrane life. In addition, some polymeric membranes have limited mechanical stability, leading to a reduction in permeability at high pressures and possible membrane failure, particularly in systems employing rapid high-pressure backpulsing.

These limitations have motivated the development of a variety of inorganic membranes with greatly enhanced chemical, thermal, and mechanical stability compared to polymeric ones (Shah et al., 2007). With inorganic membranes, processing at high temperature (up to

°500 C) and extreme pH value (1-14) is possible. In addition, inorganic membranes can be cleaned with aggressive chemical, organic solvent or hot water stream. The base materials for the preparation of ceramic membranes are alumina ($A1_2O_3$), titania (TiO_2), silica ($Si0_2$), and zirconia ($Zr0_2$) (Van der Bruggen et al., 2010). A number of different techniques is available to prepare synthetic UF and MF membranes. Some of these techniques can be used to prepare organic (polymeric) as well as inorganic membranes.

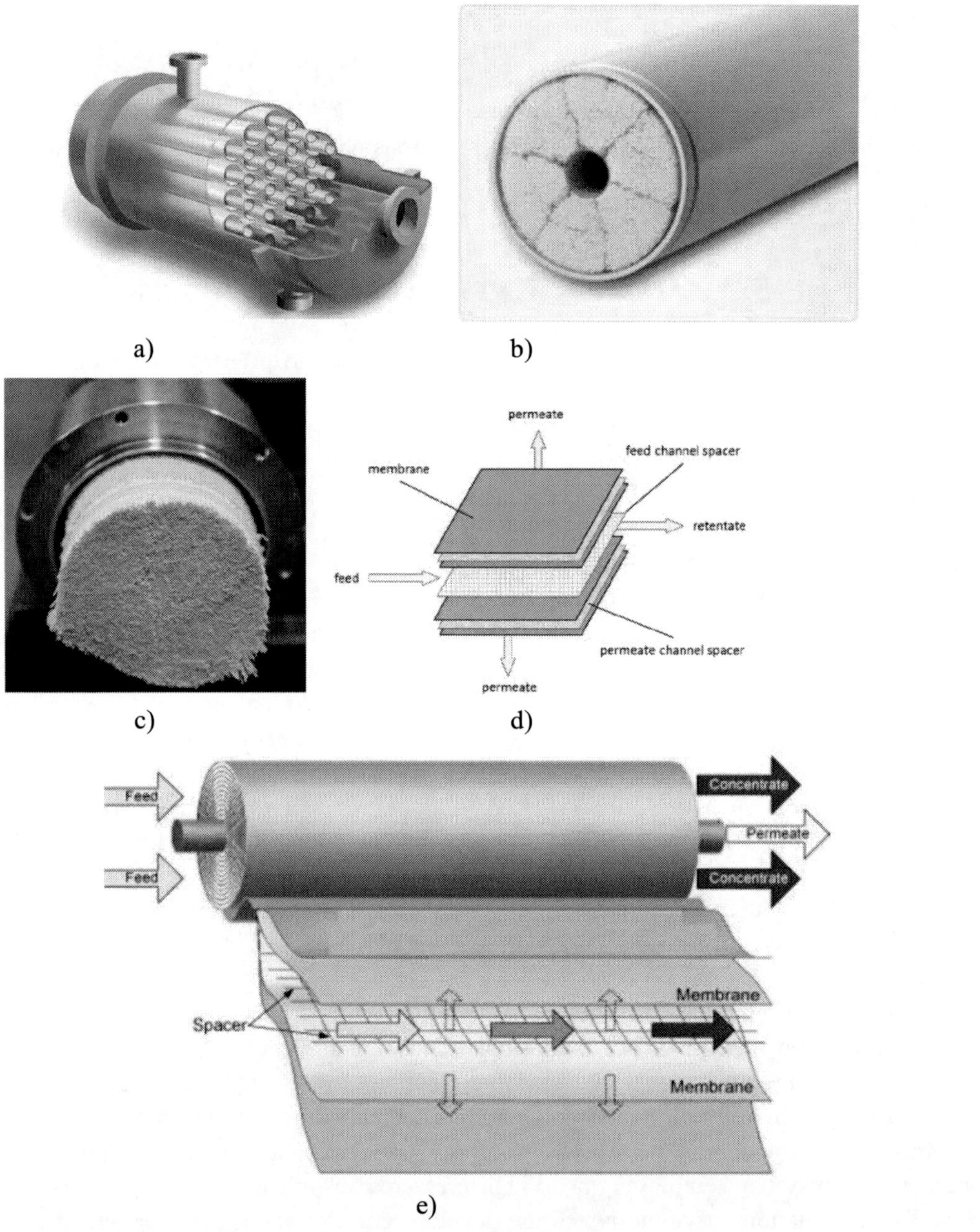

Figure 6. Schematic representation of a (a) tubular membrane module; (b) capillary membrane module; (c) hollow fiber membrane module; (d) plate and frame membrane module and (e) spiral-wound membrane module.

MF membranes are prepared by sintering, track-etching, stretching or phase inversion techniques, while UF membranes are mainly prepared by phase inversion. Sintering and the sol/gel process are the most common techniques to prepare ceramic membranes. Sintering is limited to pore sizes around 0.1 μm; practically all UF membranes are prepared with the sol-gel technique.

MF and UF membranes used at industrial level are installed in proper devices generally called *membrane modules*. On large industrial scale, membrane modules are available in five basic designs: hollow fibre, spiral wound, tubular, plate and frame and capillary. They are quite different in their design, mode of operation, production costs and energy requirement for pumping the feed solution through the module.

The choice of the module configuration, as well as the arrangement of modules in a system, is based on economic considerations, the type of separation, ease of cleaning, the control of concentration polarization and membrane fouling. Each design has its own special applications, advantages, and disadvantages.

In tubular membrane modules (Figure 6a) several polymeric membranes are cast on the inside of porous paper or plastic tubular inserts. Membrane tubes are assembled in a housing according to a shell-and-tube configuration. The feed solution is pressurized inside the tubes and the permeate is collected in the shell side of the housing. Tube diameters are typically in the range 10-25 mm.

Tubular membrane modules are suitable for feed solutions with high concentration of suspended solids offering an easy control of concentration polarization and membrane fouling phenomena; they operate under turbulent flow conditions with Reynolds numbers greater than 10,000 and can be cleaned with standard CIP techniques; however, this configuration offers the lowest packing density resulting in high floor space requirement and the highest energy consumption in comparison with other configurations. The tubular design with ceramic inorganic membranes is frequently used in the food processing industries.

In capillary membrane modules, membranes (Figure 6b) are in the form of self-supporting tubes with diameters from 0.2 to 3 mm arranged in parallel in a shell tube. The feed is pumped in the lumen side of membranes while the permeate is collected in the shell side.

The high membrane area per module volume and the low production cost are typical advantages of capillary membrane modules; in addition, in these modules concentration polarization and membrane fouling can be controlled through a physical process called back flushing in which the permeate flow is reversed allowing to dislodge the fouling material from the membrane surface. The main drawbacks are the required low operating pressure that cannot exceed values of 4-6 bar due to the limited stability of the membranes.

Hollow fibre membrane modules (Figure 6c) consist of a cylindrical housing or shell containing a bundle of several membrane fibres with outer diameters from 200 to 500 μm. The feed solution is commonly applied in the lumen side while the permeate is removed from the shell side.

The hollow fibre membrane module is characterised by the highest packing density if compared with other configurations and its production is very cost effective. Drawbacks are related to the difficult control of concentration polarization and membrane fouling. Consequently, an extensive pretreatment of the solution is needed in order to remove particles, macromolecules or other materials which can precipitate at the membrane surface.

In the plate-and-frame membrane module, a feed channel spacer is sandwiched between two flat-sheet membranes separated by permeate channel spacers (Figure 6d). Commercial units are usually horizontal with the membranes mounted vertically. These units are mainly used in small-scale applications concerning the production of pharmaceuticals or fine chemicals. The low packing density restricts their use to clear feed streams containing only fine suspended solids; in addition, these modules are quite expensive and the exchange of membranes is labor intensive.

Spiral-wound membrane modules allow the efficient packaging of flat-sheet membranes in a convenient cylindrical form. These modules consist of a sandwich of flat-sheet membranes, spacers and porous permeate flow material wrapped around a central permeate collecting tube (Figure 6e). The feed solution is pumped axially through the feed channels, while the permeate is forced into the permeate channel and spiral towards the central collecting tube. Commercial modules are about 1 meter long with diameters between 10 and 60 cm. Membrane areas can be in the range 3-60 m^2. In general two or more modules are fitted in series and suitably sealed into a pressure housing.

This configuration provides a relatively large membrane area per unit volume and, consequently, it is characterized by low capital costs. By contrast, spiral wound membrane modules cannot be mechanically cleaned and a feed pretreatment is required.

In Table 1 characteristics of all the modules which must be considered in a system design, including packing density, operating cost, fouling tendency and ease of cleaning, are reported.

Table 1. Properties of different module configurations

Type of module	Operating cost	Fouling tendency	Ease of cleaning	Pre-filtration	Packing density
Tubular	very high	very good	good	no	low
Capillary	low	very good	poor	yes	high
Hollow fibre	very low	very poor	very poor	yes	very high
Spiral wound	low	good	medium	yes	medium
Plate and frame	medium	good	good	yes	low

6. Industrial Applications of MF and UF Processes

UF and MF processes are employed in both production and analytical applications for concentration, purification and fractionation of various products in diverse fields such as food, medical and biotechnological industries, paper industry, dairy industry, etc. Basically, the most common relevant UF and MF applications can be summarised as follows:

- removal of particles from liquid and gas streams coming from chemical, biological and food industry;
- clarification and sterile filtration of heat sensitive solutions and beverages;
- production of pure water in the electronic industries;

- products purification, gas filtration, process solvent recovery in the chemical industries;
- wastewater treatment.

An overview of different MF and UF applications on a large technical scale is depicted in Figure 7. In the food and beverage industry the use of MF for the retention of cellular components, microorganisms and other solids from alcoholic or non-alcoholic beverages, as well as clarification and concentration, is widely employed; different solutions can be processed including milk, beer, wine, whiskies, drinking water, syrups, and fruit juices. A more exhaustive description of MF and UF applications in this field will be given in the next case studies. Almost every manufacturing industry (food, steel, textile, animal, etc.) and service establishment generate a huge quantity of wastewater daily.

The need for stringent pollution control provides tremendous opportunities for membrane technology. Cross-flow MF and UF processes offer interesting alternatives in the treatment of wastewaters if compared to conventional treatments for their high removal efficiencies allowing reuse of the treated water and/or recovery of some valuable waste constituents (Scholz and Lucas, 2003; Benìtez et al., 2006).

UF and MF membranes have emerged as one of the best suitable alternatives for the management of highly polluted effluents generated in the textile industry processes (Marcucci et al., 2001). These technologies are capable either of being used as a pre-treatment or to achieve complete water reclamation, depending on the process involved.While MF works out for removing suspended matter, UF allows the removal of particles and macromolecules but does not eliminate color completely. However, UF permeate may meet the requirements for allowing wastewater reuse in the minor processes (such as rinsing and washing) of the textile industry (Barredo-Damas et al., 2012).

UF is widely applied in the pulp and paper industry. It is a suitable method for increasing the reuse of process water, reducing fresh water consumption and diminishing the amount of discharged effluent.

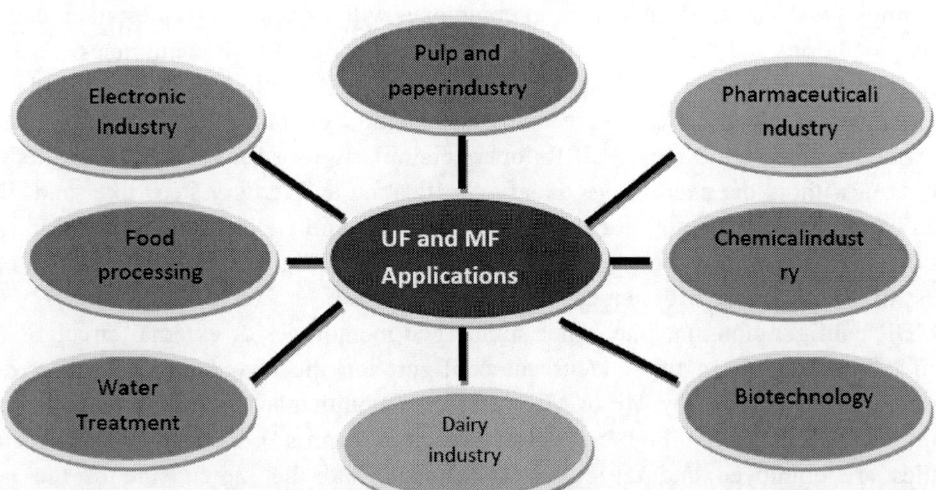

Figure 7. Applications of UF and MF processes.

In addition, UF is easy to apply in the recycling of valuable raw materials and in the separation of compounds for further processing. For example, hemicelluloses can be separated from process water in thermomechanical pulping for the preparation of value-added products. Other applications include colour removal from kraft mill bleaching effluents, recovery of lignin from kraft black liquor and recovery of paper coatings (Puro et al., 2010; Jönsson and Wimmerstedt, 1985; Jönsson et al., 1996).

In the leather industry UF can be used to remove organic compounds from spent soaking and deliming effluents, to recover sulphides from spent dehairing baths, to remove fat substances from spent degreasing baths and as pre-treatment step in an integrated membrane process devoted to the recovery of chromium salts from exhausted tanning liquors (Cassano et al., 2001).

The wastewaters generated in fish and poultry processing industry contain a large amount of organic materials. These wastewaters are usually discharged into the sea without any treatment. The discovery of potentially valuable proteins in the wastewaters, in recent years, has drawn much attention from several researchers to recover proteins by membrane technology. UF ceramic membranes have been used to reduce the organic load from fish meal wastewaters and to allow the recovery of valuable raw materials and proteins (Afonso et al., 2004). Chabeaud et al. (2009) used UF membranes with MWCO lower than 7 kDa to improve the bioactivity of a saithe protein hydrolysate containing peptides, by fractionation or concentration of some specific molecular weight peptide classes.

Oily wastewaters can be treated by UF in order to produce a water phase which can be discharged into a sewer avoiding post-treatments and a concentrated oil phase that can be incinerated. Other interesting applications in the field of wastewaters are related to the removal of flexographic ink in wash water from the printing process (Chabot et al., 1998), the treatment of latex whitewater (Breslau et al., 1992) for the removal of latex emulsions, the recovery of paint from aqueous solutions employed in electrocoating processes (Porter, 1979).

In micellar-enhanced UF (MEUF) low molecular weight compounds permeable to UF membranes can be removed thanks to their association with larger impermeable molecules. For example, surfactants are able to form complexes with organic molecules and metals under specific conditions and the formed micelles can be separated by UF membranes (Puasa et al., 2011). The combination of the activated sludge process with MF or UF membranes in membrane bioreactors (MBRs) has received considerable attention in the last two decades for wastewater treatment and reuse. MBRs operate similarly to a conventional activated sludge process but without the need for secondary clarification and tertiary steps like sand filtration (Melin et al., 2006). MF and UF membranes assure a complete retention of the biomass: this leads to a high microbial concentration in the reactor and a highly efficient biological reaction process in the bioreactor with reduced sludge production (Lee, 2003).

MBR configurations include either submerged membranes or external circulation (side-stream configuration). In the side-stream configuration the mixed liquor is filtered under pressure through a cross-flow UF or MF membrane module placed outside the bioreactor and the biomass is circulated at high cross-flow velocity (2-4 m s^{-1}). Basically tubular membrane modules are employed and the liquor is pumped into the lumen side of the module. Transmembrane pressures are in the range of 1-4 bar. This configuration allows to obtain high permeate fluxes (generally in the range 50-120 L $m^{-2}h^{-1}$); however, the application on large scale is limited by the high energy consumption of the recirculating pump. In the submerged

MBR process the membrane is placed inside the bioreactor and the filtration is performed by suction removal of the effluent. The submerged MBR process operates at low transmembrane pressure values (about 0.5 bar) producing permeate fluxes lower than side-stream systems (from 15 to 50 L $m^{-2}h^{-1}$); significant advantages result in terms of reduced fouling phenomena and energy consumption.

Integrated systems including water softening, MF, UF, RO, UV-sterilization and mixed bed ion-exchange (or electrodeionization) have been realised for the production of ultrapure water from surface water. Typical applications include the production of ultrapure water to be used in production of electronic devices or analytical laboratories (Tansel, 2008).

Table 2. Industrial applications of MF and UF processes

Industrial sector	Application
Pharmaceutical	Sterile filtration: parenterals, ophthalmic solutions, fermentation products (antibiotics, vaccines and bioengineered proteins), tissue culture media; removal of microorganisms and particulates from air and other gases (vent filters, filtration of air or nitrogen used in fermentors)
Electronic	Sterile filtration: final point-of-use filters for ultrapure water
Nuclear	Removal of corrosion products in the primary coolant loop of boiling water reactors
Food/Beverage	Sterile filtration: wine, beer, bottled water, gases, sugar solutions, edible oil, syrups, vinegar, whisky, brandy; Clarification and fractionation of cheese whey, defatting and reduction of microbial load in milk; clarification of wine, beer, fruit juices, vinegar, bottled water, beet and sugar cane solutions, purification of dextrose stream; preconcentration of milk for cheese manufacture
Medical	Hemodialysis, biohybrid organs, analytical and diagnostic devices
Biotechnology	Downstream processing (concentration, clarification); marine biotechnology; biological conversions by membrane bioreactors (membrane recycle bioreactors, plug flow bioreactors)
Water treatment	Process water, treatment of grinding and polishing waters, finely dispersed suspension; retention of activated carbons, recovery of process water, filtration of process chemicals, solvents; removal of heavy metals as hydroxides, removal of lignin, removal of oil water effluents
Bulk chemicals/ Petrochemical	Sterile filtration: solvents and reagents, inorganic solutions, fatty acids, waxes, polymer fibres and films
Paints/coatings	Painting solutions, hydraulic fluid inks, foam, plating solutions, wastewaters
Sugar refining	Removal of colloidal and macromolecular impurities from raw juice, clarification of thin juice
Vegetable oils	Removal of waxes from sunflower oil
Corns and other grains	Dextrose clarification
Animal products	Treatment of wastewaters from meat processing plants, reduction in low molecular weight components from gelatin, removal of glucose from egg white, recovery of proteins from effluents of fish processing

A wide range of MF and UF applications in the pharmaceutical industry, including antibiotic processing, production of therapeutic proteins, ophthalmic preparations, fermentations, injectable drugs (parenterals), etc., are already consolidated (Scott, 1995). Antibiotics processing involving the use of MF membranes includes a variety of treatments such as the bulk clarification after the fermentation step, the filtration of solvents, the processing of the bulk final solution in the solvent based process. In the production of biologicals from mammalian cells, applications of microporous systems include: cell-harvest, perfusion of continuous cell culture vessels, clarification of biologicals, pre-column clarification.

In the fermentation process used in the production of antibiotics, hormones, aminoacids and enzymes, the inlet and outlet air is filtered to remove airborne microorganic contaminants from the fermentor. MF applications include also the sterile filtration of the nutrient feed and liquid product and 0.22-µm-rated filters are usually used. These filters are also used for the sterilization of heat-sensitive ophthalmic components.

In the processing of parenterals, MF is primarily used for final sterilisation of heat labile products, the reduction of bacterial burden before the final sterilization by autoclave and the removal of particles from parenteral solutions and aerosols generated during their processing.

Physiologically and anatomically, a microporous membrane is an ideal artificial surface for growing mammalian cells. A variety of polymeric membranes is currently used for cell harvesting processes. With a proper selection of membranes, modules and operating parameters high cell recoveries (up to 100%) can be reached.

Main industrial applications of MF and UF processes are summarised in Table 2.

7. Case Study: MF and UF in Food and Beverage Industry

The food industry represents a significant part of the turnover of the membrane manufacturing industry world-wide. Membrane science and technology is an expanding field in the preservation, nutritional performance and safety of foods.

Membrane processes have contributed to the revision of traditional flow-sheets in the processing of different products with the major issue of simplification, improvement of competitiveness, process or product novelty and environmental friendliness. In particular, MF and UF processes have been applied widely in food processing industry in the last two decades. Compared to conventional competitive separation operations (decantation, filtration, centrifugation, chromatography, etc.) UF and MF processes offer different advantages in terms of:

- flexibility in equipment design and operations;
- easy scale-up;
- separation without phase change;
- minimum space requirements for plant operation;
- reduction of mechanical and thermal stress hence retaining chemical properties of the feed components;
- possibility to separate compounds in a wide range of molecular weights;

- low operating costs;
- low energy requirement;
- improvement of the final product quality;
- eco-friendly processes.

In the food and beverage industry the use of MF and UF is dominated by the application in the treatment of milk and whey followed by fruit juices, wine and beer. A general overview related to these applications will be provided in the following sections.

7.1. Milk and Whey

In milk processing technological innovations are referred basically to the treatment of milk in cowshed, the storage of raw material, the reduction of the total bacterial charge and packaging. Further, from the cheese manufacture result huge quantities of by-products (such as whey) containing sugars, proteins and mineral salts which contribute to environmental pollution. Substances such as lactose and proteins, with different purity degree, can find interesting and gainful uses not only in the modern food industry but also in the pharmaceutical sector. Membrane operations as mild technologies offer different advantages over traditional techniques (such as thermal evaporation) in terms of efficiency and control of the process.

By referring to milk processing, consolidate applications concern the removal of bacteria, the milk standardization, the production of concentrated milk and caseins fractionation (Atra et al., 2005; Daufin et al., 2001).

Traditionally the removal of microorganisms from fluids of the dairy industry (milk, whey, brine) is based on thermal treatment. MF membranes with pores of 0.2 μm should be able to separate fat and bacteria from the rest of milk components. Early studies performed with ceramic membranes suggested that the removal of fat substances before MF is necessary in order to increase the concentration factor and to increase permeation flux. The formation of dynamic or secondary membranes in the use of polymeric membranes operating in cross-flow configuration was also observed as a drawback. The uniform transmembrane pressure (UTP) concept, introduced commercially by Alfa(Tetra)-Laval and US Filter, aimed at minimizing the formation of secondary membranes operating at extremely high velocities and low transmembrane pressures. This system produced a bacterial retention of 99% and a substantial retention of fat substances without a significant change in concentration of other components (Saboya and Maubois, 2000; Vadi and Rizvi, 2001).

A process combining the thermal treatment with MF for the removal of bacteria from raw milk was introduced and implemented on industrial scale in 1989 by Alfa Laval. This process, known as "Bactocatch", leads to a removal of 99.8-99.9% of total bacteria, a 99.95% reduction of *Bacillus cereus* and a reduction of anaerobic spores up to values lower than 3 spores/L. The process allows to remove completely somatic cells improving the shelf-life of milk from 12 to 18 days without negative effects on its sensorial qualities (Maubois, 1990). A schematic representation of the Bactocatch process is depicted in Figure 8.

The UF process can be used to concentrate milk with the aim of reducing refrigeration and transportation costs. UF can also be used to standardize milk proteins offering a valid solution to natural variations of its composition (Korolczuc et al., 1986). UF membranes retain proteins, fats, insoluble and bound salts allowing the permeation of lactose and soluble salts. The final content of proteins, lactose and minerals in the retentate depends on the duration of the process and operating conditions (pH and temperature). The use of UF to standardize the level of proteins and total solids of milk employed in the manufacture of fermented products (i.e. yogurts) allows to obtain high quality products if compared with those prepared from milk power-added or evaporated milk. Some quality attributes of fermented products, such as firmness and acidity, can be easily controlled by using membrane technology. Further, the selection of appropriate membranes, the separation conditions and the thermal treatment level can contribute to the production of a wide range of products with new characteristics.

PES and PS membranes are generally employed in milk fractionation. Although these membranes are characterised by a higher fouling tendency if compared with cellulosic membranes they are more thermally stable (50-55°C) and can be cleaned within a wide pH range. This should be considered in relation to the conditions required for the sanitization of the plant and for membrane cleaning procedures. The required operating conditions are characterised by high feed velocities in order to minimize the polarization concentration phenomenon which, in the case of milk fractionation, is quite significant.

The production of specialty milk-based beverages represents today one of the most promising activities in the dairy industry. The UF of skim milk offers interesting perspectives in the production of beverages with high calcium and protein content but relatively low content of fat and cholesterol.

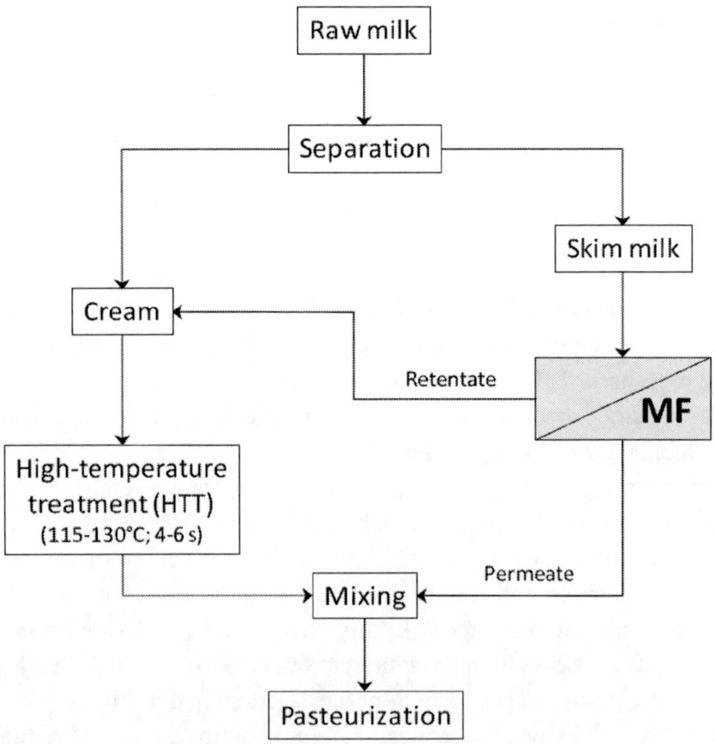

Figure 8. The "Bactocatch"process for milk treatment developed by Alfa-Laval.

Calcium in milk is mostly in the insoluble or bound form or is associated with casein micelles which are retained by UF membranes, while the concentration of sodium, potassium and lactose remains unchanged. The separation of milk fat globules from whole milk, fat enriched milk and cream has been realised using special MF membranes. Fat globules are separated in large globules (>2μm) and small globules (<2 μm) fractions (Goudédranche et al., 2000). The use of small globules makes the final product (drinking milk, yoghurt, sour cream, camembert) more unctuous and with finer textural characteristics if compared with products made from large globules.

Protein concentrates prepared with membrane processes have gained a considerable interest over the last years. Terms such as milk protein (MP) and milk protein concentrate (MPC) are referred to products containing 50-85% and more than 85% of proteins, respectively. Through a combination of UF, MF and diafitration and a precise control of pH and temperature, chemical, physical and functional properties of MP and MPC products can be modified for different applications.

Some MF applications in milk processing concern the fractionation of caseins. In particular, the treatment of skim milk with 0.2 μm membranes produces a permeate containing most whey proteins and lipids free and a retentate containing native caseins. The retentate can be standardized up to obtain the desired caseins/total protein ratio. Products containing more than 90% of caseins can be obtained through a further treatment of the retentate by diafiltration.

The integration of MF with other membrane processes, enzymatic processes and/or chromatographic techniques offers interesting perspectives for the purification of caseins and bioactive peptides from skim milk which can be of interest for pharmaceutical applications.

The principal use of MF and UF processes in the dairy industry concerns the treatment of milk for cheese manufacture.

Cheese manufacture is a fractionation process in which casein and fat are concentrated in the curd while lactose, soluble proteins, minerals and other minor components are lost in the whey. The history of cheesemaking using membranes started in the late 1960s with the invention of the MMV process (named after its inventors Maubois, Mocquot and Vassal) (Maubois and Mocquot, 1975; Zeman and Zidney, 1996). In this approach milk is preconcentrated by UF in a liquid pre-cheese with a protein/fat/solids level similar to that normally found in cheese. Then pre-cheese is converted to cheese according to the conventional cheesemaking method. This approach reduces the whey drainage and there is no need for cheese vats. This principle was firstly used for the Camembert cheese manufacture but many applications have been developed successfully for the manufacture of other cheese varieties including fresh, soft, semi-hard and hard cheese.

The advantages of UF in cheese making are mainly in terms of reduction in the use of enzymes (rennet), increased yields of cheese of 10-30%, reduced volume of milk to handle (with a consequent reduction in the number of cheesemaking vats), reduction of produced whey since water and lactose are removed in the process, uniformity in the quality with minimization of seasonal variations, continuous and automated process (Cheryan, 1998).

Thanks to the introduction of ceramic and PS membranes the UF process can be applied to the fermentative process of milk in order to produce fresh cheeses with superior qualities if compared to those prepared with traditional methodologies. Savings of skim milk of 13-14%

result in the application of UF in the manufacture of quarg in comparison with the traditional method.

Applications of UF in cheesemaking differ mainly in the concentration degree of milk before or during the coagulation process. Employing a concentration factor of 1.2-2:1 milk can be standardised in order to minimise variations of its composition and increasing the protein content up to 4.5-5%. The retentate, known as low concentrated retentate (LCR), can be used in the manufacture of different cheeses. Also medium concentrated retentates (MCR) produced by UF up to a concentration factor of 2-6:1 are used in the production of different cheeses (Kosikowski, 1986 a and b).

Potential applications of membrane processes in cheesemaking are not only limited to milk concentration. The capacity of MF to control the α-casein/β-casein ratio offers new possibilities of controlling cheese quality. Further, the application of the MF/HTT process in cheesemaking is an interesting perspective to improve cheese quality and to control the cheese proteolytic potential.

A multistage process to maximize the removal of serum proteins from skim milk before cheese making was developed by Nelson and Barbano (2005). The process removed 94-96% of serum proteins from skim milk producing a low concentration factor retentate with concentrations of soluble minerals, nonprotein nitrogen and lactose similar to the original skim milk.

In the first stage skim milk is microfiltered and the permeate is then submitted to an UF step. In the second stage the retentate from MF is diluted back to the original weight of skim milk with the permeate from UF. Finally, the mixture is diafiltered with the UF permeate of a second diafiltration step. The result after the second diafiltration is a skim milk with a lower serum protein content. The retentate from MF could be blended with cream to standardize milk for traditional Cheddar cheese making.

The processing of cheese whey by UF is one of the most successful applications in the dairy industry. Whey is a by-product of the cheese making and casein industry. Typically, 10-20 kg of cheese and about 80-90 kg of liquid whey are produced from 100 kg of milk. The whey composition depends on the production process: it has a low solids content and a lactose:protein ratio which makes it difficult to use as it is. The biological oxygen demand (BOD) ranges between 32 and 60 g/L with consequent disposal problems.

UF is largely used for the production of whey protein concentrates (WPC): the initial protein content of 10-12% (dry bases) can be increased up to 80% with a simultaneous decreasing in lactose and some salts. Diafiltration can be used after the concentration step to purify the WPC above 50% protein (dry bases). WPC can be further fractionated into β-lactoglobulin and α-lactoglobulin fractions or be used for the manufacture of macropeptides with pharmacoterapeutic value (Zidney, 1998).

The MF process is generally used in the treatment of whey as pre-treatment for the UF step. Whey usually contains small quantities of fat and casein which prevent the attainment of high purity WPC. These compounds are effectively removed by MF membranes in addition to some precipitated salts and bacteria.

The fractionation of whey proteins by MF and/or UF is another interesting application of membrane processes in the dairy sector. The separation of α- and β-lactalbumin from a binary mixture of these proteins was investigated by Cheang and Zidney (2003) by using a 30 kDa cellulose membrane and a 50 kDa PES membrane at different pH values.

According to the obtained results, the separation of these proteins can be achieved with a high selectivity and a high yield rate by optimal selection of pH, ionic strength and MWCO. Variations of physicochemical and hydrodynamic conditions could induce large differences in protein transmission through MF and UF membranes (Muller et al., 2003).

7.2. Fruit and Vegetables Juices

The introduction of membrane technologies in the industrial transformation cycle of the fruit juices represents one of the technological answers to the problem of the production of juices with high quality, natural fresh taste and additive free. Juice clarification, stabilisation, depectinization and pre-concentration are typical steps where membrane processes such as MF and UF have been successfully utilized.

The traditional clarification of fruit juices is based on the use of several batch operations that are labor- and time-consuming. Typically, after preliminary operations (sorting, washing and peeling steps) the fruit is crushed and then submitted to a pressing step to remove large particulates. The press juice is pasteurized and then treated with pectinases to reduce cloudiness and viscosity making the subsequent filtration easier. A preliminary clarification step is generally performed by using fining agents such as gelatin. After decanting the juice is submitted to a precoat filtration by using diatomaceous earth as filter aids to remove suspended solids, colloids, proteins and condensed polyphenols.The final clear juice, having a turbidity of 1.5-3 nephelometric turbidity units (NTU), can be pasteurized before packaging.

MF and UF can be used successfully to replace several of these operations. These processes offer different advantages over conventional clarification methodologies in terms of: low energy requirements and costs; possibility of avoiding the use of gelatines, adsorbents and other filtration aids (avoiding dust formation and sludge deposition); possibility of lower temperature processing (hence reduction of thermal damage to food during processing); simpler process design;reduction of waste products; possibility of avoiding use of chemical agents; increased juice yield (juice recovery of 96-98% can be obtained in contrast with conventional processes); reduction in enzyme utilization (depending on the juice enzyme usage is reduced from 50 to 75%); easy cleaning and maintenance of the equipment; better product quality; possibility of operating in a single step reducing working times (in membrane cross-flow filtration a clear juice can be obtained in 2-4h); possibility of avoiding pasteurization (at 60–65°C) and sterilization (at about 110°C) (MF and UF can be considered as cold-sterilization, as they remove the bacteria and other live organisms present in the juice) (Fukumoto, et al., 1998; Koseoglu et al., 1990).

Fruit juices contain typically 75-90% water, 9-25% sugars, 1-5% organic acid, 0.1-0.2% dietary fiber, and 0.2-0.6% proteins. Other constituents, such as minerals (potassium, sodium calcium, magnesium, iron, copper, phosphorous, zinc), vitamins (A, B, C), flavor constituents (esters, aldehydes, alcohols, volatile essential oils), pigments (carotenoids, anthocyanins andpolyphenolics), lipids (simple waxes, steroids and aliphatic compounds), nucleotides, starch, pectin and microorganisms are present in smaller amount (Girard and Fukumoto, 2000).

Some compounds, including sugars, organic acids, aminoacids, phenolic compounds, pigments, essential oils, volatiles (including esters, aldehydes, alcohols), vitamins and minerals have molecular weight less than 1 kDa. Proteins, starch, enzymes, pectic substances, lipids, microorganisms and other soluble and insoluble complexes are larger entities. Various enzymes are found in fruits and their juices; most of them have molecular weight in the range 10-90 kDa. Starch appears as insoluble granules with diameters of 1 to 16 μm in freshly pressed apple juice. Pectins have a wide molecular weight range (10-500 kDa), with a global average weight of about 100 kDa. Colloids are composed mainly of polysaccharides with diameters between 0.001 and 0.01 μm: they confer an opalescent appearance to fruit juices (Girard and Fukumoto, 2000).

UF and MF membranes are typically used to separate juices into a fibrous concentrated pulp (retentate) and a clarified fraction free of spoilage microorganisms (permeate). These membranes retain high molecular weight compounds (pectin or proteins) and allow low molecular weight solutes (sucrose, acids, salts, aroma and flavor compounds) to permeate through the membrane.

The selection of MF or UF membranes for fruit juice clarification should take in consideration the viscosity of the raw juice, the sterilization ability and the investment costs of the membranes, the retention of specific compounds and the membrane performance. For pulpy juices, with high solids content and viscosity, large-bore tubular modules or plate and frame modules with large spacers are preferred. On the other hand, particle clogging can occur in the narrow flow path of hollow fiber or spiral-wound membrane modules reducing permeate flux and interfering with subsequent chemical and mechanical cleaning. Consequently, the use of thin channel devices requires additional pre-treatments of the raw juice to reduce the overall content of solids.

PS membranes have been used extensively for juice UF; in comparison with cellulose acetate membranes they can withstand short exposures to hypochlorites during periodic cleaning cycles. PVDF and polyamide (PA) membranes are also largely used in some juice UF systems. Ceramic membranes have greater resistance to chemical degradation, much longer life and higher costs if compared with organic membranes. Retention and performance should be tested on pilot plant.

The most used configurations for the clarification of fruit juices at industrial level are tubular (inner diameter 5-10 mm), capillary (1-1.5 mm) and plate-and-frame membrane modules. The tubular configuration is associated with low packing density and high membrane replacement costs. A new concept, recently developed, combines a high-speed separator with spiral-wound UF modules to overcome these limitations (Lipinzki, 2010). Vibrating membrane systems are particularly appropriate for juice with high pulp content since they develop a higher shear rate at the membrane surface reducing membrane fouling.

Permeate fluxes in the clarification of fruit juice by MF and UF are strongly affected by operating conditions such as transmembrane pressure, cross-flow velocity, temperature and volume reduction factor (VRF), nature of the membrane and nature of the feed solution (Cheryan, 1998).

In general, the permeate flux increases with membrane pore size while the retention of sugars decreases as membrane pore size is increased (Girard and Fukumoto, 2000). The adsorption of colloidal material is strongly affected by the membrane material. Besides, the surface structure and flux resistance of the fouling layer is affected by the membrane morphology. Dense fouling layers were detected on smooth surfaced membranes. On the contrary, rough membranes produced a looser surface fouling layer with a lower flow resistance per unit thickness of foulant (Riedl et al., 1998).

A limiting flux is generally reached when the transmembrane pressure is increased up to a specific value; after that any further pressure increase determines no significant increase of the permeate flux. The existence of a limiting flux can be related to the formation of a viscous and gelatinous-type layer which is responsible for an additional resistance to the permeate flux in addition to that of the membrane.

An increase in cross-flow velocity enhances wall shear stress on the membrane surface reducing concentration polarization and reversible fouling on the membrane surface; consequently, the mass transfer coefficient is enhanced leading to an increasing of permeation flux (Vadi and Rivzi, 2001).

When the operating temperature is raised, the feed viscosity is reduced and diffusion coefficients of macromolecules increase. According to the film model, the effect of these two factors is to enhance mass transfer and to increasethe permeation rate (Cassano et al., 2007a).

A decreasing in permeate flux is observed when the juice concentration is increased.

VRF is another parameter influencing the productivity of MF and UF membranes in the clarification of fruit juices. High VRF values (>10) can be easily reached with juices with low pulp content such as grape and apple juices. For very pulpy juices, such as tropical fruits, optimum VRF values are less than 5. The yield (1-1/VRF) of recovered juice is a crucial item in the process economics.

Several Authors investigated the effect of UF and MF membrane properties and operating conditions on physicochemical properties of fruit juices (Cardoso de Oliveira et al., 2012; Carneiro et al., 2002; Cassano et al., 2004, 2007a, 2007b and 2008; de Bruijn et al., 2002 and 2003; Hernandez et al., 1992; Laorko et al., 2010; Matta et al., 2004; Rai et al., 2006; Vladisavljević et al., 2003). Basically, pectic materials and colloidal particles are removed by UF membranes resulting in lower viscosity and improvement in color and clarity of the filtered juice. Almost all soluble solids are recovered in the permeate. pH and density are generally not affected by juice filtration. Membrane filtration with a pore size of 0.2 μm could serve as a cold sterilization process which could achieve the preservation of phytochemical compounds.

Table 3. Rejection of UF and MF membranes in the clarification of cactus pear juice

	R(%)	
	MF	UF
suspended solids	100	100
TAA	12.7	14
Polyphenols	0	6.2
Flavonoids	36.7	41.5
Proteins	76.9	78.3
ascorbic acid	22.5	28.5
Betacyanins	100	100
Betaxanthins	72.3	77

Recently, Cassano et al. (2010) studied the influence of UF and MF processes on the physic-chemical composition of cactus pear juice, in order to evaluate the effect of the clarification treatment on the content of main parameters which characterize the nutritional and functional properties of the fruit. The cactus pear juice was clarified by using PVDF flat–sheet membranes with pore size and MWCO of 0.20 μm and 200 kDa, respectively. The analytical results showed that the physicochemical and nutritional properties of cactus pear juice are comparable with those of fresh cactus pear except for the absence of suspended solids and betalains which remained concentrated in the retentate. The permeate of both processes is a clear solution with a low proteiccontent, enriched of antioxidant compounds (polyphenols, Vitamin C, etc.), sugars, amino acids and minerals, due to the low rejection of both membranes towards these compounds (Table 3).

Clarified fruit juices can be used in the manufacture of differentproducts such as cocktails, liqueurs, fizzy beverages, flavoured mineral water, etc. Moreover, natural antioxidants occurring in clarified juices may be used as components of composite food formulations in order to stabilize them or may be extracted and added to foods.

Finally, clarified juices can be submitted to a concentration step by using thermal evaporation or membrane operations such as reverse osmosis (RO), membrane distillation (MD) and osmotic distillation (OD).

High-strength citrus juice concentrates can be produced by using UF or MF as a pre-treatment to separate the pulp from the serum, followed by a combination of high- and low-retention RO membranes (Cheryan and Alvarez, 1995). A pilot scale system investigating an integrated process for the clarification (by UF), concentration (by thermal evaporation) and aroma compounds recovery (by pervaporation, PV) from apple juice has been exploited by Alvarez et al. (2000). Integrated membrane processes for the clarification (by UF), preconcentration (by RO) and concentration (by osmotic distillation, OD) of different fruit and vegetable juices (orange, carrot, lemon, kiwifruit and cactus-pear juices) have been also proposed by Cassano et al. (2003, 2004 and 2007c).

7.3. Wine and Beer

Wine clarification aimed at removing suspended solids and undesirable substances from wine is one of the most important steps in the wine-making process.

Conventional separation techniques such as static defecation, centrifugation and depth-filtration are often unsuccessful and difficult to realize. These processes, being often discontinuous, require extensive spaces, due to the intrinsic limitations imposed by the dimensions of vats, the high energy consumption for the temperature control and the huge quantities of coadiuvants and additives (bentonite, gelatines, activated carbon, casein, silica sol, diatomaceous earth, etc.) with consequent problems of disposal and environmental impact.

Cross-flow MF and UF operations offer different advantages over the use of diatomaceous-earth filtration in wine clarification (Güell, 1999; Salazar et al., 2007) in terms of low operating temperatures, reduction of energy consumption, removal of bacteria, cells and spores, reduction of SO_2 addition, reduction of the pectin content, proteins and colloids, sterilized and clear product in one single continuous operation (Rayess et al., 2011).

UF can be used either before the fermentation (for clarifying the must) or after (for treating the finished wine). UF of the must is used for the removal of colloids, high molecular weight tannins, polysaccharides, proteins and undesirable microorganisms. After the fermentation process and before storage, MF can be used to remove the yeast or UF to improve stability of the finished wine. Wine turbidity is caused mainly by suspended solids material like yeast residues and colloidal substances. MF membranes with average pore diameter of 0.2 μm can ensure microbiological limpidity and stability in a single operation: they yield products with low turbidity (less than 1 NTU), low in germs, without affecting their organoleptic characteristics (Daufin et al., 2001). A lower polysaccharide content was observed in microfiltered wines in comparison with the traditional filtration (Serrano et al., 1992). These compounds, together with polyphenols, have been considered as the main cause of fouling in both MF mineral and organic membranes (Belleville et al. 1992; Vernhet et al., 1999). In particular, the adsorption of polyphenols on polymeric membranes is stronger on more polar membranes, while the polarity of the membrane surface is a limiting factor in polysaccharide adsorption (Vernhet et al., 1997). The differences in fouling behavior between wine samples cannot be explained completely by the different initial polyphenol and polysaccharide concentrations but complex interactions between foulant molecules and foulant-membrane must be considered (Cameira-dos-Santos et al., 1994).

Goncalves et al. (2001) evaluated the effect of operating parameters (transmembrane pressure and concentration factor) on the permeate flux and the removal of polysaccharides and polyphenols from white and red wines by using MF and UF fluoropolymer membranes (0.1 μm pore size and 100 kDa MWCO, respectively). They found that permeate fluxes in the treatment of red wines were much lower than those obtained with white wines. At the same transmembrane pressure (1.0 bar) final permeate fluxes of 34 and 18 L/m^2h were obtained in the MF and UF of red wine, respectively. The wine clarification by MF was associated with a small removal of polysaccharides and polyphenols in the case of white wine and a slightly higher removal in the case of red wine.

The UF process permits also to reduce haze or deposit formation in bottled wines due to protein aggregation during storage. Haze is a common defect of commercial wines making them unacceptable for consumers. Numerous works have been devoted to the identification of protein structures implied in colloidal instabilities, as well as to the elucidation of factors that trigger or prevent haze formation. The traditional method of haze removing is based on the use of fining agents, such as bentonite, which cause also a retention of color and some of flavor compounds. UF permits to remove the haze-causing proteins preserving flavor and color of the wine.

Considerations of the brewing process indicate two areas where MF might play a useful role: (i) loss reduction in the brewing process; (ii) as a technological alternative to the conventional solid-liquid separation (Daufin et al., 2001). Loss reduction concerns the recovery of extract during the wort clarification and beer recovery from fermentation and maturation tank bottoms. Extract recovery after trub separation with MF allows the production of a permeate with acceptable quality, a significant trub volume reduction (over 50%) and the possibility to operate at 75°C.

After fermentation yeast is settling at the bottom of the fermentation vessels. The settled tank bottoms account for 1.5-2% of the total beer volume and contain a high proportion of beer that is lost if not recovered. Natural sedimentation, centrifugation and filter-press can be used in order to recover "green beer" or "rough beer" from tank bottoms. However, these processes are expensive, provide a relatively low moisture solid discharge and the filtrate can be not completely clear. MF of fermentation tank bottoms reduces the beer loss: the permeate recovered by MF can be recycled in the wort or in the maturation vessels (Nielsen, 1989).

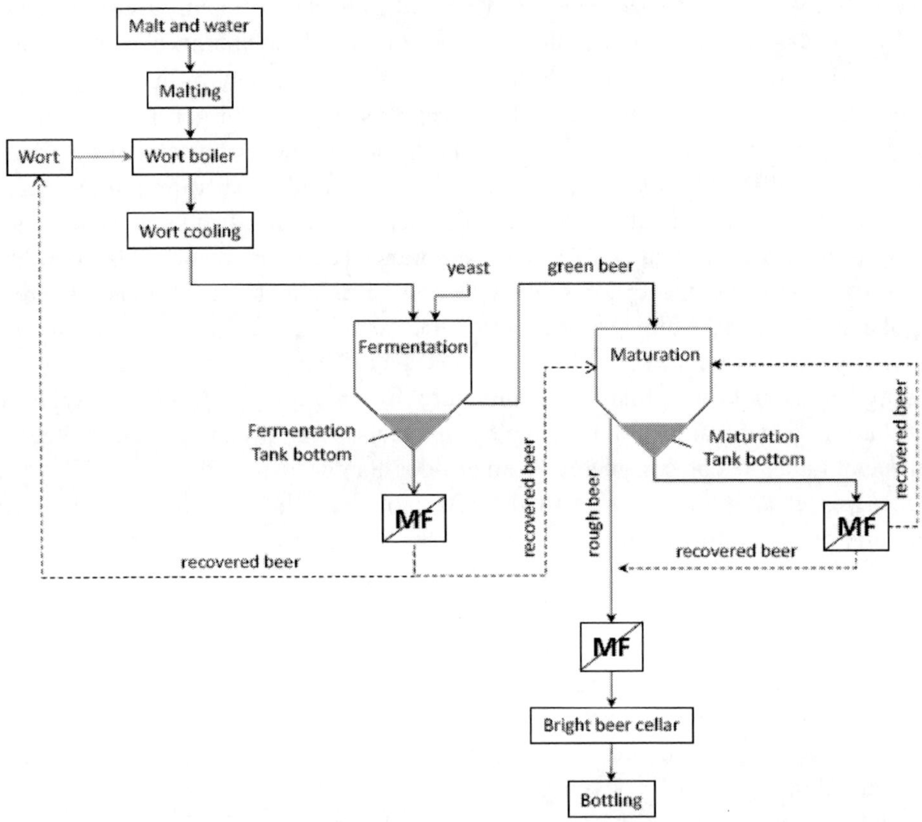

Figure 9. Application of MF in beer production.

The beer recovered from the maturation tank bottom can be returned to the maturation vessel or sent to a final clarification (Figure 9). MF produces a permeate of acceptable quality with minimal loss of original gravity, color and bitterness. It can be operated at low temperatures (close to 0°C) achieving an economic flux and hygienic beer recovery.

Conventional beer clarification process employs filter press or pressure vessel filters, which are commonly pre-coated with porous particles of diatomaceous earth (DE) as filter aids. A number of advantages have been suggested for replacement of the conventional filtration process with cross-flow MF. These include the elimination of filter aids and associated handling and disposal problems, reduced beer losses, high solids handling capacity, substitution of heat pasteurisation and a better product quality.

Beer clarification by MF process demands a fine balanced retention of large particles and a transmission of soluble macromolecules including carbohydrates, proteins, flavour, and colour compounds which give the'wholesome' quality of a beer (Gan et al., 2001).

Conclusion and Future Trends

MF and UF processes show a high efficiency over conventional separation technologies in different fields. They offer new opportunities in the design, rationalization and optimization of industrial processes. The development of new applications of established MF and UF processes will be driven by economical and environmental targets.

The possibility of integrating these processes together with new membrane operations, such as pervaporation, electrodialysis and membrane contactors, or with conventional processes (such as centrifugation, evaporation, liquid-liquid extraction and adsorption) appears an attractive way to achieve important benefits in the logic of the process intensification strategy.

REFERENCES

Afonso, M.D., Ferrer, J. and Borquez, R. (2004) An economic assessment of proteins recovery from fish meal effluents by ultrafiltration. *Trends in Food Science and Technology*, 15, 506-512.

Alvarez, S., Riera, F.A., Alvarez, R., Coca, J., Cuperus, F.P., Bouwer, S., Boswinkel, G., van Gemert, R.W., Veldsink, J.W., Giorno, L., Donato, L., Todisco, S., Drioli, E., Olsson, J., Trägårdh, G., Gaeta, S.N. and Panyor, L. (2000) A new integrated membrane process for producing clarified apple juice and apple juice aroma concentrate. *Journal of Food Engineering*, 46, 109–125.

Atra, R., Vatai, G., Bekassy-Molnar, E. and Balint, A. (2005). Investigation of ultra and nanofiltration for utilization of whey protein and lactose, *Journal of Food Engineering*, 67, 325–332.

Baker, W.R. (2000) *Membrane technologies and applications*, Mc-Graw-Hill Companies, USA.

Barredo-Damas, S., Alcaina-Miranda, M.I., Iborra-Clar, M.I. and Mendoza-Roca, J.A.. (2012) Application of tubular ceramic ultrafiltration membranes for the treatment ofintegrated textile wastewaters. *Chemical Engineering Journal*,192, 211-218.

Belleville, M.P., Brillouet, J.M., Tarodo de la Fuente, B. and Montounet, M. (1992) Fouling colloids during micropurous alumina membrane filtration of wine. *Journal of Food Science*, 57, 396-400.

Benìtez, F.J., Acero, J.L. and Leal, A.I. (2006) Application of microfiltration and ultrafiltration processesto cork processing wastewaters and assessmentof the membrane fouling. *Separation and Purification Technology*,50, 354–364.

Bertran, C.D., Hoogland, M.R., Li, H., Odell, R.A. and Fane, A.G. (1993) Flux enhancement in cross-flow microfiltration using a collapsible-tube pulsation generator. *Journal of Membrane Science*, 84, 279-292.

Bird, R.B., Stewart, W.E. and Lightfoot, E.N. (1965) *Transport Phenomena*, John Wiley and Sons, New York.

Breslau, B.R., Buckley, R.G., Eastburn, G.L., Schmitz, M.C. and Tansey, S.P. (1992) Ultrafiltration processes for the recovery of polymeric lattices from white water, U.S. Patent 5,171,767.

Cameira-dos-Santos, P.J., Cheynier, V., Brillouet, J.M., Cadot, Y., Escudier, J.L., Roesink, H.D.V. and Montounet, M. (1994) Cross-flow filtration of wines with polymeric membranes. The influence of wine making technology and backpulse on the fouling process. Prehrambeno-tehnološka i Biotehnološka Revija, 32, 77-73.

Cardoso de Oliveira, R., Caleffi Docê, R. and Davanted de Barros, S.T. (2012) Clarification of passion fruit juice by microfiltration: Analyses of operating parameters, study of membrane fouling and juice quality. *Journal of Food Engineering*, 111, 432-439.

Carneiro, L., Sa, I.D.S., Gomes, F.D.S., Matta, V.M. and Cabral, L.M.C. (2002) Cold sterilization and clarification of pineapple juice by tangential microfiltration. *Desalination*, 148, 93–98.

Cassano, A., Molinari, R., Romano, M. and Drioli, E. (2001) Treatment of aqueous effluents of the leather industry by membrane processes. A review. *Journal of Membrane Science*,181, 111–126.

Cassano, A., Drioli, E., Galaverna, G., Marchelli, R., Di Silvestro, G. and Cagnasso, P. (2003) Clarification and concentration of citrus and carrot juices by integrated membrane processes. *Journal of Food Engineering*, 57, 153-163.

Cassano, A., Jiao, B. and Drioli, E. (2004) Production of concentrated kiwifruit juice by integrated membrane processes. *Food Research International*, 37, 139–148.

Cassano A., Marchio, M. and Drioli, E. (2007a) Clarification of blood orange juice by ultrafiltration: analyses of operating parameters, membrane fouling and juice quality. *Desalination*, 212, 15-27.

Cassano, A., Donato, L. and Drioli, E. (2007b) Ultrafiltration of kiwifruit juice: Operating parameters, juice quality and membrane fouling. *Journal of Food Engineering*, 79, 613–621.

Cassano, A., Conidi, C., Timpone, R., D'Avella, M. and Drioli, E. (2007c) A membrane-based process for the clarification and the concentration of the cactus pear juice. *Journal of Food Engineering*, 80, 914-921.

Cassano, A., Donato, L., Conidi, C. and Drioli, E. (2008) Recovery of bioactive compounds in kiwifruit juice by ultrafiltration. *Innovative Food Science and Emerging Technologies*, 9, 556-562.

Cassano, A., Conidi, C. and Drioli E. (2010) Physico-chemical parameters of cactus pear (Opuntia ficus-indica) juice clarified by microfiltration and ultrafiltration processes. *Desalination,* 250, 1101–1104.

Chabeaud, A., Vandanjion, L., Bourseau, P., Jaouen, P., Chaplain-Derouiniot, M. and Guerard, F. (2009) Performance of ultrafiltration membranes for fractionating a fish protein hydrolysate: application to the refining of bioactive peptidic fractions. *Separation and Purification Technology*, 66, 463-471.

Chabot, B., Krishnagopalan, G.A. and Abubakr, S. (1998) Effect of ultrafiltration permeate recycling on deinking efficiency of flexo-printed newspapers. *Progress in Paper Recycling*, August, 28-38.

Cheang, B. and Zidney, A.L. (2003) Separation of α-lactalbumin and β–lactalbumin using membrane ultrafiltration. *Biotechnology and Bioengineering*, 83, 201-209.

Cheryan, M. and Alvarez, J.R. (1995) Food and Beverage industry applications, in: Noble, R.D. and Stern, S.A. (Eds.), *Membrane Separation Technology-Principles and Applications*, Elsevier, Amsterdam, 341.

Cheryan, M. (1998) *Ultrafiltration and Microfiltration Handbook*, Technomic Publishing Company, Lancaster.

Daufin, G., Escudier, J.E., Carrère, Bérot, S., Fillaudeau, L. and Decloux, M. (2001) Recent and emerging applications of membrane processes in the food and dairy industry. *Transaction of the Institution of Chemical Engineers*, 79, 89-102.

Dave, Y.G. and Reddy, A.V.R (2011) Preparation, characterization, acid stability and organic fouling of poly(acrylonitrile-co-methacrylic acid) ultrafiltration membranes. *Desalination,* 282, 9-18.

de Bruijn, J., Venegas, A. and Borquez, R. (2002) Influence of crossflow ultrafiltration on membrane fouling and apple juice quality. *Desalination*, 148, 131-136.

de Bruijn, J.P.F., Venegas, A., Martinez, J.A. and Bórquez, R. (2003) Ultrafiltration performance of Carbosep membranes for the clarification of apple juice. *LWT-Food Science and Technology*, 36, 397-406.

Drioli, E. and Romano, M. (2001) Progress and New Perspectives on Integrated Membrane Operations for Sustainable Industrial Growth. *Industrial Engineering Chemical Resource*, 40, 1277-1300.

Field, R.W., Wu, D., Howell, J.A. and Gupta, B.B. (1995) Critical flux concept for microfiltration fouling. *Journal of Membrane Science*, 100, 259-272.

Field, R.W., Wu, D., Howell, J.A. and Gupta, B.B. (1996) Mass transport and design of membrane system, in: Scott, K. and Hughes, R. (Eds.), *Industrial Membrane Separation Technology*, Chapman and Hall, London, 93.

Fukumoto, L.R., Delaquis, P. and Girard B. (1998) Microfiltration and ultrafiltration ceramic membranes for apple juice clarification. *Journal of Food Science*, 63, 845–850.

Gan, Q., Howell, J.A., Fiel, R.W., England, R., Bird, M.R., O'Shaughnessy, C.L. and McKechnie, M. T. (2001) Beer clarification by microfiltration-product quality control andfractionation of particles and macromolecules. *Journal of Membrane Science,* 194, 185–196.

Girard B. and Fukumoto L. (2000) Membrane processing of fruit juices and beverages: a review. *Critical Reviews in Food Science and Nutrition*, 40, 91-157.

Goldsmith, R.L., de Filippi, R.P., Hossain, S. and Timmins, R.S. (1971) Industrial Ultrafiltration in: Bier, M., *Membrane Processes in Industry and Biomedicine*, Plenum Press, New York.

Gonçalves, F., Fernandes, C. and de Pinho, M.N. (2001) White wine clarification by micro/ultrafiltration: effect of removed colloids in tartaric stability. *Separation and Purification Technology*, 22-23, 423-429.

Goudédranche, H., Maubois, J.L. and Fauquant, J. (2000) Fractionation of globular milk fat by membrane microfiltration. *Lait*, 80, 93-98.

Güell, C. (1999) Membrane separation techniques in wine and beer production. *Environment Protection Engineering,* 25, 87-101.

Hérmia, J. (1982) Constant pressure blocking filtration laws. Application to power-law non-Newtonian fluids, *Transaction of the Institution ofChemical Engineers*, 60, 183-187.

Hernandez, E., Chen, C.S., Shaw, P.E., Carter, R.D. and Barros, S.M. (1992). Ultrafiltration of orange juices: effecton soluble solids, suspended solids and aroma. *Journal of Agriculture and Food Chemistry*, 40, 986–988.

Jönsson, A.S. and Wimmerstedt, R. (1985) The application of membrane technology in the pulp and paper industry. *Desalination*, 53, 181–196.

Jönsson, A.S., Jönsson, C., Teppler, M., Tomani, P. and Wännström, S. (1996) Treatment of paper coating colour effluents by membrane filtration. *Desalination*, 105, 263–273.

Korolczuc, J. Maubois, J.L. and Fauquant, J. (1986) Mécanisation en formagerie de pâtes molles. *Proc. XXII International Dairy Congress*. Reidel D. Publishing Company, The Hague, The Netherlands, 123-128.

Koseoglu, S.S., Lawhon, J.T. and Lusas, E.W. (1990) Use of membranes in citrus juice processing. *Food Technology*, 44, 90–97.

Kosikowski, F.V. (1986a) Membrane separations in food processing, in: McGregor, W.C. (Ed.) *Membrane. Separations in Biotechnology*, Marcel Dekker, New York, 201-254.

Kosikowski, F.V. (1986b) New cheesemaking procedures utilizing ultrafiltration. *Food Technology*, 40, 71-77.

Laorko, A., Zhenyu, L., Tongchitpakdee, S., Chantachum, S. and Youravong, W. (2010) Effect of membrane property and operating conditions on phytochemical properties and permeate flux during clarification of pineapple juice. *Journal of Food Engineering*, 100, 514-521.

Lee, W. (2003) Sludge characteristics and their contribution to microfiltration in submerged membrane bioreactors. *Journal of Membrane Science*, 216, 217-227.

Lipinzki, F. (2010) Cross-flow membrane applications in the food industry,in: Peinemann, K.V., Nunes, S.P. and Giorno, L. (Eds.), *Membrane Technology, Volume 3: Membranes for Food Applications*, Weinheim, Wiley-VCH Verlag GmbH and Co. KGaA, 1-24.

Loeb, S. and Sourirajan, S. (1962). Seawater demineralization by means of a semipermeable membrane, in:Gould R., *Advances in Chemistry, ACS Series No. 38*, American Chemical Society,Washington, 117-132.

Marcucci, M.G., Nosenzo, G., Capannelli, G., Ciabatti, I., Corrieri, D. and Ciardelli, G. (2001) Treatment and reuse of textile effluents based on new ultrafiltration and other membranes technologies. *Desalination*, 138, 75-82.

Matta, V.M., Moretti, R.H. and Cabral, L.M.C. (2004) Microfiltration and reverse osmosis for clarification and concentration of acerola juice. *Journal of Food Engineering*, 61, 477-482.

Maubois, J.L. and Mocquot, G. (1975) Application of membrane ultrafiltration to preparation of various types of cheese. *Journal of Dairy Science*, 58, 1001-1007.

Maubois, J.L. (1990) Nouvelles applications des technologies à membrane dans l'industrie laitière. *Proc. XXIII Congrès International de Laiterie*, 2, Montrèal, 1775-1790.

Melin, T., Jefferson, B., Bixio, D., Thoeye, C., De Wilde, W., De Koning, J., van der Graaf, J. and Wintgens, T. (2006) Membrane bioreactor technology for wastewater treatment and reuse. *Desalination*, 187, 271-282.

Mulder, M. (1991) *Basis principles of membrane technology*, Kluwer Academic Publishers, Dordrecht.

Muller, A., Chaufer, B., Merin, U. and Daufin, G. (2003) Purification of α-lactalbumin from a prepurified acid whey: Ultrafiltration or precipitation. *Lait*, 83, 439-451.

Nelson, B.K. and Barbano, D.M. (2005) A microfiltration process to maximize removal of serum proteins from skim milk before cheese making. *Journal of Dairy Science*, 88, 1891-1900.

Nielsen, C.E. (1989) Microfiltration route to recovering beer from tank bottoms. *Brewing and Distilling International*, September, 20-21.

Puasa, S.W., Ruzitah, M.S. and Sharifah, A.S.A.K. (2011) An overview of micellar-enhanced ultrafiltration in wastewater treatment process, *2011 International Conference on Environment and Industrial Innovation IPCBEE,* 12,167-172.

Porter, M. C. (1979) Membrane Filtration, in: Schweizter, P.A. (Ed), *Handbook of Separation Techniques forChemical Engineers*, McGraw-Hill, New York.

Puro, L., Kallioinen, M., Mänttäri, M., Natarajan, G., Cameron, D.G. and Nyström, M. (2010) Performance of RC and PES ultrafiltration membranes in filtration of pulp mill process waters. *Desalination,* 264, 249-255.

Rai, P., Majumdar, G.C., Sharma, S., Das Gupta, S. and De, S. (2006) Effect of various cut off membrane on permeate flux and quality during filtration of mosambi (Citrussinensis (L) Osbeck) juice. *Food and Bioproduct Processing*, 84, 213–219.

Rayess, Y.E., Bacchin, P., Taillandier, P., Raynal, J., Mietton-Peuchot, M. and Devatine, A. (2011) Cross-flow microfiltration applied to oenology: A review. *Journal of Membrane Science*, 382, 1-19.

Reddy, A.V.R., Mohan, D.J., Bhattacharya, A., Shah, V.J. and Ghosh P.K. (2003) Surface modification of ultrafiltration membranes by preadsorption of a negatively charged polymer I. Permeation of water soluble polymers and inorganic salt solutions and fouling resistance properties. *Journal of Membrane Science,* 214, 211-221.

Reddy, A.V.R., Trivedi, J.J., Devmurari, C.V., Mohan, D.J., Singh, P., Rao, A.P., Joshi,S.V. and Ghosh, P.K., (2005) Fouling resistant membranes in desalination and water recovery. *Desalination*, 183 , 301-306.

Riedl, K., Girard, B. and Lencki, R.W. (1998) Influence of membrane structure on fouling layer morphology during apple juice clarification. *Journal of Membrane Science*, 139, 155-166.

Ripperger, S. and Almann, J. (2002) Crossflow microfiltration- state of the art. *Separation and Purification Technology*, 26, 19-31.

Saboya, L.V. and Maubois, J.L. (2000) Current developments of microfiltration technology in the dairy industry. *Lait*, 80, 541-553.

Salazar, F.N., De Bruijn, J.P.F., Seminario, L., Güell, C. and López, F. (2007) Improvement of wine cross-flow microfiltration by a new hybrid process. *Journal of Food Engineering,* 79, 1329-1336.

Scholz, W. and Lucas, M., (2003) Techno-economic evaluation of membrane filtrationfor the recovery and re-use of tanning chemicals.*Water Research*, 3, 1859–1867.

Scott, K. (1995) *Handbook of industrial membranes*, Elsevier Advanced Technology,Oxford.

Serrano, M., Pontens, B. and Ribéreau-Gayon, P. (1992) Étude de différents membranes de microfiltration tangentielle. Comparaison avec la filtration sur précouche de diatomées. *Journal International des Sciences de la Vigne et du Vin*, 26, 97-116.

Shah, T.N., Foley, H.C. and Zydney, A.L. (2007) Development and characterization of nanoporous carbon membranes for protein ultrafiltration. *Journal of Membrane Science*, 295, 40-49.

Sherwood, R.K., Pigford, R.L. and Wilke, C.R. (1975) *Mass transfer*, McGraw Hill, New York.

Strathmann, H., Giorno, L. and Drioli E. (2006) *An introduction to membrane sciences and technology,* Consiglio Nazionale delle Ricerche, Roma.

Tansel, B. (2008) New Technologies for Water and Wastewater Treatment: A Survey of Recent Patents. *Recent Patents on Chemical Engineering*, 1, 17-26.

Vadi, P.K. and Rizvi, S.S.H. (2001) Experimental evaluation of a uniform transmembrane pressure cross-flow microfiltration unit for the concentration of micellar casein from skim milk. *Journal of Membrane Science*, 189, 69-82.

Van der Bruggen, B., Vandecasteele, C., Van Gestel, T., Doyenb, W. and Leysenb, R. (2003). A review of pressure-drivenmembrane processes in wastewater treatment and drinking water production. *Environmental Progress*, 22, 46-56.

Vernhet, A., Bellon-Fontaine, N.M., Brillouet, J.M., Roesink, E. and Montounet, M. (1997) Wetting properties of microfiltration membrane: determination by means of the capillary rise technique and incidence on the adsorption of wine polysaccharide and tannins. *Journal of Membrane Science*, 128, 163-174.

Vernhet, A., Pellerin, P., Belleville, M.P., Planque, J. and Montounet, M. (1999) relative impact of major polysaccharides on the performance of an organic microfiltration membrane. *American Journal of Enology and Viticulture*, 50, 51-56.

Vladisavljević, G.T., Vukosavljević, P. and Bukvić, B. (2003) Permeate flux and fouling resistance in ultrafiltration of depectinized apple juice using ceramic membranes. *Journal of Food Engineering*, 60, 241-247.

Zeman, L.J. and Zidney A.L. (1996) *Microfiltration and Ultrafiltration*, Marcel Dekker Inc., New York.

Zydney, A.L. (1998) Protein separations using membrane filtration: new opportunitiesfor whey fractionation. *International Dairy Journal*, 8, 243–250.

In: Membrane Processes for Sustainable Growth
Editors: Angelo Basile and Alfredo Cassano
ISBN: 978-1-62618-446-6

Chapter 7

APPLICATION OF MEMBRANE TECHNOLOGY FOR THE DEVELOPMENT OF SUSTAINABLE BIOTECHNOLOGY PROCESSES

***Faisal I. Hai*[1]* *and Oskar Modin*[2]**
[1]Strategic Water Infrastructure Laboratory,
School of Civil Mining and Environmental Engineering,
The University of Wollongong, Australia
[2]Division of Water Environment Technology,
Department of Civil and Environmental Engineering,
Chalmers University of Technology, Göteborg, Sweden

ABSTRACT

The scope of application of membranes in biotechnology has widened significantly in the recent years. Although many of the membrane options are yet to achieve wide industrial applications, they show tremendous potential for the transformation and synthesis of value-added products, energy production, therapeutic applications and environmental remediation. This chapter provides an overview of membrane applications in selected established and emerging biotechnology processes. Approaches to overcoming the technology bottlenecks that impede the scale-up of such systems have been discussed in this chapter.

Keywords: Biotechnology; membrane reactors; enzyme; fine chemical purification and synthesis; emerging pollutants; bioelectrochemical systems

LIST OF ABBREVIATIONS

ADH	Alcohol dehydrogenease
ABTS	2,2'-azino-bis(3-ethylbenzothiazoline-6-sulphonic acid)

* Corresponding author; e-mail: faisal@uow.edu.au

BES	Bioelectrochemical system
DEA	Diethylamine
EMR	Enzymatic membrane reactor
GLU	Glutaraldehyde
GMA	Glycidyl metacrylate
HMDA	Hexamethylenediamine
HRP	Horseradish peroxidase
MBR	Membrane bioreactor
MBfR	Membrane biofilm reactor
MEC	Microbial electrolysis cell
MF	Microfiltration
MFC	Microbial fuel cell
MnP	Manganese-dependant peroxidase
PE	Polyethylene
PEG	Poly (ethyleneglycol)
PEMFC	Proton exchange membrane fuel cell
PVA	Polyvinyl alcohol
PVDF	Polyvinylidene fluoride
RED	Reverse electrodialysis
rBDNF	Recombinant Brain-Derived Neurotrophic Factor
UF	Ultrafiltration

1. Introduction-Membrane Application in Biotechnology

A typical membrane process step involves passage of feed through the membrane such that the molecules are separated in retentate and permeate on the basis of their size, charge, or other differential interaction (Rathore and Shirke, 2011). Membranes have long been an integral part of biotechnology processes (Charcosset, 2006).

For instance, membrane filtration has successfully replaced separation methods such as rotary vacuum filters or centrifugation in many plants by significantly improving product yields and reducing operator involvement and maintenance costs. Apart from the well established methods such as sterile filtration of fermentation media and purification of buffers and proteins, other applications of membrane processes have been introduced more recently. Examples of such emerging processes include biocatalytic membrane reactors for fine chemical synthesis, bioelectricity production, or pollution remediation, membrane chromatography, and membrane contactors. Table 1 lists examples of applications of membrane in various biotechnological processes.

Early membrane systems were adopted from technology originally developed for other industrial applications. During the last two decades, new membranes and modules have been developed specifically to meet the requirements of the biotechnology industry (van Reis and Zydney, 2007).

Table 1. Various applications of membrane in biotechnology

Category	Application example	Reference
Separation/Recovery/ Concentration/	Clarify suspensions for cell harvesting and recovery of macromolecules	(Nabais and Cardoso, 1999; Rathore and Shirke, 2011)
	Buffer exchange and desalting	(Phillips and Signs, 2001)
	Sterilize liquids to remove bacteria/ virus	(Jornitz et al., 2002)
Membrane contactor	Membrane emulsification	(Charcosset et al., 2004)
	Preparation of precipitates	(Chen et al., 2004)
Adsorptive membranes	Membrane chromatography	(Ghosh, 2002)
Enzymatic membrane reactors	Synthesis of food and pharmaceutical products	(Giorno and Drioli, 2000)
	Wastewater treatment	(Hai et al., 2012 ; Hai and Yamamoto, 2011)
Membrane biofilm reactors	Wastewater treatment	(Modin et al., 2008a; Modin et al., 2008b; Modin et al., 2010)
Membrane bioelectrochemical systems	Microbial fuel cell for electrical energy recovery	(Rabaey and Verstraete, 2005)
Biomedical applications	Tissue engineering, haemodialysis, membrane therapeutic treatments	(Catapano and Vienken, 2008)

Advances in the design and application of membrane technology for microfiltration and dead-end filtration have contributed to significant improvements in process performance and integration, in some cases allowing for a combination of multiple unit operations in a given step (Roush and Lu, 2008).

Membrane systems exploit the inherent properties of high selectivity, high surface-area-per-unit-volume, and their potential for controlling the level of contact and/or mixing between two phases. The potential advantages of membrane reactor technology over more conventional approaches include its higher efficiency and reduced costs owing to the integration of bioconversion and product purification, thus reducing equipment costs and the number of processing steps (Giorno and Drioli, 2000).

2. Membrane Application in Upstream and Downstream Processing

Ultrafiltration and microfiltration are commonly used to recover macromolecules and retain suspended colloids and particles, and are being integrated into both upstream and downstream biotechnology processes (Charcosset, 2006). The process for a typical product from a microbial or mammalian cell culture can involve several membrane filtration steps including separation of impurities from the process stream (e.g., viral clearance and sterile filtration), concentration of the feed stream to reduce total volume of the process stream or to reach a targeted product concentration (e.g., ultrafiltration), and finally, buffer exchange to deliver the product in the desired buffer system (e.g., diafiltration) (Rathore and Shirke, 2011). A few recent articles (Jornitz et al., 2002; Rathore and Shirke, 2011; Saxena et al., 2009; van Reis and Zydney, 2007) have comprehensively reviewed the application of membrane technology in upstream and downstream biotechnology processes. This section intends to provide a succinct overview of such applications.

2.1. *Clarification* and Sterilization

Membranes have traditionally been used for size-based separations with high-throughput. A large range of ultrafiltration and microfiltration applications are used to clarify suspensions for cell harvesting, and sterilize liquids to remove viruses and bacteria (Charcosset, 2006; Jornitz et al., 2002; Kuriyel and Zydney, 2000; Phillips and Signs, 2001; Rios et al., 2007; Saxena et al., 2009; Van Reis and Zydney, 2001; van Reis and Zydney, 2007). For the initial harvest of therapeutic products from mammalian, yeast and bacterial cell cultures, tangential flow microfiltration (MF) can be preferable to other options such as centrifugation, depth filtration or expanded-bed chromatography. In contrast to centrifugation, MF using 0.2 μm-rated membranes generates a particle-free harvest solution that requires no additional clarification before subsequent purification (van Reis and Zydney, 2007). Sterile filtration is performed in a normal flow, or dead-end, configuration using 0.2 μm pore size membranes that have been validated for the absolute removal of *Brevundimonas diminuta* (van Reis and Zydney, 2007). However, there is evidence that sterilizing filters can pass very small microorganisms under some process conditions, causing some users to employ 0.1 μm-rated filters to provide enhanced sterility assurance in pharmaceutical processes (Sundaram et al., 1999; Wickramasinghe, 2008).

2.2. Recovery, Purification and Concentration of Fine Chemicals

Membrane filtration is also particularly well suited in production of various fine chemicals including pharmaceuticals (Charcosset, 2006; Ghosh, 2008; Lipnizki, 2010). Pharmaceuticals are produced by fermentation and recovered from their broths by membrane filtration.

Numerous applications of membrane technologies for the recovery of valuable "bioproducts" from fermentation broth have been reported in the literature (Saxena et al., 2009). Table 2 lists some selected examples. The processes of fine chemical production include two key stages, namely fermentation and product recovery, and the biggest challenge lies in the recovery. Originally most of the separation of microorganisms from fermentation broth was performed by centrifugation.

Recently, cross-flow MF has created much interest in applications such as separation of cells in continuous fermentation processes (Jeantet et al., 1996). Often fermentation product is present as a component of a complex solution from which it needs to be separated and purified via two stage downstream filtration (Giorno and Drioli, 2000). For example, the production of L-lactic acid can be obtained by continuous fermentation in a membrane fermentor which consists of a traditional fermentor combined with an ultrafiltration unit.

Table 2. Membrane application for recovery and purification of fine chemicals from fermentation broths

Product	Membrane type	Purpose of membrane use	Reference
Antibiotics benzylpenicillin (Pen G), erythromycin and medmycin	Ultrafiltration	Removal of bioemulsifiers in antibiotic fermentation broth to improve subsequent solvent extraction of antibiotics	(Li et al., 2004)
Heterogeneous immunoglobulins (IgG)	Microfiltration	Recovery from transgenic goat milk	(Baruah and Belfort, 2004)
Recombinant Brain-Derived Neurotrophic Factor (rBDNF) inclusion bodies	Microfiltration and diafiltration	Concentration and purification from *E. coli* cell suspensions	(Schutyser et al., 2002)
Naturally glycosylated therapeutic proteins	Microfiltration	Recovery from animal cell cultures	(Vogel and Kroner, 1999)
Yeast alcohol dehydrogenease (ADH)	Microfiltration	Recovery and purification from bakers' yeast for the extraction of an intracellular enzyme product	(Levesley and Hoare, 1999)

The solution recovered as permeate contains L-lactic acid together with other small molecules that are not retained by the membrane, while the cells and macromolecules are recycled back to the bioreactor. The product then needs to be further purified and concentrated. The purification process can take place by membrane-based solvent extraction carried out through two membrane modules (Giorno and Drioli, 2000). Examples of other value-added product recovery by membrane filtration include recovery of heterogeneous immunoglobulins (IgG) (Baruah and Belfort, 2004), concentration and purification of recombinant Brain-Derived Neurotrophic Factor (rBDNF) inclusion bodies (Schutyser et al., 2002), recovery of naturally glycosylated therapeutic proteins (Vogel and Kroner, 1999) and recovery and purification of yeast alcohol dehydrogenease (ADH) (Levesley and Hoare, 1999). In order to keep pace with the new purification and process economic challenges posed by the advent of high-dose chronic therapies using recombinant DNA derived antibodies, current research and development efforts are directed toward drastic improvements in selectivity while maintaining the inherent high-throughput characteristics of membranes (van Reis and Zydney, 2007).

2.3. Membrane Fouling

The trend towards high-density cell cultures, and the corresponding increase in the level of cell debris, can cause unacceptable fouling of membrane systems, which require extensive cleaning protocols (Levesley and Hoare, 1999; van Reis and Zydney, 2007; Vogel and Kroner, 1999). Much effort is still being devoted to developing new membrane modules with improved mass-transfer characteristics for ultrafiltration and microfiltration processes

(Charcosset, 2006; Saxena et al., 2009). For example, dean vortex devices have very high mass-transfer rates owing to the presence of centrifugal flow instabilities (Luque et al., 1999; Schutyser et al., 2002). These devices show significant increases in protein transmission and capacity, although fouling remains a problem in many applications. An alternative approach is to use high frequency back-pulsing to continually clean the membrane surface (Levesley and Hoare, 1999). High-frequency back-pulsing was shown to improve flux, reduce fouling and increase protein transmission in the purification of conjugated vaccine products (Meacle et al., 1999). Other examples include rotating disk filters (Jaffrin et al., 2004; Lee et al., 1995), cylindrical Taylor vortex devices (Parnham Iii and Davis, 1995) and conically shaped rotors (Vogel and Kroner, 1999). New membrane materials are also being developed that are more resistant to fouling and/or have better chemical stability in harsh cleaning environments (Van Reis and Zydney, 2001).

3. Enzymatic Membrane Reactors (EMRs)

3.1. Basics of EMRs

EMRs are one type of membrane bioreactors (MBRs) which combine selective mass transport with enzyme-catalyzed biochemical reactions (Hai et al., 2012; Hai and Yamamoto, 2011). The commercialized form of MBRs i.e., the biosolid separation MBR differs from EMR in that the former involves whole-cell microbes in suspended form, and the membrane there acts mainly as a solid-liquid separator. In contrast, as the name suggests, an EMR utilizes enzymes harvested from a culture, but not the whole cell. An EMR accomplishes separation of enzymes from products and/or substrates by a semipermeable membrane (Jochems et al., 2011; Lopez et al., 2002). Permeable substrates and products can be selectively separated from the reaction mixture, and the selective removal of products from the reaction site increases the conversion of product-inhibited or thermodynamically unfavorable reactions (Giorno and Drioli, 2000). EMRs are particularly attractive in the context of the recent trend towards environmentally friendly technologies, as they do not require additives, are able to function at moderate temperature and pressure, and reduce the formation of by-products. Compared to chemical catalysts, the catalytic action of enzymes is extremely efficient and selective. Enzyme-catalyzed reactions can demonstrate higher reaction rates and greater stereo-specificity under milder reaction conditions (Giorno and Drioli, 2000).

An EMR can be simply a reactor containing enzyme in suspension and combined with a membrane operation unit. However, EMRs can be designed in a way that the membrane acts both as a catalytic and separation unit (Figure 1). Enzymes can be flushed along a membrane module, segregated within a membrane module, or immobilized in or on the membrane by various methods such as entrapment, gelification, physical adsorption, ionic binding, covalent binding or cross-linking. The choice of the reactor configuration depends on the properties of the reaction system (Giorno and Drioli, 2000). The membrane can have a flat-sheet shape, assembled in a plate-and-frame module or a spiral wound module, or tubular-like, assembled in tube-and-shell modules. Varieties of polymeric, ultrafiltration (predominantly) membranes have been used in EMRs (Akay et al., 2002; Bohdziewicz, 1998; Lante et al., 2000).

Membranes of both symmetric and asymmetric structure have been used (Giorno and Drioli, 2000). Usually commercially available membranes are modified to make them suitable for enzyme immobilization (Jochems et al., 2011).

2. Application of EMRs in Fine Chemical Synthesis

The application of membrane processes such as microfiltration, ultrafiltration, reverse osmosis, membrane extraction and so on for recovery and purification of fermentation products has been discussed in Section 2.2. In addition to such sequential application of bioreactors and membranes, integrated EMRs can be applied for synthesis of fine chemicals (Charcosset, 2006; Giorno and Drioli, 2000). This section discusses the application of EMRs in the agro-food and pharmaceutical industries.

3.2.1. Agro-Food Industry

Several applications of EMRs in the agro-food sector have been documented in the literature (Table 3). The main applications of EMRs include reducing the viscosity of juices by hydrolyzing pectins, reducing the lactose content in milk and whey by its conversion into digestible sugar, the treatment of musts and wines by the conversion of polyphenolic compounds and anthocyanes, and the removal of peroxides from dairy products (Giorno et al., 2010; Giorno and Drioli, 2000; Mazzei et al., 2010).

One of the first cases of the use of EMRs in the food sector was the production of milk with low lactose content (Illanes et al., 1990). Nowadays, the hydrolysis of lactose (present in whole milk or cheese whey) by the use of EMRs is an effective technique running on a large scale (Mazzei et al., 2010). EMRs have been also used in reduction of viscosity of fruit juice by hydrolyzing pectins (Alvarez et al., 1996), treatments of musts and wine by the conversion of polyphenolic compounds and antocyanes (Cantarelli et al., 1989), hydrolysis of oil (Pronk et al., 1988), production of sugars (Silman et al., 1980) and the removal of peroxides from dairy products. More recently EMRs have been successfully applied in the production of functional food and nutraceuticals, and alternative food or food ingredients that confer health benefit (Giorno and Drioli, 2000; Mazzei et al., 2010).

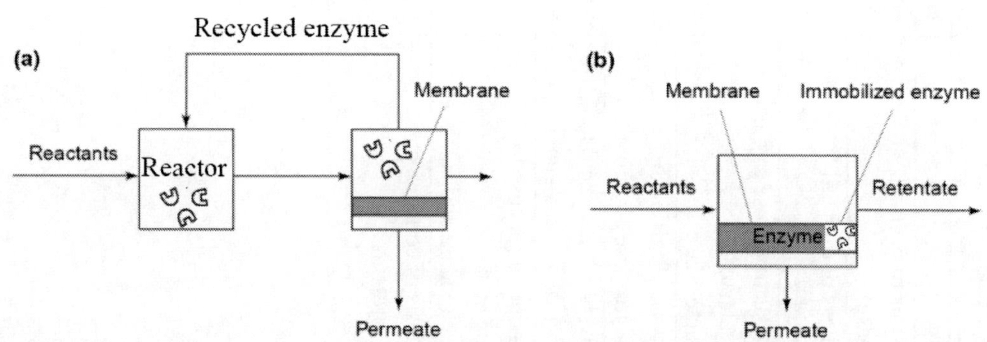

Figure 1. Simplified schematic of configurations of EMRs: (a) a reactor combined with a membrane operation unit, (b) a reactor with the membrane active as a catalytic and separation unit (Modified after Giorno and Drioli, (2000)).

Table 3. Application of EMRs in the agro-food industry (modified after (Giorno and Drioli, 2000))

Reaction	Membrane bioreactor	Purpose	References
Hydrolysis of pectines (pectinase)	CSTR with UF membrane	Production of bitterness and clarification of fruit juice and wine	(Alvarez et al., 1996)
Hydrolysis of olive oil triglycerides (lipase)	Hydrophobic plate-and-frame membrane reactor	Treatment of oils	(Pronk et al., 1988)
Hydrolysis of lactose to glucose and β-galactose (β -galactosidase)	Axial-annular flow reactor	Delactosization of milk or whey for human consumption	(Illanes et al., 1990)
Hydrolysis of raffinose (α-galactosidase and invertase)	Hollow fibre reactor with segregated enzyme	Production of monomeric sugars	(Silman et al., 1980)
Hydrolysis of malic acid to lactic acid (*Lactobacillus oenos*)	MF capillary membranes with entrapped cells	Improve taste in white wine	(Cantarelli et al., 1989)
Conversion of glucose to gluconic acid (glucose oxidase and catalase)	Packed bed reactor	Prevention of discolouration and off-flavour of egg products during storage	(Ramachandran et al., 2006; Tomotani and Vitolo, 2007)
Hydrolysis of collagen and muscle proteins (protease, papain)	CSTR with UF membrane	Meat tenderization	(Giorno and Drioli, 2000; Nakajima et al., 1992)

Table 4. Application of EMRs in the synthesis of pharmaceutical products

Synthesized material	Enzyme/Source of enzyme	EMR type	Reference
3-phenylglycidic acid ester	Lipase (*Serratia marcescens*)	Hydrophilic hollow-fiber membrane reactor	(Matsumae et al., 1994)
(S)-ibuprofen acid	Lipase	Hollow fiber membrane reactor	(Bhatia et al., 2004)
L-aspartic acid	Aspartase (*Escherichia coli*)	Entrapment in polyacrylamide gel	(Shi et al., 1997)
(S)-Naproxen	Lipase(*Candida rugosa*)	Tubular silicone rubber membrane	(Xin et al., 2001)
(R)-isomer and (S)-isomer of 2-hydroxy octanoic acid	Lipase (*Candida rugosa* or *Pseudomonas cepacia*)	Biphasic membrane reactor	(Sakaki et al., 2002)
Lovastatin	Lipase (*Candida rugosa*)	Nylon support	(Yang et al., 1997)
Diltiazem	Lipase	Multiphase hollow-fiber enzyme membrane reactor	(Lopez and Matson, 1997)
A derivative of kyotorphin	α-chymotrypsin (Bovine pancreas)	α-alumina mesoporous tubular support	(Belleville et al., 2001)

3.2.2. Pharmaceutical Products

Several studies have demonstrated the technical feasibility of EMRs for the production of amino acids, arylpropionic acids, amines, carboxylic acids, antibiotics, anticancer and anti-inflammatory drugs, vitamins, and optically pure enantiomers and antioxidants (Charcosset, 2006; Mazzei et al., 2010). Table 4 lists some important examples of application of EMRs for the synthesis of pharmaceutical compounds.

3.3. Application of EMRs in Pollutant Removal

Membrane reactors are being increasingly used in enzyme-catalyzed transformations. However, fewer studies have investigated the application of enzyme-based wastewater treatment systems. This section provides an overview of the studies which have tested various forms of EMRs for problematic compounds which are not well removed in conventional reactors. Table 5 lists a few selected examples.

Four sequential batch reactors containing raw soybean seed-hulls (source of soybean peroxidase) achieved greater than 96% removal of phenol with a retention time of 20 min in each reactor (Flock et al., 1999). While exploring soybean, horseradish and artichoke peroxidases for 4-chlorophenol conversions, Gomez et al. (2012) observed that conversion values increased when higher substrate concentrations were used and that the influence of the molar ratio H_2O_2:chlorophenol and the spatial time was negligible. In all cases, higher 4-chlorophenol conversions were attained with soybean peroxidases, reaching almost 100% in the permeate. In another study, oxidation of polyphenols was achieved by means of tyrosinase -catalyzed reaction carried out in a membrane bioreactor where the enzyme was confined upstream the membrane (Calabrò et al., 2009). A highly effective ultrafiltration biodegradation process for phenol was demonstrated with a complex of enzymes isolated from a bacterial strain *Pseudomonas* immobilized on the asymmetric structure of ultrafiltration membrane from the casting solution (Bohdziewicz, 1998).

Burton et al. (1998) described that polyphenol oxidase (PPO), obtained from *Agaricus bisporus*, can be used in hydroxylating a range of phenolic substrates to yield catechols which are then oxidised by the enzyme to give o-quinone products. In bioreactors containing customised novel asymmetric capillary membranes as the enzyme support, high concentrations of phenolic substrates were converted (Burton et al., 1998). A capillary membrane bioreactor with polyphenol oxidase immobilized on single capillary membranes was developed and tested for the removal of phenolic compounds from synthetic and industrial effluents (Edwards et al., 1999a). No precipitated melanoid products were observed in the permeate from the capillary membrane bioreactor, and in order to remove the colored quinone-type products of the reaction, a packed column containing chitosan was integrated into the system. Almost complete removal of the colored quinones and associated polymers from the permeate was observed. In another study, internally skinned polysulphone capillary membranes were coated with a viscous chitosan gel and used as an immobilization matrix for polyphenol oxidase (Edwards et al., 1999b). Bench-scale, single-capillary membrane bioreactors were then used to determine the influence of the chitosan coating on product removal after substrate conversion by immobilized polyphenol oxidase during the treatment of industrial phenolic effluents (Edwards et al., 1999b).

Microporous polypropylene hollow fiber membranes impregnated with horseradish peroxidase (HRP) and laccase were used to study the degradation of selected hydroxylated aromatic compounds (3,4-dimethylphenols, 4-ethylphenol, 2-hydroxy-1,2,3,4-tetrahydro-naphthalene, 2-hydroxy-decahydronaphthalene and 4-hydroxy-biphenyl) (Moeder et al., 2004). It was found that, with the exception of 2-hydroxy-decahydronaphthalene, all substrates were efficiently degraded (50-100% within 48 h). In preliminary investigations, the transport behavior of the analytes through the membranes before and after enzyme immobilization was studied in batch experiments. In addition, a continuous-flow hollow fiber membrane device was used to examine the degradation performance of the immobilized enzymes on the substrates. It was successfully demonstrated that freshly prepared radish juice can be used to replace the commercially available horseradish peroxidase. The method of overall monitoring used was a combination of solid-phase microextraction/gas chromatography/mass spectrometry.

The results of a study of immobilization of a commercial laccase onto a spiral-wound asymmetric polyethersulphone membrane were described by Lante et al. (2000). The immobilised enzyme system displayed a promising half-life of more than 150 h for oxidation of syringaldazine. The laccase membrane reactor was applied to the biodegradation of a model phenol solution containing 18 phenolic substrates, including chlorophenols, cresols and methoxyphenols. The type and/or the position of substituent group were observed to affect the level of substrate oxidation. A cross-flow enzyme-immobilized membrane reactor was used to investigate the removal of phenol and catechol from water (Akay et al., 2002). A strain coded as AEK-1, identified as a species of *Pseudomonas syringae* was used to obtain the crude enzyme extract. Crude enzyme extract was chemically immobilized onto a flat polyamide membrane with nominal pore size of 0.2 μm. The degradation process was carried out in a cross-flow filtration module. The highest phenol-degradation rate was 1.3 $\mu g/cm^2.s$ when the initial concentration of phenol was 500 μg/mL at the flux rate of 4.60×10^{-3} $mL/cm^2.s$.

Laccase from *Rhus vernicifera* was immobilized on a nylon membrane chemically grafted with glycidyl methacrylate (GMA) (Durante et al., 2004). Hexamethylenediamine (HMDA) and glutaraldehyde (GLU) were used as spacer and bifunctional coupling agent, respectively. Quinol was used as substrate. To know how the immobilization procedures affected the enzyme reaction rate the catalytic behavior of soluble and insoluble laccase was studied under isothermal conditions as a function of pH, temperature and substrate concentration. From these studies, two main singularities emerged from the experimental data: (i) the narrower pH-activity profile of the insoluble enzyme in comparison to that of the soluble counterpart; (ii) the increase of the affinity of the immobilised enzyme for its substrate. The behaviour of the catalytic membrane was also studied in a non-isothermal bioreactor as a function of substrate concentration and size of the applied transmembrane temperature difference. It was found that under non-isothermal conditions and keeping constant the average temperature of the bioreactor, the enzyme reaction rate linearly increases with the increase of the temperature difference. These results have been discussed in the frame of reference of the process of thermodialysis driving thermodiffusive transmembrane substrate fluxes, which add to the diffusive ones. The advantages of the catalytic process carried out under non-isothermal conditions have been thrown in relief through the evaluation of the reduction of the production times and of the percentage increases of the enzyme activity (Durante et al., 2004).

The application of laccase and manganese-dependant peroxidase (MnP) from *Trametes versicolor* to facilitate removal of aromatic hydrocarbons from a petrochemical industrial effluent was investigated. The enzymes were immobilized onto polysulphone ultrafiltration membranes thus facilitating enzyme-substrate contact and the activity and duration of activity was determined with respect to the enzymatic degradation of the aromatics within the effluent (Edwards et al., 2002).

Laccase immobilization onto a hydrophilic PVDF microfiltration membrane and its application for removing a herbicide derivative, N',N'-(dimethyl)-N-(2-hydroxyphenyl)urea (2-HF), from wastewater were studied (Jolivalt et al., 2000). 2-HF was transformed via an oxidation reaction catalyzed by laccase mostly into an insoluble product, which was simultaneously separated by filtration through the membrane.

Performance of various laccase-membrane reactor configurations including direct enzyme contact, enzyme impregnated, immobilized enzyme and a reactor system based on laccase immobilization in chitosan membranes for decolorization of azo dye (acid black 10 BX) were examined using laccase enzyme purified from white rot fungi *Pleurotus ostreatus* 1804 (Katuri et al., 2009). Experimental data showed that laccase has great potential for color removal without addition of external redox mediators. Various process parameters viz. aqueous phase of pH 6.0, enzyme concentration of 1.75 U/ml, dye concentration of 20 mg/L, temperature of 30 °C and reaction time of 120 min were optimized to achieve maximum decolorization efficiencies. Moreover, different laccase-membrane reactor configurations were tested to determine the efficacy of repeated application of laccase on dye decolorization process. Among the different reactor configurations employed, laccase encapsulated in chitosan membrane showed advantages such as short-term contact period and reusability of enzyme for a number of cycles (Katuri et al., 2009).

An enzymatic membrane reactor (EMR) for the oxidation of azo dyes by manganese peroxidase (MnP) was developed (López et al., 2004). At best conditions, a continuous operation with a dye decolorization higher than 85% and minimal enzymatic deactivation was feasible for 18 days. An ultrafiltration polyethersulfone synthetic membrane (Prep/Scale-TFF Millipore) with a nominal molecular weight cut-off of 10 kDa was used in this study. In another study, the treatment of a textile dye solution using laccase and a mediator, namely, ABTS was demonstrated in an enzyme membrane reactor. At a hydraulic retention time of 6 h, the process was operated for a period of 15 days with nearly 95% decolorization, 10% reduction in flux and 70% recovery of active ABTS (Chhabra et al., 2009).

Cyanidase, an immobilized enzyme preparation for hydrolyzing cyanide to ammonia and formate, was applied for the treatment of cyanide-containing waste waters from the food industry (Basheer et al., 1993). Apricot seed extract was chosen as a model effluent. The enzymatic hydrolysis of pure amygdalin, the main cyanogenic glycoside in the extract, and the degradation of the cyanide formed was investigated and compared with the behavior of the real extract in a batch slurry reactor.

A diffusional-type, flat-membrane reactor with immobilized cyanidase was developed, where the enzyme is effectively protected from adverse effects of high molecular components contained in the extract. For monitoring continuous-membrane reactor operation, a new unsegmented ammonia measurement system was developed and applied. In continuous operation the cyanidase retained its original activity for more than 400 hours on steam.

Table 5. Selected examples of removal of recalcitrant pollutants by enzymatic membrane reactors

Target pollutant	Type of membrane reactor	Enzyme	Reference
4-chlorophenol	Enzyme in suspension	Soybean, horseradish and artichoke-peroxidases	(Gómez et al., 2012)
Dye	Enzyme in suspension	Laccase	(Chhabra et al., 2009)
Dye	Enzyme immobilized onto membrane	Laccase	(Katuri et al., 2009)
Phenol and cyanide	Enzyme immobilized onto membrane	Crude mixed enzyme extract (from a *Pseudomonas* strain)	(Bodzek et al., 1996)
Phenylurea pesticide	Enzyme immobilized onto membrane	Laccase	(Jolivalt et al., 2000)
Xylan	Enzyme immobilized onto membrane	Xylanase	(Edward et al., 2003)

In another study (Bodzek et al., 1996), an attempt at simultaneous biodegradation of the highly toxic groups of substances, namely phenols and cyanides, formed in the coke industry was described. An ultrafiltration process was carried out in the batch mode of operation using a flat polyacrylonitrile membrane with an immobilized enzymatic fraction isolated from a bacterial strain of *Pseudomonas*.

The optimal ultrafiltration process parameters, i.e. transmembrane pressure and stirring rate in the bioreactor, as well as the influence of the concentration of xenobiotics on the effectiveness of the process, were determined. The dynamics of the decomposition of phenol and cyanides were investigated. Edward et al. (Edward et al., 2003) built a novel membrane bioreactor for the degradation of synthetic xylan effluent.

The reactor contains 30 internally skinned polysulphone membranes as an immobilization matrix for xylanase, the degrading agent, and was constructed with stainless steel to withstand high temperatures, as *Thermomyces lanuginosus* SSBP xylanase has an optimum temperature of 50°C. Overall, 85.1% of the xylanase was immobilized onto the polysulphone membranes by adsorption. Preliminary results showed that the immobilized enzyme was capable of degrading the xylan effluent. Continuous degradation of phenol in an enzymatic membrane reactor with immobilized enzyme involved the lowest consumption of enzyme (López et al., 2011). However, the immobilization process increased the enzyme cost 100-fold. In comparison, a continuous enzymatic membrane reactor with suspended enzyme gave high degradation efficiency, and, therefore, may be a viable technology for phenol removal (López et al., 2011).

3.4. Barriers against Large Scale Application of EMRs

Compared to chemical catalysts, enzymes are particularly suitable for certain applications, specifically in terms of energy consumption, safety, pollution prevention and the high quality of products produced. However, the use of enzymes on an industrial scale is not yet fully established (Giorno and Drioli, 2000; Mazzei et al., 2010). Only a few references are available regarding industrial application of EMRs for fine chemical synthesis. In relation to enzymes, the major technological difficulties of using EMRs on an industrial level are (Giorno et al., 2010; Giorno and Drioli, 2000; Mazzei et al., 2010): (1) the availability of pure enzyme at an acceptable cost (often the commercial enzymes are mixtures of several proteins); (2) difficulties in immobilizing enzymes that often need expensive cofactors; (3) the necessity for biocatalysts to operate at low substrate concentrations; and (4) microbial contaminations. In addition, as in all membrane reactors, concentration-polarization phenol-mena and membrane fouling strongly affect the performance of this type of reactors. Appropriate fluid dynamic conditions and reactor design are necessary to control performance at a steady state. Additional impediments exist against application of EMRs for wastewater treatment. Gradual loss of enzymatic activity due to various physical, chemical and biological inhibitors under wastewater conditions is inevitable. Thus, in comparison to other applications, the application of enzyme-based wastewater treatment systems remains a less studied topic (Katuri et al., 2009; Majeau et al., 2010), and no full scale EMR plants for wastewater treatment could be identified in the literature.

4. Membrane Biofilm Reactors (MBfRs)

4.1 Basics of MBfRs

In a membrane biofilm reactor (MBfR), a membrane separates a gas phase from a liquid phase and a biofilm is attached on the liquid-facing side of the membrane. Compounds present in the liquid and other compounds diffusing from the gas phase through the membrane are biochemically converted by the biofilm. An MBfR can be used to degrade liquid or gas phase contaminants. The most common type is the oxygen-based MBfR (Figure 2 A). Air, pure oxygen, or oxygen-enriched air is fed through the membrane to a biofilm that oxidizes substrates in the liquid (Syron and Casey, 2008a). Feeding oxygen through a membrane is advantageous compared to conventional sparging method. Using an MBfR, the oxygen is fed directly to the biofilm with near 100% efficiency. At elevated intramembrane oxygen pressures, high oxygen transfer rates can be obtained. Other types of MBfRs also exist, for example, hydrogen-based MBfR and biofilter MBfR.

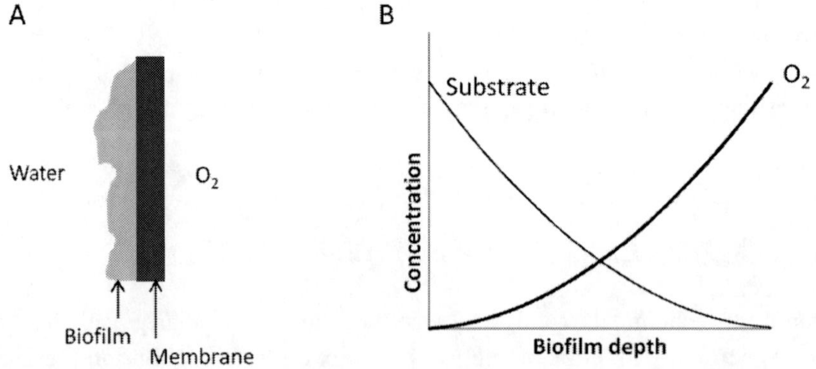

Figure 2. (A) Schematic of oxygen-based membrane biofilm reactor, (B) substrate concentration profiles in membrane-attached biofilm.

4.2. Pollutant Removal by MBfRs

Most oxygen-based MBfR studies have focused on removal of organics and nitrogen compounds from wastewater. Because of the counter-diffusion of oxygen and liquid substrate, membrane-attached biofilms are especially well-suited for simultaneous removal of organics and nitrogen (Figure 2 B). Timberlake et al. (1998) designed a reactor for simultaneous nitrification, denitrification, and organics oxidation. Nitrifiers will thrive near the membrane where the oxygen concentration is high. Denitrifiers will grow near the bulk liquid where the conditions are anoxic and organic substrate is plentiful. Aerobic organics oxidation may take place in an intermediate biofilm section. Using molecular techniques, researchers have shown that nitrifiers and denitrifiers indeed are spatially separated in membrane-attached biofilms (Cole et al., 2004; LaPara et al., 2006).

The hydrogen-based MBfR feeds hydrogen gas to a membrane-attached biofilm. Hydrogen acts as electron donor for reduction of oxidized contaminants in the liquid, for example nitrate and perchlorate (Ergas and Reuss, 2001; Lee and Rittmann, 2000; Nerenberg and Rittmann, 2004). In the methane-based MBfR, methane gas is supplied to the biofilm. Such reactors have been investigated for denitrification (Modin et al., 2008a; Modin et al., 2008b; Modin et al., 2010), pesticide removal (Modin et al., 2008c), and cometabolic degradation of trichloroethylene (Clapp et al., 1999). The biofilter MBfR targets contaminants in the gas phase rather than the liquid phase. A contaminated gas stream flows past the membrane. The contaminants diffuse through the membrane and are degraded by the biofilm growing on the liquid-facing side. The liquid consists of a nutrient medium that supports degradation of the target contaminant. Biofilter MBfR have been investigated for a range of contaminants, but toluene is often used as a model compound (Ergas and McGrath, 1997; Hartmans et al., 1992; Parvatiyar et al., 1996). Other target compounds include for example propene (Reij et al., 1995), dichloroethane (Freitas dos Santos et al., 1995), dimethyl sulfide (De Bo et al., 2002), nitric oxide (Min et al., 2002), and carbon dioxide (Kumar et al., 2010).

4.3. MBfR Membranes

Membranes used in MBfRs can be classified as porous, dense, or composite. In a porous membrane, the gas passes through the pores and the intramembrane gas pressure must be low to avoid bubble formation on the liquid side (Ahmed and Semmens, 1992). A dense membrane, on the other hand, allows high gas pressures without bubble formation. The gas dissolves into the membrane polymer, e.g. silicone rubber, diffuses through the membrane and desorbs on the liquid-facing side (Yasuda and Lamaze, 1972). Composite membranes consist of both porous and dense layers and have been used frequently in MBfR studies. By adding porous layers as structural support, a very thin dense membrane layer can be used. Since the thickness of the layer determines its gas transfer resistance, using a thin layer is beneficial.

For a MBfR to function properly, a biofilm of appropriate thickness should grow on the membrane surface. Too thick biofilms can results in substrate diffusion limitation of the removal rate (Syron and Casey, 2008b). Catastrophic sloughing, when large parts of the biofilm are suddenly detached, is another problem that should be addressed (Rishell et al., 2004). Researchers have modified membranes to improve attachment and retention of biofilms. There are two types of modifications: chemical alteration of the membrane and addition of support material onto the membrane surface.

Hibiya et al. (2000) showed that a chemically altered polyethylene (PE) membrane had higher adhesiveness to nitrifying biofilms than an unaltered PE membrane. The membrane was altered using radiation induced grafting in a process developed by Tsuneda et al. (1995) to capture proteins. They used radiation induced grafting to attach chains of the epoxy-group-containing monomer glycidyl metacrylate (GMA) to PE membranes. The graft chains form a tentacle-like structure on the membrane surface. The epoxy groups were converted to diethylamine (DEA) groups, which are anion exchange groups, by submerging the membrane in a diethylamine solution. The remaining epoxy groups were hydrophilized by introducing ethanolamino (EA) groups through soaking in ethanolamine solution. The density of DEA groups on the membrane was positively correlated with its protein binding capacity (Tsuneda et al., 1995). Lee et al. (1997) further compared three types of GMA-grafted PE membranes. The first contained only the GMA graft chain; in the second membrane, parts of the epoxy groups had been converted to DEA groups; in the third membrane, the remaining epoxy groups had been converted to EA groups (Figure 3). The second and third membrane types were also prepared in replicates with varying DEA density. They tested the various types of membranes for capture of gram-positive and gram-negative bacteria. The second type with high DEA density performed best, particularly in capturing gram-positive bacteria. Terada et al. (2003) used a DEA-membrane in an oxygen-based MBfR accomplishing simultaneous nitrification and denitrification. In another study Terada et al. (2004) confirmed that the DEA membrane performed better than an unmodified PE membrane for initial attachment of nitrifying bacteria.

Terada et al. (2005) investigated GMA-grafted membranes with either the anion exchanging DEA groups or cation exchanging sulfonic acid groups (SO_3H) introduced. Adhesion of five gram-negative bacteria on the DEA membranes occurred faster than on the membrane containing sulfonic acid groups.

Their results also indicated that surface roughness and increased membrane potential were beneficial for bacterial attachment. Surface roughness increases with degree of grafting and the membrane potential increases with the introduction of anion exchange groups.

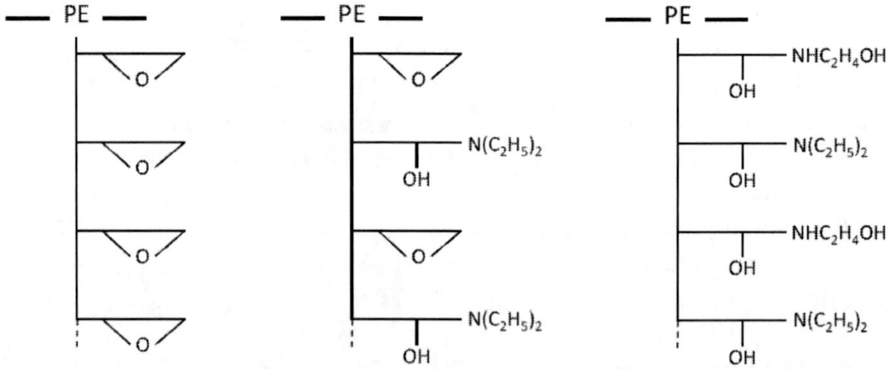

Figure 3. Polyethylene (PE) membranes with GMA graft chains (left), graft chains containing DEA groups (middle), and graft chains containing DEA and EA groups (right).

Further investigation of the membrane potential showed that although higher membrane potential promotes adhesion, too high potential may negatively affect bacterial viability.

The gram-negative *Escherichia coli* showed significantly reduced viability at a membrane potential above -8 mV. The gram-positive *Bacillus subtilis* was less affected (Terada et al., 2006a). Lackner et al. (2009) used radiation induced grafting to modify PE and polypropylene (PP) membranes. They introduced poly (ethyleneglycol) (PEG) chains with either NH_2 groups or CH_3 groups. The combination of PEG chains and NH_2 functional groups improved initial attachment as well as shear resistance of a nitrifying biofilm. PEG chains or NH_2 groups alone did not have the same positive effect. Neither did PEG chains and CH_3 groups. The base polymer used did not affect performance.

Rather than chemical modification of the membrane polymer itself, another option to retain biofilm is to dress the membrane with a special support material. A ferro-nickel fibrous slag composed of mixed oxides of SiO_2, MgO, and Al_2O_3 covering a silicone membrane has been used in a series of studies on denitrification using hydrogen-based MBfR (Terada et al., 2006b), combined nitrogen and phosphorous removal (Terada et al., 2006c) and nitrification (Terada et al., 2009) using oxygen-based MBfR. The fibrous support layer allowed a larger amount of biomass to be immobilized on the membrane compared to bare silicone tubing. The biomass loss after high water flow rate was also lower with fibrous support (Terada et al., 2009). Hirasa et al. (1991) prepared a fibrous support material of <1μm fibers of poly(vinyl alcohol) treated with dimethylaminoacetaldehyde. The fibers were woven around a silicone tubing membrane fed with air. Yamagiwa et al. (1994) used the fibrous support oxygen-based MBfR for combined nitrification, denitrification, and carbon removal from wastewater. They showed that nitrifiers were dominant within the fibrous support while denitrifiers grew in the surface biofilm attached to the support. Polyurethane foam sheets placed on a Teflon membrane (Yamagiwa et al., 1998) and porous fabric of pyridinium-type polymer covering microporous carbon tube membrane (Gong et al., 2008) have also been used. In the latter study, they investigated partial ammonium oxidation coupled with anammox. They found that aerobic ammonium-oxidizing bacteria appeared in the membrane-support interface while anammox bacteria were present in the outer portion of the fibrous support, near the bulk liquid (Gong et al., 2008). Hsieh et al. (2002) immobilized nitrifying bacteria in polyvinyl alcohol (PVA), which was coated as a solid matrix around a silicone membrane. The biofilm-PVA matrix was robust over an experimental period of 20 weeks and the author suggested that the PVA strengthened the biofilm against shear stress.

5. Membrane Bioelectrochemical Systems (BESs)

5.1. Basics of Membrane BESs

Bioelectrochemical systems (BES) are devices in which living microbes interact with solid electrodes. The microbes either use an anode as electron acceptor for oxidation reactions or a cathode as electron donor for reduction reactions (Rabaey et al., 2007). Thus far, the most studied type of BES is the microbial fuel cell (MFC), which is used to recover electrical energy from dissolved organic matter in e.g. wastewater. In a MFC, microbes oxidize organic matter and transfer electrons to the anode. The electrons flow through an external circuit to the cathode, where oxygen is reduced to water (Figure 4). The oxygen reduction reaction can also be catalyzed by microbes using the cathode as electron donor and oxygen as electron acceptor (Clauwaert et al., 2007; Rabaey et al., 2008). The flow of electrons from the anode to the cathode is balanced by an equal but opposite flow of ionic charge through the liquid. Electrical energy can be recovered from the system since the overall reaction, oxidation of organics and reduction of oxygen, is thermodynamically favorable. Another type of BES is the microbial electrolysis cell (MEC) for hydrogen production (Liu et al., 2005; Rozendal et al., 2006a). Instead of oxygen reduction, protons are reduced to hydrogen gas at the cathode. The proton reduction can be either abiotic on e.g. platinum or catalyzed by microbes (Rozendal et al., 2008a). The overall reaction, oxidation of organics and reduction of protons to hydrogen gas, is thermodynamically unfavorable, so a small voltage input is needed to drive the reaction.

Membranes play a crucial role in many types of BESs. In MFCs and MECs, ion exchange membranes are often used to separate the anode and cathode compartments.

The membrane should prevent crossover of oxygen or hydrogen from the cathode compartment to the anode compartment and organic substrate in the opposite direction. At the same time, the membrane should be as permeable as possible to the transfer of ionic charge to minimize the internal resistance of the system. The use of membranes in BESs results in many challenges but also leads to several opportunities.

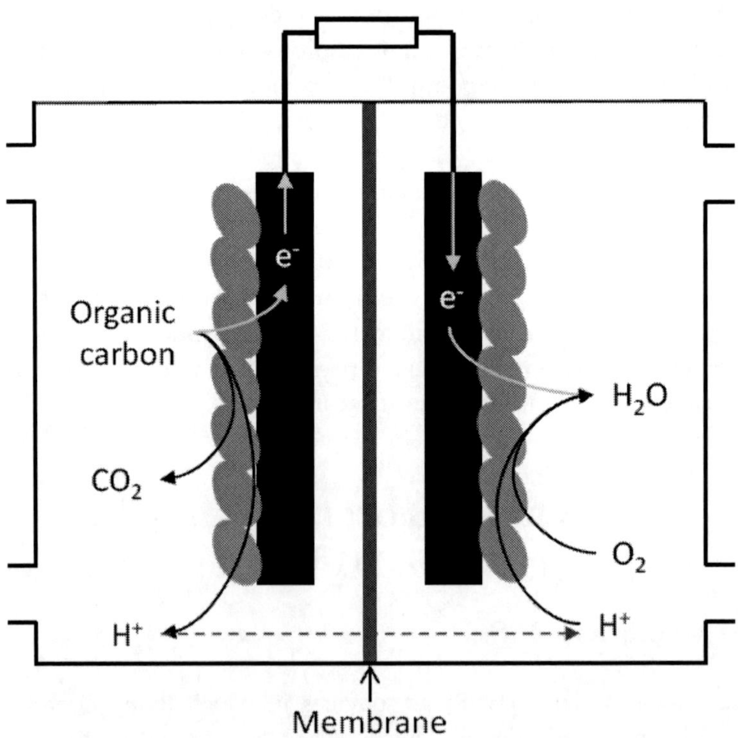

Figure 4. Schematic of a microbial fuel cell.

5.2. Membrane-Associated Challenges

Most BESs have been operated with various types of ion exchange membranes as separator between anode and cathode. Early on, the Nafion proton exchange membrane was a popular choice (Bond and Lovley, 2003; Liu et al., 2004), probably because it is widely used in the chlor-alkali industry and in proton exchange membrane fuel cells (PEMFC). Nafion consists of a tetrafluoroethylene (Teflon) backbone with perfluoroether side chains terminated by sulfonate groups (SO_3^-). Protons move through the membrane either by "hopping" between hydrolyzed SO_3^- groups or by electroosmotic drag where protons bound to water move through the membrane (Peighambardoust et al., 2010). The use of Nafion membranes have led to operational problems in BESs. Gil et al. (2003) investigated an MFC consisting of an anaerobic anode compartment fed with wastewater and an aerobic cathode compartment fed with oxygenated phosphate buffer. A Nafion membrane separated the two compartments. They observed that the pH increased in the cathode compartment and decreased in the anode compartment, leading to deteriorating performance. In a MFC, organic compounds are oxidized at the anode and oxygen reduced at the cathode according to the following equations:

$$CH_2O + H_2O \rightarrow CO_2 + 4H^+ + 4e^- \tag{1}$$

$$O_2 + 4H^+ + 4e^- \rightarrow 2H_2O \tag{2}$$

As shown in equations 1 and 2, the anode reaction releases protons while the cathode reaction consumes protons. Thus, to maintain the pH in the anode and cathode compartments, the transfer of one electron through the external circuit must be accompanied by the transfer of one proton through the proton exchange membrane. BESs depend on the activities of living microbes and therefore work best at a pH near 7, which means that the proton concentration in BESs is around 10^{-7} M. The concentration of other cations such as Na^+ and K^+ is usually 10^{-2}-10^{-3} M (Harnisch and Schröder, 2009). Since the concentration of these other cations is up to a hundred thousand times larger than the proton concentration, the other cations will account for most of the charge transfer through the membrane. Rozendal et al. (2006b) demonstrated this phenomenon and showed that Na^+, K^+, NH_4^+, Ca^{2+}, and Mg^{2+} ions were responsible for practically the entire charge transfer through the Nafion membrane. We can expect that other cation exchange membranes than Nafion will show a similar behavior.

Other types of membranes have also been investigated in BESs. Harnisch et al. (2008) compared two cation exchange membranes, an anion exchange membrane, and a bipolar membrane. Anion exchange membranes consist of positively charged groups (such as quartenary ammonium) immobilized on a polymer backbone.

They allow the transfer anions through the membrane. Bipolar membranes consist of a cation and anion exchange membrane pressed together. Ideally, charge transfer through a bipolar membrane is accomplished by a water splitting reaction between the membranes and migration of protons through cation exchange layer and hydroxide ions through the anion exchange layer. Water splitting requires energy input of approximately 22 W-h/mol corresponding to a membrane polarization of about 0.83 V (Hurwitz and Dibiani, 2001).

Theoretically, a bipolar membrane should prevent pH shifts in BESs at the expense of higher energy consumption. However, diffusion of electrolyte salts into the membrane and the migrations of the electrolyte ions through the membrane layers still causes pH shifts (Harnisch et al., 2008). Rozendal et al. (2008b) compared anion exchange-, cation exchange-, bipolar-, and charge mosaic membranes. The latter consists of both positively and negatively charged groups in the membrane allowing transfer of both cations and anions. In terms of pH shifts, the bipolar membrane performed better than the other three, although pH shifts still occurred. The charge mosaic membrane appeared to function as a cation exchange membrane, transferring mostly cations. Unless a membrane with extremely high specificity for proton or hydroxide ion transfer can be developed, the application of ion exchange membranes in BESs will likely always lead to pH shifts between anode and cathode compartments.

To avoid the problems associated with ion exchange membranes, BESs have been constructed with porous separators such as ultrafiltration membranes (Kim et al., 2007), microfiltration membranes (Sun et al., 2009), and course pore separators (Fan et al., 2007).

BESs have also been operated completely without membranes (Liu and Logan, 2004) and with liquid recirculation between anode and cathode compartments (Freguia et al., 2008). All these strategies typically lead to increased crossover of substrates between the anode and cathode compartments, which results in lower coulombic efficiencies (Li et al., 2011).

5.3. Membrane-Associated Opportunities

The pH shift that takes place when ion exchange membranes are applied in BESs does not necessarily have to be a problem, but could also be viewed as an opportunity. Rabaey et al. (2010) used this phenomenon to produce caustic from wastewater in a bioelectrochemical reactor. They constructed a lamellar type bioelectrochemical reactor containing graphite felt anode and stainless steel cathode separated by cation exchange membranes. The anode was first fed with an acetate-containing nutrient medium and later with real brewery wastewater. With acetate feed, they were able to produce a caustic concentration equivalent to 3.4% (weight) NaOH. With real wastewater the current generation and caustic concentration were lower. The key to using a bioelectrochemical reactor for generation of caustic is to have a much larger anode flow rate than cathode flow rate. The NaOH generation at the cathode is accompanied by an equal H^+ generation at the anode (Figure 5 A). While it is desirable to generate a concentrated caustic solution at the cathode, the conditions at the anode should be kept stable and near neutral pH to allow for biological activity.

The caustic could also be produced in combination with hydrogen peroxide in BESs. The pulp and paper industry uses hydrogen peroxide and caustic soda for bleaching and brightening processes (Foller and Bombard, 1995). BESs could be used to produce hydrogen peroxide by the cathodic reduction of oxygen (Rozendal et al., 2009). Since the pulp and paper generates a wastewater rich in organics, this wastewater could potentially be used to power the production of a mixture of hydrogen peroxide and caustic, which can be used onsite.

In wastewater treatment plants, the alkalinity rise in the catholyte could be used directly to support nitrification (Modin et al., 2011). Wastewater treatment plants often consist of a treatment train for the wastewater and another treatment train for the generated sludge. The sludge treatment processes – thickening, anaerobic digestion, and dewatering – generate reject water that is fed back to the wastewater treatment train. The reject water contains high ammonium concentration and constitutes a major ammonium load to the plant; thus, it is often treated in a separate reactor before being fed back to the plant. The ammonium oxidation reaction is acidifying (equation 3), so chemicals like NaOH or $Ca(OH)_2$ have to be added to the reactor to maintain a near neutral pH.

$$NH_4^+ + 1.5O_2 \rightarrow NO_2^- + H_2O + 2H^+ \quad (3)$$

Modin et al. (2011) proposed that a BES could be used to extract alkalinity from the influent wastewater to support nitrification of reject water. The influent wastewater would flow through the anode compartment and the reject water through the cathode compartment. The influent wastewater flow rate is typically 100 times larger than the reject water flow rate, which means that a small alkalinity drop in the anode compartment would be accompanied by a 100 times larger alkalinity concentration increase in the cathode compartment.

Cao et al. (2009) developed a MFC for desalination of saline water. By placing an anion exchange membrane facing the anode and a cation exchange membrane facing the cathode, they created a central compartment in the MFC, between the anode and cathode compartments. Saline water is fed to the central compartment. When current is flowing, cations migrate toward the cathode and anions migrate toward the anode, desalinating the central compartment (Figure 5 B). The desalination rate was further improved by stacking several desalinating compartments between the anode and cathode (Chen et al., 2011; Kim and Logan, 2011). The system does not necessarily have to be operated as a MFC; with a voltage input, hydrogen gas can be recovered at the cathode while water is desalinated (Mehanna et al., 2011).

Kim and Logan (2011a) combined BES with reverse electrodialysis (RED) systems (Figure 5 C). RED recovers energy from salinity gradients between salt water and fresh water, which occur e.g. where river water and sea water meet. RED systems consist of stacks of alternating cation- and anion exchange membranes. The spaces between the membranes are filled with alternating salt water and fresh water. A voltage develops over each membrane because of the salinity gradient. Conventional RED systems are operated with abiotic anode and cathode electrodes and may suffer from large overpotentials, which are voltage losses associated with the electrode reactions.

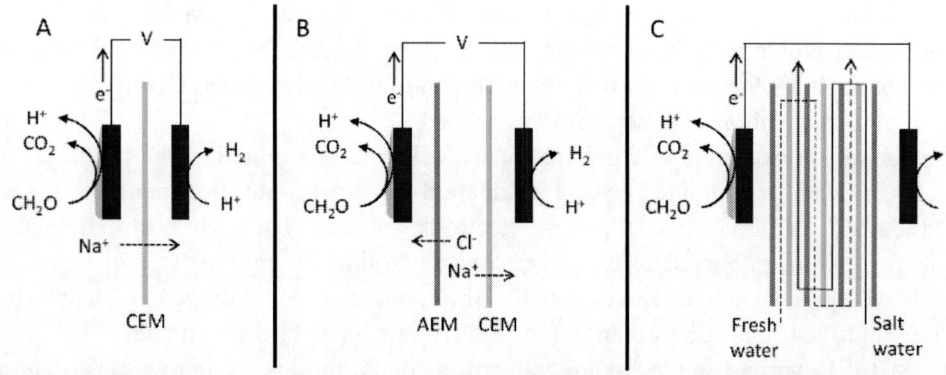

Figure 5. Integration of ion exchange membranes with microbial bioelectrochemical systems. (A) A single cation exchange membrane (CEM) is used to generate a caustic solution at the cathode. (B) An anion exchange membranes (AEM) and CEM allows desalination of the central compartment. (C) A stack of AEM and CEM allows energy recovery from salinity gradients.

A biological anode oxidizing organics has relatively low overpotential. Thus, combining a MFC with a RED stack boosts the electrical energy output (Kim and Logan, 2011). A RED stack can also allow hydrogen production in a MEC without an external voltage input.

The combination of porous membranes and BES also offers opportunities. In membrane bioreactors (MBR), porous membranes are combined with biological treatment for direct solid-liquid separation. Wang et al. (2011) constructed a combined MBR-MFC system. The cathode was made of a 40-μm stainless steel mesh that also acted as filtration material for the effluent. Although the steel mesh had a large pore size, biofilm development on the mesh improved the filtration performance and rejected small flocs from escaping with the effluent. The effluent turbidity was about 0.8 NTU. The biofilm also improved the catalytic properties for reduction of oxygen.

Conclusion and Future Trends

The application of ultrafiltration and microfiltration membranes in processes including initial harvesting of fermentation media, sterile filtration of fermentation media and purification of buffers and proteins are now standard practices.

Other applications of membrane processes such as membrane chromatography, membrane contactors and MBRs have been introduced more recently. Although the applications of ultrafiltration and microfiltration membranes in upstream/downstream processing of fermentation media have been well-established, processes such as MBRs and membrane chromatography are yet to gain the expected levels of success. Several forms of MBRs have been introduced. Among them, only the biosolid-separation MBRs for wastewater treatment have been commercialized. The other forms of MBRs, such as the EMRs for fine chemical synthesis or pollutant degradation, MBfRs for pollutant degradation, and BESs for e.g. bioelectricity production still await large-scale application and/or commercialization. The main technological difficulties in using the emerging MBRs on an industrial level are related with rate-limiting aspects and scale-up difficulties of this technology, together with the stability of the biocatalysts (e.g. enzyme in an EMR).

Membranes are important in bioelectrochemical systems. For some applications, such as energy recovery from wastewater using MFC or MEC, currently available ion exchange membranes are problematic because they do not have high enough specificity for proton transfer. For such applications, further research about alternative process configurations and the use of porous membranes as separators is needed.

The use of ion exchange in combination with BES have resulted in the development of novel systems such as combined caustic production and wastewater treatment, the microbial desalination cell, and the microbial reverse electrodialysis cell. This field is still in its infancy with the first set of publications based on research trying to exploit the properties of ion exchange membranes in combination with BES appearing in 2009-2011. More work is needed to improve reactor performance and identify suitable application niches.

The MBfR is a promising technology that offers opportunities to supply gaseous substrate to biofilms at high rate and efficiency. This could make it possible to accomplish high rate oxidation of high strength wastewater. By manipulating the intramembrane gas pressure we could potentially also control the microenvironments within membrane-attached biofilms to achieve complex biodegradation processes that require the involvement of several microbial groups, such as nitrogen removal from wastewater. It is challenging to maintain biofilms of appropriate thickness on membranes. Too thick biofilms lead to substrate diffusion limitation whereas too thin biofilms can lead to incomplete removal of the target contaminant and loss of the supplied gas. The majority of MBfR studies have utilized unmodified membrane obtained directly from manufacturers. However, modification of the membrane surface can improve adhesion and retention of biofilm. Several promising modification methods have been tested. Now we need to know which method is most appropriate for which treatment process. Here, mathematical modeling can be a useful tool.

The varieties of physical and chemical mechanisms (e.g., mixing, reaction and permeation), which a membrane can facilitate, appear to be one of the most attractive aspects of application of membrane processes in biotechnology. Therefore, successful new membrane applications will undoubtedly continue to appear in the near future. Theoretical understanding of the physical and chemical phenomena governing these processes will be instrumental to ensure their penetration into the biotechnology industry.

REFERENCES

Ahmed, T., Semmens, M.J., 1992. Use of sealed end hollow fibers for bubbleless membrane aeration: experimental studies. *Journal of Membrane Science*, 69, 1-10.

Akay, G., Erhan, E., Keskinler, B., Algur, O.F., 2002. Removal of phenol from wastewater using membrane-immobilized enzymes: Part II. Cross-flow filtration. *Journal of Membrane Science*, 206, 61-68.

Alvarez, V., Andres, L.J., Riera, F.A., Alvarez, R., 1996. Microfiltration of Apple Juice using Inorganic Membranes: Process Optimization and Juice Stability. *Canadian Journal of Chemical Engineering*, 74, 156-162.

Baruah, G.L., Belfort, G., 2004. Optimized recovery of monoclonal antibodies from transgenic goat milk by microfiltration. *Biotechnology and Bioengineering*, 87, 274-285.

Basheer, S., Kut, Ö.M., Prenosil, J.E., Bourne, J.R., 1993. Development of an enzyme membrane reactor for treatment of cyanide-containing wastewaters from the food industry. *Biotechnology and Bioengineering*, 41, 465-473.

Belleville, M.P., Lozano, P., Iborra, J.L., Rios, G.M., 2001. Preparation of hybrid membranes for enzymatic reaction. *Separation and Purification Technology*, 25, 229-233.

Bhatia, S., Long, W.S., Kamaruddin, A.H., 2004. Enzymatic membrane reactor for the kinetic resolution of racemic ibuprofen ester: modeling and experimental studies. *Chemical Engineering Science*, 59, 5061-5068.

Bodzek, M., Bohdziewicz, J., Kowalska, M., 1996. Immobilized enzyme membranes for phenol and cyanide decomposition. *Journal of Membrane Science*, 113, 373-384.

Bohdziewicz, J., 1998. Biodegradation of phenol by enzymes from Pseudomonas sp. immobilized onto ultrafiltration membranes. *Process Biochemistry*, 33, 811-818.

Bond, D.R., Lovley, D.R., 2003. Electricity Production by Geobacter sulfurreducens Attached to Electrodes. *Applied and Environmental Microbiology*, 69, 1548-1555.

Burton, S.G., Boshoff, A., Edwards, W., Rose, P.D., 1998. Biotransformation of phenols using immobilised polyphenol oxidase. *Journal of Molecular Catalysis B: Enzymatic*, 5, 411-416.

Calabrò, V., Curcio, S., De Paola, M.G., Iorio, G., 2009. Optimization of membrane bioreactor performances during enzymatic oxidation of waste bio-polyphenols. *Desalination*, 236, 30-38.

Cantarelli, C., Brenna, O., Giovanelli, G., Rossi, M., 1989. Beverage stabilization through enzymatic removal of phenolics. *Food Biotechnol*, 3, 203-213.

Cao, X., Huang, X., Liang, P., Xiao, K., Zhou, Y., Zhang, X., Logan, B.E., 2009. A new method for water desalination using microbial desalination cells. *Environmental Science and Technology*, 43, 7148-7152.

Catapano, G., Vienken, J., 2008. Biomedical Applications of Membranes Advanced Membrane Technology and Applications. John Wiley and Sons, Inc., pp. 489-517.

Charcosset, C., 2006. Membrane processes in biotechnology: An overview. *Biotechnology Advances*, 24, 482-492.

Charcosset, C., Limayem, I., Fessi, H., 2004. The membrane emulsification process - A review. *Journal of Chemical Technology and Biotechnology*, 79, 209-218.

Chen, G.G., Luo, G.S., Xu, J.H., Wang, J.D., 2004. Membrane dispersion precipitation method to prepare nanopartials. *Powder Technology*, 139, 180-185.

Chen, X., Xia, X., Liang, P., Cao, X., Sun, H., Huang, X., 2011. Stacked Microbial Desalination Cells to Enhance Water Desalination Efficiency. *Environmental Science and Technology*, 45, 2465-2470.

Chhabra, M., Mishra, S., Sreekrishnan, T.R., 2009. Laccase/mediator assisted degradation of triarylmethane dyes in a continuous membrane reactor. *Journal of Biotechnology*, 143, 69-78.

Clapp, L.W., Regan, J.M., Ali, F., Newman, J.D., Park, J.K., Noguera, D.L., 1999. Activity, structure, and stratification of membrane-attached methanotrophic biofilms cometabolically degrading trichloroethylene. *Water Science and Technology*, 39, 153-161.

Clauwaert, P., van der Ha, D., Boon, N., Verbeken, K., Verhaege, M., Rabaey, K., Verstraete, W., 2007. Open air biocathode enables effective electricity generation with microbial fuel cells. *Environmental Science and Technology*, 41, 7564-7569.

Cole, A.C., Semmens, M.J., LaPara, T.M., 2004. Stratification of Activity and Bacterial Community Structure in Biofilms Grown on Membranes Transferring Oxygen. *Applied and Environmental Microbiology*, 70, 1982-1989.

De Bo, I., Van Langenhove, H., Heyman, J., 2002. Removal of dimethyl sulfide from waste air in a membrane bioreactor. *Desalination*, 148, 281-287.

Durante, D., Casadio, R., Martelli, L., Tasco, G., Portaccio, M., De Luca, P., Bencivenga, U., Rossi, S., Di Martino, S., Grano, V., Diano, N., Mita, D.G., 2004. Isothermal and non-isothermal bioreactors in the detoxification of waste waters polluted by aromatic compounds by means of immobilised laccase from Rhus vernicifera. *Journal of Molecular Catalysis B: Enzymatic*, 27, 191-206.

Edward, V.A., Pillay, V.L., Swart, P., Jacobs, E., Singh, S., 2003. Degradation of synthetic xylan effluent using a membrane bioreactor. *South African Journal of Science*, 99, 315-317.

Edwards, W., Bownes, R., Leukes, W.D., Jacobs, E.P., Sanderson, R., Rose, P.D., Burton, S.G., 1999 a. A capillary membrane bioreactor using immobilized polyphenol oxidase for the removal of phenols from industrial effluents. *Enzyme and Microbial Technology*, 24, 209-217.

Edwards, W., Leukes, W.D., Bezuidenhout, J.J., 2002. Ultrafiltration of petrochemical industrial wastewater using immobilised manganese peroxidase and laccase: Application in the defouling of polysulphone membranes. *Desalination*, 149, 275-278.

Edwards, W., Leukes, W.D., Rose, P.D., Burton, S.G., 1999 b. Immobilization of polyphenol oxidase on chitosan-coated polysulphone capillary membranes for improved phenolic effluent bioremediation. *Enzyme and Microbial Technology*, 25, 769-773.

Ergas, S.J., McGrath, M.S., 1997. Membrane bioreactor for control of volatile organic compound emissions. *Journal of Environmental Engineering*, 123, 593-598.

Ergas, S.J., Reuss, A.F., 2001. Hydrogenotrophic denitrification of drinking water using a hollow fibre membrane bioreactor. *Journal of Water Supply: Research and Technology - AQUA*, 50, 161-171.

Fan, Y., Hu, H., Liu, H., 2007. Enhanced Coulombic efficiency and power density of air-cathode microbial fuel cells with an improved cell configuration. *Journal of Power Sources*, 171, 348-354.

Flock, C., Bassi, A., Gijzen, M., 1999. Removal of aqueous phenol and 2-chlorophenol with purified soybean peroxidase and raw soybean hulls. *Journal of Chemical Technology and Biotechnology*, 74, 303-309.

Foller, P.C., Bombard, R.T., 1995. Processes for the production of mixtures of caustic soda and hydrogen peroxide via the reduction of oxygen. *Journal of Applied Electrochemistry*, 25, 613-627.

Freguia, S., Rabaey, K., Yuan, Z., Keller, J., 2008. Sequential anode–cathode configuration improves cathodic oxygen reduction and effluent quality of microbial fuel cells. *Water Research*, 42, 1387-1396.

Freitas dos Santos, L.M., Hömmerich, U., Livingston, A.G., 1995. Dichloroethane removal from gas streams by an extractive membrane bioreactor. *Biotechnology Progress*, 11, 194-201.

Ghosh, R., 2002. Protein separation using membrane chromatography: Opportunities and challenges. *Journal of Chromatography A*, 952, 13-27.

Ghosh, R., 2008. Biopharmaceutical Separations by Ultrafiltration Advanced Membrane Technology and Applications. John Wiley and Sons, Inc., pp. 435-450.

Gil, G.-C., Chang, I.-S., Kim, B.H., Kim, M., Jang, J.-K., Park, H.S., Kim, H.J., 2003. Operational parameters affecting the performannce of a mediator-less microbial fuel cell. *Biosensors and Bioelectronics*, 18, 327-334.

Giorno, L., De Bartolo, L., Drioli, E., 2010. Membrane Bioreactors. in: E. Drioli, Giorno, L. (Ed.) Comprehensive Biotechnology (Second Edition). Academic Press, Burlington, pp. 263-288.

Giorno, L., Drioli, E., 2000. Biocatalytic membrane reactors: applications and perspectives. *Trends in Biotechnology*, 18, 339-349.

Gómez, M., Murcia, M.D., Ortega, S., Barbosa, D.S., Vayá, G., Hidalgo, A.M., 2012. Removal of 4-chlorophenol in a continuous membrane bioreactor using different commercial peroxidases. *Desalination and Water Treatment*, 37, 97-107.

Gong, Z., Liu, S., Yang, F., Bao, H., Furukawa, K., 2008. Characterization of functional microbial community in a membrane-aerated biofilm reactor operated for completely autotrophic nitrogen removal. *Bioresource Technology*, 99, 2749-2756.

Hai, F.I., Nghiem, L.D., Modin, O., 2012 Biocatalytic membrane reactors: Removal of recalcitrant and emerging pollutants. in: A. Basile (Ed.) Handbook of membrane reactors (volume 2). Woodhead publishing, Cambridge.

Hai, F.I., Yamamoto, K., 2011. Membrane Biological Reactors. in: P. Wilderer (Ed.) Treatise on Water Science. Elsevier, Oxford, pp. 571-613.

Harnisch, F., Schröder, U., 2009. Selectivity versus Mobility: Separation of Anode and Cathode in Microbial Bioelectrochemical Systems. *ChemSusChem*, 2, 921-926.

Harnisch, F., Schröder, U., Scholz, F., 2008. The Suitability of Monopolar and Bipolar Ion Exchange Membranes as Separators for Biological Fuel Cells. *Environmental Science and Technology*, 42, 1740-1746.

Hartmans, S., Leenen, E., Voskuilen, G., 1992. Membrane bioreactor with porous hydrophobic membranes for waste-gas treatment. *Studies in Environmental Science*, 51, 103-106.

Hibiya, K., Tsuneda, S., Hirata, A., 2000. Formation and characteristics of nitrifying biofilm on a membrane modified with positively-charged polymer chains. *Colloids and Surfaces B: Biointerfaces*, 18, 105-112.

Hirasa, O., Ichijo, H., Yamauchi, A., 1991. Preparation of new support for immobilization of activated sludge. *Journal of Fermentation and Bioengineering*, 71, 376-378.

Hsieh, Y.L., Tseng, S.K., Chang, Y.J., 2002. Nitrification using polyvinyl alcohol-immobilized nitrifying biofilm on an O2-enriching membrane. *Biotechnology Letters*, 24, 315-319.

Hurwitz, H.D., Dibiani, R., 2001. Investigation of electrical properties of bipolar membranes at steady state and with transient methods. *Electrochimica Acta*, 47, 759-773.

Illanes, A., Ruiz, A., ZÃºÃ±iga, M.E., Aguirre, C., O'Reilly, S., Curotto, E., 1990. Immobilization of lactase for the continuous hydrolysis of whey permaete. *Bioprocess Engineering*, 5, 257-262.

Jaffrin, M.Y., Ding, L.H., Akoum, O., Brou, A., 2004. A hydrodynamic comparison between rotating disk and vibratory dynamic filtration systems. *Journal of Membrane Science*, 242, 155-167.

Jeantet, R., Maubois, J.L., Boyaval, P., 1996. Semicontinuous production of lactic acid in a bioreactor coupled with nanofiltration membranes. *Enzyme and Microbial Technology*, 19, 614-619.

Jochems, P., Satyawali, Y., Diels, L., Dejonghe, W., 2011. Enzyme immobilization on/in polymeric membranes: Status, challenges and perspectives in biocatalytic membrane reactors (BMRs). *Green Chemistry*, 13, 1609-1623.

Jolivalt, C., Brenon, S., Caminade, E., Mougin, C., Pontié, M., 2000. Immobilization of laccase from Trametes versicolor on a modified PVDF microfiltration membrane: characterization of the grafted support and application in removing a phenylurea pesticide in wastewater. *Journal of Membrane Science*, 180, 103-113.

Jornitz, M., Soelkner, P., Meltzer, T., 2002. Sterile filtration--a review of the past and present technologies. *PDA Journal of Pharmaceutical Science and Technology*, 56, 192-195.

Katuri, K.P., Venkata Mohan, S., Sridhar, S., Pati, B.R., Sarma, P.N., 2009. Laccase-membrane reactors for decolorization of an acid azo dye in aqueous phase: Process optimization. *Water Research*, 43, 3647-3658.

Kim, J.R., Cheng, S., Oh, S.-E., Logan, B.E., 2007. Power Generation Using Different Cation, Anion, and Ultrafiltration Membranes in Microbial Fuel Cells. *Environmental Science and Technology*, 41, 1004-1009.

Kim, Y., Logan, B.E., 2011a. Hydrogen production from inexhaustible supplies of fresh and salt water using microbial reverse-electrodialysis electrolysis cells. *Proceedings of the National Academy of Sciences*, 108, 16176-16181.

Kim, Y., Logan, B.E., 2011b. Microbial Reverse Electrodialysis Cells for Synergistically Enhanced Power Production. *Environmental Science and Technology*, 45, 5834-5839.

Kim, Y., Logan, B.E., 2011c. Series Assembly of Microbial Desalination Cells Containing Stacked Electrodialysis Cells for Partial or Complete Seawater Desalination. *Environmental Science and Technology*, 45, 5840-5845.

Kumar, A., Yuan, X., Sahu, A.K., Dewulf, J., Ergas, S.J., Van Langenhove, H., 2010. A hollow fiber membrane photo-bioreactor for CO2 sequestration from combustion gas coupled with wastewater treatment: a process engineering approach. *Journal of Chemical Technology and Biotechnology*, 85, 387-394.

Kuriyel, R., Zydney, A.L., 2000. Sterile filtration and virus filtration. *Methods in Biotechnology Downstream Protein Processing*, 9, 185-194.

Lackner, S., Holmberg, M., Terada, A., Kingshott, P., Smets, B.F., 2009. Enhancing the formation and shear resistance of nitrifying biofilms on membranes by surface modification. *Water Research*, 43, 3469-3478.

Lante, A., Crapisi, A., Krastanov, A., Spettoli, P., 2000. Biodegradation of phenols by laccase immobilised in a membrane reactor. *Process Biochemistry*, 36, 51-58.

LaPara, T.M., Cole, A.C., Shanahan, J.W., Semmens, M.J., 2006. The effects of organic carbon, ammoniacal-nitrogen, and oxygen partial pressure on the stratification of membrane-aerated biofilms. *Journal of Industrial Microbiology and Biotechnology*, 33, 315-323.

Lee, K.-C., Rittmann, B.E., 2000. A novel hollow-fibre membrane biofilm reactor for autohydrogenotrophic denitrification of drinking water. *Water Science and Technology*, 41, 219-226.

Lee, S.S., Burt, A., Russotti, G., Buckland, B., 1995. Microfiltration of recombinant yeast cells using a rotating disk dynamic filtration system. *Biotechnology and Bioengineering*, 48, 386-400.

Lee, W., Saito, K., Furusaki, S., Sugo, T., 1997. Capture of microbial cells on brush-type polymeric materials bearing different functional groups. *Biotechnology and Bioengineering*, 53, 523-528.

Levesley, J.A., Hoare, M., 1999. The effect of high frequency backflushing on the microfiltration of yeast homogenate suspensions for the recovery of soluble proteins. *Journal of Membrane Science*, 158, 29-39.

Li, S.Z., Li, X.Y., Cui, Z.F., Wang, D.Z., 2004. Application of ultrafiltration to improve the extraction of antibiotics. *Separation and Purification Technology*, 34, 115-123.

Li, W.-W., Sheng, G.-P., Liu, X.-W., Yu, H.-Q., 2011. Recent advances in the separators for microbial fuel cells. *Bioresource Technology*, 102, 244-252.

Lipnizki, F., 2010. Chapter 7 - Membrane Processes for the Production of Bulk Fermentation Products. in: Z.F. Cui, H.S. Muralidhara (Eds.), Membrane Technology. Butterworth-Heinemann, Oxford, pp. 121-153.

Liu, H., Grot, S., Logan, B.E., 2005. Electrochemically assisted microbial production of hydrogen from acetate. *Environmental Science and Technology*, 39, 4317-4320.

Liu, H., Logan, B.E., 2004. Electricity generation using an air-cathode single chamber microbial fuel cell in the presence and absence of a proton exchange membrane. *Environmental Science and Technology*, 38, 4040-4046.

Liu, H., Ramnarayanan, R., Logan, B.E., 2004. Production of electricity during wastewater treatment using a single chamber microbial fuel cell. *Environmental Science and Technology*, 38, 2281-2285.

Lopez, C., Mielgo, I., Moreira, M.T., Feijoo, G., Lema, J.M., 2002. Enzymatic membrane reactors for biodegradation of recalcitrant compounds. Application to dye decolourisation. *Journal of Biotechnology*, 99, 249-257.

López, C., Moreira, M.T., Feijoo, G., Lema, J.M., 2004. Dye Decolorization by Manganese Peroxidase in an Enzymatic Membrane Bioreactor. *Biotechnology Progress*, 20, 74-81.

López, C., Moreira, M.T., Feijoo, G., Lema, J.M., 2011. Economic comparison of enzymatic reactors and advanced oxidation processes applied to the degradation of phenol as a model compound. *Biocatalysis and Biotransformation*, 29, 344-353.

Lopez, J.L., Matson, S.L., 1997. A multiphase/extractive enzyme membrane reactor for production of diltiazem chiral intermediate. *Journal of Membrane Science*, 125, 189-211.

Luque, S., Mallubhotla, H., Gehlert, G., Kuriyel, R., Dzengeleski, S., Pearl, S., Belfort, G., 1999. A new coiled hollow-fiber module design for enhanced microfiltration performance in biotechnology. *Biotechnology and Bioengineering*, 65, 247-257.

Majeau, J.A., Brar, S.K., Tyagi, R.D., 2010. Laccases for removal of recalcitrant and emerging pollutants. *Bioresource Technology*, 101, 2331-2350.

Matsumae, H., Furui, M., Shibatani, T., Tosa, T., 1994. Production of optically active 3-phenylglycidic acid ester by the lipase from Serratia marcescens on a hollow-fiber membrane reactor. *Journal of Fermentation and Bioengineering*, 78, 59-63.

Mazzei, R., Drioli, E., Giorno, L., 2010. Biocatalytic Membranes and Membrane Bioreactors. in: E. Drioli, Giorno, L. (Ed.) Comprehensive Membrane Science and Engineering. Elsevier, Oxford, pp. 195-212.

Meacle, F., Aunins, A., Thornton, R., Lee, A., 1999. Optimization of the membrane purification of a polysaccharide-protein conjugate vaccine using backpulsing. *Journal of Membrane Science*, 161, 171-184.

Mehanna, M., Kiely, P.D., Call, D.F., Logan, B.E., 2011. Microbial electrodialysis cell for simultaneous water desalination and hydrogen gas production. *Environmental Science and Technology*, 44, 9578-9583.

Min, K.-N., Ergas, S.J., Harrison, J.M., 2002. Hollow-fiber membrane bioreactor for nitric oxide removal. *Environmental Engineering Science*, 19, 575-583.

Modin, O., Fukushi, K., Nakajima, F., Yamamoto, K., 2008a. A membrane biofilm reactor achieves aerobic methane oxidation coupled to denitrification (AME-D) with high efficiency. *Water Science and Technology*, 58, 83-87.

Modin, O., Fukushi, K., Nakajima, F., Yamamoto, K., 2008b. Performance of a membrane biofilm reactor for denitrification with methane. *Bioresource Technology*, 99, 8054-8060.

Modin, O., Fukushi, K., Nakajima, F., Yamamoto, K., 2010. Nitrate removal and biofilm characteristics in methanotrophic membrane biofilm reactors with various gas supply regimes. *Water Research*, 44, 85-96.

Modin, O., Fukushi, K., Rabaey, K., Rozendal, R.A., Yamamoto, K., 2011. Redistribution of wastewater alkalinity with a microbial fuel cell to support nitrification of reject water. *Water Research*, 45, 2691-2699.

Modin, O., Fukushi, K., Yamamoto, K., 2008c. Simultaneous removal of nitrate and pesticides from groundwater using a methane-fed membrane biofilm reactor. *Water Science and Technology*, 58, 1273-1279.

Moeder, M., Martin, C., Koeller, G., 2004. Degradation of hydroxylated compounds using laccase and horseradish peroxidase immobilized on microporous polypropylene hollow fiber membranes. *Journal of Membrane Science*, 245, 183-190.

Nabais, A.M.A., Cardoso, J.P., 1999. Purification of benzylpenicillin filtered broths by ultrafiltration and effect on solvent extraction. *Bioprocess Engineering*, 21, 157-163.

Nakajima, M., Shoji, T., Nabetani, H., 1992. Protease hydrolysis of water soluble fish proteins using a free enzyme membrane reactor. *Process Biochemistry*, 27, 155-160.

Nerenberg, R., Rittmann, B.E., 2004. Hydrogen-based, hollow-fiber membrane biofilm reactor for reduction of perchlorate and other oxidized contaminants. *Water Science and Technology*, 49, 223-230.

Parnham Iii, C.S., Davis, R.H., 1995. Protein recovery from cell debris using rotary and tangential crossflow microfiltration. *Biotechnology and Bioengineering*, 47, 155-164.

Parvatiyar, M.G., Govind, R., Bishop, D.F., 1996. Biodegradation of toluene in a membrane biofilter. *Journal of Membrane Science*, 119, 17-24.

Peighambardoust, S.J., Rowshanzamir, S., Amjadi, M., 2010. Review of the proton exchange membranes for fuel cell applications. *International Journal of Hydrogen Energy*, 35, 9349-9384.

Phillips, A.T., Signs, M.W., 2001. Desalting, Concentration, and Buffer Exchange by Dialysis and Ultrafiltration Current Protocols in Protein Science. John Wiley and Sons, Inc.

Pronk, W., Kerkhof, P.J.A.M., van Helden, C., van't Riet, K., 1988. Hydrolysis of triglycerides by immobilized lipase in a hydrophilic membrane reactor. *Biotechnology and Bioengineering*, 32, 512-518.

Rabaey, K., Bützer, S., Brown, S., Keller, J., Rozendal, R.A., 2010. High current generation coupled to caustic production using a lamellar bioelectrochemical system. *Environmental Science and Technology*, 44, 4315-4321.

Rabaey, K., Read, S.T., Clauwaert, P., Freguia, S., Bond, P.L., Blackall, L.L., Keller, J., 2008. Cathodic oxygen reduction catalyzed by bacteria in microbial fuel cells. *The ISME Journal*, 2, 519-527.

Rabaey, K., Rodríguez, J., Blackall, L.L., Keller, J., Gross, P., Batstone, D., Verstraete, W., Nealson, K.H., 2007. Microbial ecology meets electrochemistry: electricity-driven and driving communities. *The ISME Journal*, 1, 9-18.

Rabaey, K., Verstraete, W., 2005. Microbial fuel cells: novel biotechnology for energy generation. *Trends in Biotechnology*, 23, 291-298.

Ramachandran, S., Fontanille, P., Pandey, A., Larroche, C., 2006. Gluconic Acid: Properties, Applications and Microbial Production. *Food Technology and Biotechnology*, 44, 185-195.

Rathore, A.S., Shirke, A., 2011. Recent developments in membrane-based separations in biotechnology processes: Review. *Preparative Biochemistry and Biotechnology*, 41, 398-421.

Reij, M.W., de Gooijer, K.D., de Bont, J.A.M., Hartmans, S., 1995. Membrane bioreactor with a porous hydrophobic membrane as a gas-liquid contactor for waste gas treatment. *Biotechnology and Bioengineering*, 45, 107-115.

Rios, G.M., Belleville, M.-P., Paolucci-Jeanjean, D., 2007. Membrane engineering in biotechnology: quo vamus? *Trends in Biotechnology*, 25, 242-246.

Rishell, S., Casey, E., Glennon, B., Hamer, G., 2004. Characteristics of a methanotrophic culture in a membrane-aerated biofilm reactor. *Biotechnology Progress*, 20, 1082-1090.

Roush, D.J., Lu, Y., 2008. Advances in Primary Recovery: Centrifugation and Membrane Technology. *Biotechnology Progress*, 24, 488-495.

Rozendal, R., Hamelers, H., Euverink, G., Metz, S., Buisman, C., 2006a. Principle and perspectives of hydrogen production through biocatalyzed electrolysis. *International Journal of Hydrogen Energy*, 31, 1632-1640.

Rozendal, R.A., Hamelers, H.V.M., Buisman, C.J.N., 2006b. Effects of membrane cation transport on pH and microbial fuel cell performance. *Environmental Science and Technology*, 40, 5206-5211.

Rozendal, R.A., Jeremiasse, A.W., Hamelers, H.V.M., Buisman, C.J.N., 2008a. Hydrogen production with a microbial biocathode. *Environmental Science and Technology*, 42, 629-634.

Rozendal, R.A., Leone, E., Keller, J., Rabaey, K., 2009. Efficient hydrogen peroxide generation from organic matter in a bioelectrochemical system. *Electrochemistry Communications*, 11, 1752-1755.

Rozendal, R.A., Sleutels, T.H.J.A., Hamelers, H.V.M., Buisman, C.J.N., 2008b. Effect of the type of ion exchange membrane on performance, ion transport, and pH in biocatalyzed electrolysis of wastewater. *Water Science and Technology*, 57, 1757-1762.

Sakaki, K., Hara, S., Itoh, N., 2002. Optical resolution of racemic 2-hydroxy octanoic acid using biphasic enzyme membrane reactor. *Desalination*, 149, 247-252.

Saxena, A., Tripathi, B.P., Kumar, M., Shahi, V.K., 2009. Membrane-based techniques for the separation and purification of proteins: An overview. *Advances in Colloid and Interface Science*, 145, 1-22.

Schutyser, M., Rupp, R., Wideman, J., Belfort, G., 2002. Dean vortex membrane microfiltration and diafiltration of rBDNF E. coli inclusion bodies. *Biotechnology Progress*, 18, 322-329.

Shi, Y.F., Jin, F.X., Wu, Y.Y., 1997. Microfiltration membrane bioreactor in stirred backflush operation for biotransformation using intact cells. *Process Biochemistry*, 32, 387-390.

Silman, R.W., Black, L.T., McGhee, J.E., Bangley, E.B., 1980. Hydrolysis of raffinose in a hollow-fiber reactor using an unrefined mixture of α-galactosidase and invertase. *Biotechnology and Bioengineering*, 22, 533-541.

Sun, J., Hu, Y., Bi, Z., Cao, Y., 2009. Improved performance of air-cathode single-chamber microbial fuel cell for wastewater treatment using microfiltration membranes and multiple sludge inoculation. *Journal of Power Sources*, 187, 471-479.

Sundaram, S., Auriemma, M., Howard Jr, G., Brandwein, H., Leo, F., 1999. Application of membrane filtration for removal of diminutive bioburden organisms in pharmaceutical products and processes. *PDA Journal of Pharmaceutical Science and Technology*, 53, 186-201.

Syron, E., Casey, E., 2008a. Membrane-Aerated Biofilms for High Rate Biotreatment: Performance Appraisal, Engineering Principles, Scale-up, and Development Requirements. *Environmental Science and Technology*, 42, 1833-1844.

Syron, E., Casey, E., 2008b. Model-based comparative performance analysis of membrane aerated biofilm reactor configurations. *Biotechnology and Bioengineering*, 99, 1361-1373.

Terada, A., Hibiya, K., Nagai, J., Tsuneda, S., Hirata, A., 2003. Nitrogen removal characteristics and biofilm analysis of a membrane-aerated biofilm reactor applicable to high strength nitrogenous wastewater treatment. *Journal of Bioscience and Bioengineering*, 95, 170-178.

Terada, A., Ito, J., Matsumoto, S., Tsuneda, S., 2009. Fibrous Support Stabilizes Nitrification Performance of a Membrane-Aerated Biofilm: The Effect of Liquid Flow Perturbation. *Journal of Chemical Engineering of Japan*, 42, 607-615.

Terada, A., Kaku, S., Matsumoto, S., Tsuneda, S., 2006a. Rapid autohydrogenotrophic denitrification by a membrane biofilm reactor equipped with a fibrous support around a gas-permeable membrane. *Biochemical Engineering Journal*, 31, 84-91.

Terada, A., Yamamoto, T., Tsuneda, S., Hirata, A., 2004. Enhancement of biofilm formation onto surface-modified hollow-fiber membranes and its application to a membrane-aerated biofilm reactor. *Water Science and Technology*, 49, 263-268.

Terada, A., Yamamoto, T., Tsuneda, S., Hirata, A., 2006b. Sequencing batch membrane biofilm reactor for simultaneous nitrogen and phosphorus removal: Novel application of membrane-aerated biofilm. *Biotechnology and Bioengineering*, 94, 730-739.

Terada, A., Yuasa, A., Kushimoto, T., Tsuneda, S., Katakai, A., Tamada, M., 2006c. Bacterial adhesion to and viability on positively charged polymer surfaces. *Microbiology*, 152, 3575-3583.

Terada, A., Yuasa, A., Tsuneda, S., Hirata, A., Katakai, A., Tamada, M., 2005. Elucidation of dominant effect on initial bacterial adhesion onto polymer surfaces prepared by radiation-induced graft polymerization. *Colloids and Surfaces B: Biointerfaces*, 43, 99-107.

Timberlake, D.L., Strand, S.E., Williamson, K.J., 1988. Combined aerobic heterotrophic oxidation, nitrification and denitrification in a permeable-support biofilm. *Water Research*, 22, 1513-1517.

Tomotani, E.J., Vitolo, M., 2007. Immobilized glucose oxidase as a catalyst to the conversion of glucose into gluconic acid using a membrane reactor. *Enzyme and Microbial Technology*, 40, 1020-1025.

Tsuneda, S., Saito, K., Furusaki, S., Sugo, T., 1995. High-thoughput processing of proteins using a porous and tentacle ion exchange membrane. *Journal of Chromatography A*, 689, 211-218.

Van Reis, R., Zydney, A., 2001. Membrane separations in biotechnology. *Current Opinion in Biotechnology*, 12, 208-211.

Van Reis, R., Zydney, A., 2007. Bioprocess membrane technology. *Journal of Membrane Science*, 297, 16-50.

Vogel, J.H., Kroner, K.H., 1999. Controlled shear filtration: A novel technique for animal cell separation. *Biotechnology and Bioengineering*, 63, 663-674.

Wang, Y.-K., Sheng, G.-P., Li, W.-W., Huang, Y.-X., Yu, Y.-Y., Zeng, R.J., Yu, H.-Q., 2011. Development of a Novel Bioelectrochemical Membrane Reactor for Wastewater Treatment. *Environmental Science and Technology*, 45, 9256-9261.

Wickramasinghe, S.R., 2008. Tangential-Flow Filtration for Virus Capture Advanced Membrane Technology and Applications. John Wiley and Sons, Inc., pp. 541-555.

Xin, J.Y., Li, S.B., Xu, Y., Chui, J.R., Xia, C.G., 2001. Dynamic enzymatic resolution of Naproxen methyl ester in a membrane bioreactor. *Journal of Chemical Technology and Biotechnology*, 76, 579-585.

Yamagiwa, K., Ohkawa, A., Hirasa, O., 1994. Simultaneous Organic Carbon Removal and Nitrification by Biofilm Formed on Oxygen Enrichment Membrane. *Journal of Chemical Engineering of Japan*, 27, 638-643.

Yamagiwa, K., Yoshida, M., Ito, A., Ohkawa, A., 1998. A new oxygen supply method for simultaneous organic carbon removal and nitrification by a one-stage biofilm process. *Water Science and Technology*, 37, 117-124.

Yang, F., Weber, T.W., Gainer, J.L., Carta, G., 1997. Synthesis of lovastatin with immobilized Candida rugosa lipase in organic solvents: Effects of reaction conditions on initial rates. *Biotechnology and Bioengineering*, 56, 671-680.

Yasuda, H., Lamaze, C.E., 1972. Transfer of gas to dissolved oxygen in water via porous and nonporous polymer membranes. *Journal of Applied Polymer Science*, 16, 595-601.

In: Membrane Processes for Sustainable Growth
Editors: Angelo Basile and Alfredo Cassano
ISBN: 978-1-62618-446-6

Chapter 8

PRESSURE-DRIVEN MEMBRANE APPLICATIONS: NANOFILTRATION

Bart Van der Bruggen[1,*] and Yang Zhang[1]

[1]KU Leuven, Department of Chemical Engineering, Laboratory of Applied Physical Chemistry and Environmental Technology, Leuven, Belgium

ABSTRACT

This chapter describes the fundamentals of nanofiltration as a pressure driven membrane separation process to be used as a sustainable method for application in water production and recycling of wastewater for secondary use. Firstly, methods for characterization of nanofiltration membranes are briefly described. These include both direct and indirect measurements. Direct observation is often applied, but operates in fact at resolution limits. Therefore, indirect measurements that relate the separation performance to, e.g., pore sizes are also important. Based on the separation principles in nanofiltration, models for fluxes in aqueous nanofiltration are further explained. These are often phenomenological; the Hagen-Poiseuille equations are the key in understanding these models. For predicting the retention, more complex approaches are needed. Some of these that can be used for uncharged solutes are explained.

One of the main remaining challenges in nanofiltration, membrane fouling, is introduced. Finally, an overview is given of application areas for nanofiltration, which includes drinking water applications, water recycling and wastewater treatment in various industries such as the textile industry, the dairy industry, electroplating and others.

***Keywords**:* Membrane filtration; nanofiltration; pressure-driven membranes

1. INTRODUCTION

Nanofiltration (NF) is a pressure-driven membrane separation process of which the first applications started to be used on large scale in the 1990's. It is a separation process where

* E-mail: bart.vanderbruggen@cit.kuleuven.be.

low molecule weight organics and multivalent ions are retained by a membrane. Applications are found in the production of drinking water (softening, removal of NOM and color) and in industrial water treatment [1]. Pressure differences used for nanofiltration are about 5-20 bar [2]. Typical nanofiltration membranes are made of several polymer layers, with a thin selective layer (thickness to 1 μm) on top of a thicker nonselective support and usually one or more intermediate layers. Membranes are made of many different materials, mostly from polymers such as aromatic polyamides, polysulfone, polyethersulfone and substituted poly(vinyl alcohols), poly(acrylonitrile), poly(phenylene oxide) as well as from different modifications of these polymers. NF membranes can also be made of inorganic materials such as alumina, titania, hematite, and/or silica on alumina or of mixtures of organic and inorganic materials such as zirconia and polyphosphazene [3].

The properties of NF membranes lie between ultrafiltration (UF) and reverse osmosis (RO) membranes. They retain molecules with molar mass of about 200-1000; that corresponds to a pore diameter of about 0.7-1.3 nm. The surface of nanofiltration membrane is usually charged due to the presence of ionized functional groups. This leads to a high retention for multivalent ions. However, compared to RO, NF membranes have relatively low retentions for monovalent ions. This allows to clearly distinguish NF and RO membranes. NF membranes were developed originally as 'low pressure reverse osmosis' membranes, intended to have a larger salt passage and a lower energy consumption than regular RO membranes [4].

NF membranes have been on the market for more than 20 years, and have been applied industrially for almost 20 years. Today, NF has become a well-known and trusted purification method for aqueous separations; solvent nanofiltration is a younger discipline and although the scientific interest in solvent separation is booming, industrial applications are appearing at a slower rate [5].

NF has been applied for removal of hazardous substances from drinking water [6], metal recovery from effluents [7], treatment of wastewater from the textile industry [8], brewery industry [9], pulp and paper industry [10], and purification in the pharmaceutical [11], food and biotechnological industries [12].

2. Characterization of Nanofiltration Membranes by Observation and Modeling

Characterization of the membrane structure, such as pore radius, pore density, pore shape, pore length, tortuosity and morphology, is very important in view of understanding the process; therefore, characterization methods are of crucial importance for nanofiltration. Several characterization methods can be applied, both based on direct instrument observation and experimental methods. Some of these are specific for nanofiltration [13], but others can be used for a wider range of membrane structures.

Microscopic observation is the most direct method to characterize the membrane pore structure. Atomic force microscopy (AFM) and Scanning electron microscopy (SEM) have been applied routinely for the membrane observation [14]. Secondly, bubble point pressure measurements, typically used for porous membranes to measure the pore size distribution, are sometimes applied for integrity testing [15]. The third method is thermoporometry. The temperature of liquid solidification and/or solid melting is lower in smaller pores and thus by measuring the freezing and/or the melting thermodiagram, the pore size and its distribution can be determined in wet environment [16,17]. Of course, methods for determination of the membrane chemistry such as FT-IR and XPS are also widely used.

Although such direct methods are often applied, they are in fact at the very limits of their applicability, or over these limits. For nanofiltration membranes with relatively large pore size, they can possibly be applied, and it might even be possible to visualize pore openings. For membranes with smaller pore sizes, however, the concept of a 'pore' becomes rather vague since the dimensions of such assumed pores go down to the order of magnitude of the free volume of the polymer. Direct observation methods therefore do not work well anymore.

An indirect method for characterization of membrane properties is through the interpretation of molecular transport through a nanofiltration membrane, which is closely related to the separation performance of the membrane. If the relationship between the flux and rejection and the membrane structure is known, the membrane structure (thickness, tortuosity, pore size, pore density, etc.) can be characterized. Of course, models that can interpret experimental flux and rejection and translate these into information on the membrane pore structure are necessary for the characterization of the membrane.

In the past several decades three major approaches have been studied by many membranologists for describing transport phenomena through porous membranes, and each theory has led to its own way for description and modeling of the transport of solute molecules. The first approach for analysis is based on irreversible thermodynamics derived by Kedem and Katchalsky [18] and Spiegler and Kedem [19]. The second approach is the Stefan-Maxwell multicomponent diffusion equations, which has been introduced into analysis of membrane transport by Peppas and Meadows [20] and Robertson and Zydney [21]. The last approach is called the hydrodynamic model or pore model. It started from the pioneering work done by Ferry in 1936 [22].

The former two approaches treat the membrane as a black box, and thus can be applicable to both porous and non-porous membranes. The equations are derived phenomenologically, and they relate inputs and outputs of the membrane and involve the membrane transport properties. The latter is derived from the fundamental hydrodynamic equation for the trans-capillary transport of rigid spheres.

The extended Nernst–Planck model has been used to characterize membranes in terms of both structural and electrical parameters [23]. Derived models have developed this approach to include both steric and hindered transport within the NF pores, often using the concept of 'effective' pores [24]. Studies of NF membranes using atomic force microscopy (AFM) [25] and nitrogen adsorption–desorption [26] have, however, shown a significant distribution of pore sizes. Cooper and van Derveer investigated the distribution of pores at polysulfone membranes by measuring dextran rejection as a function of molecular weight and found a linear dependency when plotted on log-probability paper, suggesting a log-normal distribution of pores [27].

Mochizuki and Zydney have reviewed the geometric standard deviation (GSD) values that have been reported for many different types of membranes where values ranged considerably from 1.2 to 2.9 depending on both the membrane material and the molecular weight cut-off the membrane [28]. In contrast, Leypoldt predicted sieving characteristics from measured pore size distributions and concluded that it was not possible to obtain actual distributions from experimental data of sieving coefficient as a function of molecular weight because sieving characteristics were not uniquely dependent on the assumed pore size distribution [29]. Aimar et al. [30] proposed a method for obtaining the log-normal pore size distribution of UF membranes based on the normalization of the curves of sieving coefficients as a function of molecular weight with an experimentally measured solute rejection.

The theoretical effects of pore size distributions on uncharged solute transport were studied by Mochizuki and Zydney, who have attempted to quantify solute rejection and flux using log-normal and Gaussian distributions [28]. Saksena and Zydney continued this work to investigate pore size effects in electrokinetic quantities such as zeta potential and electro-osmotic flow [31].

Van der Bruggen et al. [32] compared the steric hindrance pore model, the model of Zeman and Wales, the log-normal model and an adapted version of the log-normal model by the retention data of a broad range of small organic molecules, and concluded that the log-normal model is the most useful model to predict reflection coefficients. Based on experimental data for retentions of uncharged molecules, Van der Bruggen and Vandecasteele [33] pointed out that the modeling with molecular weight as a size parameter is nearly as valuable as the modeling with the effective diameter as a size parameter by using the log-normal model.

3. Separation and Transport Mechanisms

Nanofiltration combines removal of uncharged components on nanoscale with charge effects between solution and the membrane. The removal of uncharged components may be a result from size exclusion, as known from ultrafiltration, or may be a result from differences in diffusion rates in a non-porous structure, which depend also on molecular size. According to the Stokes–Einstein law, expressing an inverse proportionality between the diffusion constant and the size of a component, the diffusion rate will be smaller for a larger component, resulting in an effect similar to size exclusion. The charge effect, on the other hand, results in removal of (mainly multivalent) ions, the former effect results in the removal of uncharged organic species.

It is usually accepted that the rejection of uncharged (organic) molecules is determined by the size of the dissolved molecules compared to the size of the membrane pores [34,35]. Other physicochemical effects such as dipole interactions may also play a role.

All models to describe the rejection of organic molecules that have been proposed are based on the sieving mechanism, and neglect other interactions. These models make use of a parameter representing the size of the molecule (or a related parameter such as the diffusion coefficient), and a method to account for pore size distribution or steric hindrance. Rejections can be predicted, but the accuracy can be low when components are used that interact strongly with the membrane or cause fouling. In contrast, rejection of ionic components in NF is obtained in a totally different way: ions are rejected as a result of charge interactions between the membrane surface and the ions (Donnan exclusion). The divalent ions (hardness, sulphates) are more efficiently removed. For tight NF membranes, size exclusion can provide an additional ion rejection.

The NaCl rejection of membranes decreases with nanofiltration increasing salt concentration, which is a typical phenomenon if electrostatic interactions are involved in the rejection mechanisms. The distribution between the bulk on the feed side and the pore entrance is calculated using the Donnan distribution, the transport in the pore is described with the extended Nernst-Planck equation and the Donnan distribution is again applied for the distribution at the permeate side.

When there is no osmotic pressure difference across the membrane ($\Delta\pi = 0$), the transmembrane flow occurs because of the pressure difference (ΔP). This can be described as:

$$(J_V)_{\Delta\pi=0} = L_1 \cdot \Delta P$$

Or

$$L_1 = \left(\frac{J_V}{\Delta P}\right)_{\Delta\pi=0}$$

L_1 is called the hydrodynamic permeability or water permeability of the membrane and is often referred to as L_P. When there is no hydrodynamic pressure difference across the membrane ($\Delta P = 0$), there is a transmembrane flow because of the osmotic pressure difference:

$$(J_d)_{\Delta P=0} = L_2 \cdot \Delta\pi$$

Or

$$L_2 = \left(\frac{J_d}{\Delta\pi}\right)_{\Delta P=0}$$

L_2 is called the osmotic permeability or solute permeability and is often referred to as ω.

The reflection coefficient, σ, can be derived from steady-state permeation measurements. When no volume flux occurs ($J_V = 0$) under steady state conditions then:

$$L_1 \cdot \Delta P + L_2 \cdot \Delta\pi = 0$$

Or

$$(\Delta P)_{J_V=0} = -\frac{L_2}{L_1}\Delta\pi$$

In the steady state, when the osmotic pressure difference equal to the hydrodynamic pressure difference, there is no solute transport across the membrane (L_1=L_2).

The ratio L_2/L_1 is denoted as the reflection coefficient σ. It represents the diffusive transport of a given solute by the nanofiltration membrane; in practice, this is what can be observed as the retention at a very high (in principle infinite) pressure.

The reflection coefficient σ is a measure of the selectivity of a membrane and has a value between 0 and 1.

$\sigma = 1 \Rightarrow$ ideal membrane, no solute transport;

$\sigma < 1 \Rightarrow$ not a completely semipermeable membrane: solute transport;

$\sigma = 0 \Rightarrow$ no selectivity.

The volume flux (J_V) can be described as:

$$J_V = L_P(\Delta P - \sigma\Delta\pi)$$

For the solute flux (J_S), one should consider both diffusive and convective effects:

$$J_S = \overline{C_S}(1-\sigma)J_V + \omega\Delta\pi$$

The solute transport through the membrane can thus be described by three parameters: water (hydrodynamic) permeability L_P, solute permeability ω, and the reflection coefficient σ.

If the solute is not completely retained by the membrane, the osmotic pressure difference is not $\Delta\pi$ but σ.Δπ, since only the retained fraction should be taken into account.

When measuring the pure water flux ($\Delta\pi = 0$) at different operation pressures, the volume flux as a function of the transmembrane pressure is obtained, as shown in Figure 1.

A higher L_P might indicate that the membrane has a more loose structure, although the hydrophilicity of the membrane plays a role for the water flux. A more hydrophilic membrane also yields a higher flux; thus, the membrane morphology and hydrophilicity should always be evaluated together.

The following equation can be derived from the equations above:

$$\frac{J_S}{\Delta c} = \omega + (1-\sigma)J_V \frac{\bar{c}}{\Delta c}$$

where Δc is the concentration difference between the feed and the permeate and $\bar{c}$ is the mean logarithmic concentration. $\bar{c}$ can be defined as:

$$\bar{c} = (c_f - c_p)/\ln(c_f/c_p)$$

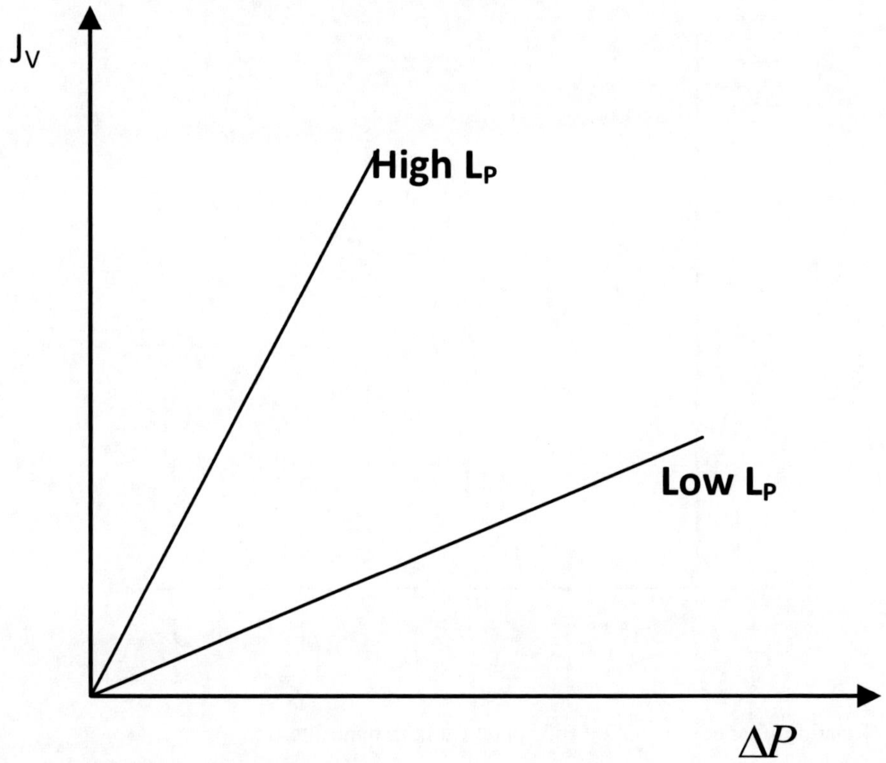

Figure 1. Schematic plot of the volumetric flux through a nanofiltration membrane as a function of the transmembrane pressure.

Thus, the relationship of the parameters can be depicted as in Figure 2.

A representation of the mass transfer process occurring in NF is shown in Figure 3.

When an external pressure ΔP is imposed on a liquid which is adjacent to a semi-permeable membrane, solvent will flow through the membrane. The general terms that are used in the description of membrane separation processes are the solvent flux (J) and the rejection (R). The solvent flux is given by:

$$J = \frac{\Delta P}{\eta R_{tot}}$$

in which ΔP is the effective transmembrane pressure [N/m^2], η the permeate viscosity [Pa.s] and R_{tot} the total resistance towards solvent flow [m^{-1}].

A neutral solute dissolved in the solvent at a concentration C_b will also flow towards the membrane. If the membrane rejects the solute, partial permeation will occur and non-permeated solute accumulates in the boundary layer, and hence a concentration profile develops in equilibrium with back diffusion to the bulk solution. This phenomenon is known as concentration polarization.

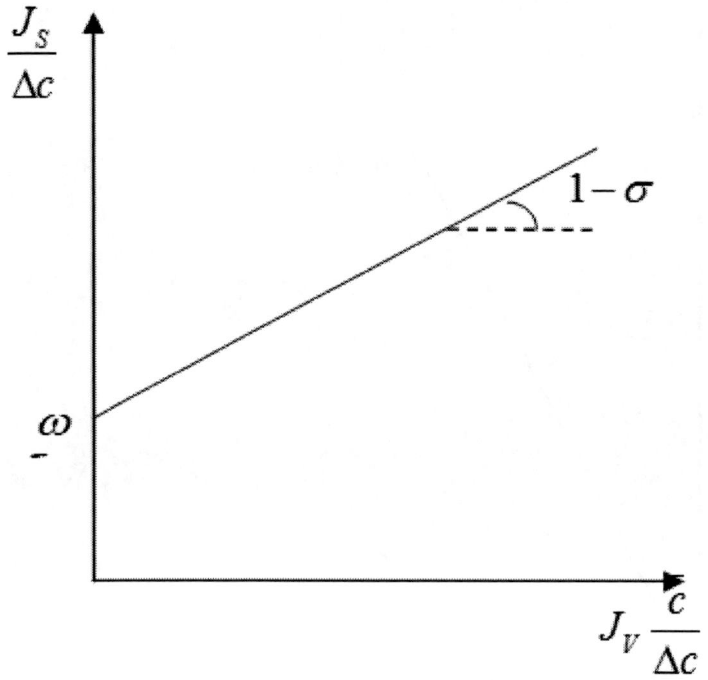

Figure 2. Relationship between solute and solvent flux in nanofiltration.

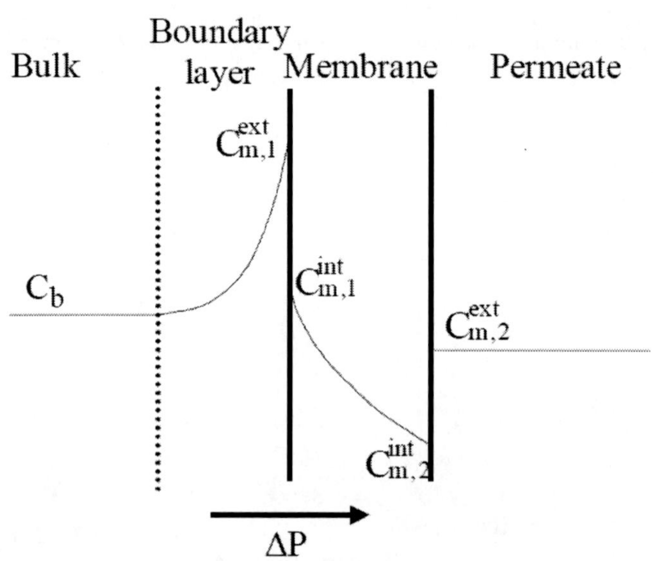

Figure 3. Mass transfer in nanofiltration.

The solute distributes at the membrane/solution interface and will be transported through the membrane by convection and diffusion. At the permeate side, a second distribution process will occur and a final concentration of solute in the permeate, $C_{m,2}^{ext}$, will be reached. For the characterization of solute behavior the retention is calculated as:

$$R = 1 - \frac{C_{m,2}^{ext}}{C_b}$$

In nanofiltration, the distribution of a non-charged solute at the boundary layer/membrane interface is considered to be determined by a steric exclusion mechanism. Steric exclusion is not typical for nanofiltration but applies to ultrafiltration and microfiltration as well. Due to its size a solute only has access to a fraction of the total surface area of a pore. This causes a geometrical exclusion of the solute from the membrane. A separation between solutes will only be accomplished when the solutes have a difference in size.

4. Models for Transport and Retention

If charge effects were not present, the equilibrium concentrations of all components would be the same inside the pores as outside provided that the ions are smaller than the pores of the membrane. However, in the case of cation exchange, the stationary phase has a large number of negatively charged groups (R) which tend to attract counterions (and repel co-ions (X)). Thus, there is a tendency for positive ions to be pulled into the stationary phase pores and for negative ions to be repelled from them. This effect is called Donnan exclusion.

Donnan exclusion has a pronounced effect on the separation in NF, compared to other pressure driven membrane processes. Due to the slightly charged nature of the membrane, solutes with an opposite charge compared to the membrane (counter-ions) are attracted, while solutes with a similar charge (co-ions) are repelled. At the membrane surface a distribution of co- and counter-ions will occur, thereby causing an additional separation.

However, electroneutrality is maintained in the solution, thus, anions permeate together with cations. The result is that typical retention sequences are obtained for mixtures containing different mono- and multivalent ions.

An interpretation of the transport mechanisms through a nanofiltration membrane is necessary for the description of the retention of uncharged molecules. Transport of uncharged molecules is a combination of diffusion and convection. This is expressed in the transport equations of Spiegler and Kedem [19] for water flux and for the flux of a dissolved component, from which the retention of a given molecule can be calculated as:

$$R = \frac{\sigma(1-F)}{1-\sigma F}$$

$$F = \exp(-\frac{1-\sigma}{P} J_V)$$

where R is the retention; J_v is the water flux(l/h m^2); P is the solute permeability(l/h m^2) and σ is the reflection coefficient. The solute retention R is given as a function of the water flux J_v and the solute permeability P.

The solute permeability P (not to be confused with the pressure) is a measure of the transport of a molecule by diffusion. The reflection coefficient σ of a given component is the maximum possible retention for that component, as explained above. The reflection coefficient can be determined by the ratio of solute radius to pore radius. The reflection coefficient can be derived from either experimentally as the retention at an infinite water flux (extrapolation to high pressures). The resulting curve for the reflection coefficient as a function of the molecular diameter (retention curve) can be used to estimate the maximal retention that can be obtained by a given membrane.

The Spiegler-Kedem equations indicate that the retention increases with increasing water flux and reaches a limiting value σ at an infinitely high water flux.

By taking only sterical effects into account and an idealized membrane structure, the Steric Hindrance Pore (SHP) model can be derived [36]. The SHP model can be used to estimate the membrane pore radius: for a solute with known radius the reflection coefficient is determined and the pore radius can be calculated with the following equation:

$$\sigma = 1 - H_F S_F$$

with:

$$H_F = 1 + (16/9)\eta^2$$

$$S_F = (1-\eta)^2\left[2-(1-\eta)^2\right]$$

$$\eta = r_s / r_p$$

in which H_F is a "wall-correction parameter" that represents the effect of the pore wall, S_F is a parameter that represents steric hindrance during transport through the pore. The solute radius and the pore radius are symbolized by r_s and r_p respectively.

This eventually leads to the following equation:

$$\sigma = 1 - \left(1 + \frac{16r_s^2}{9r_p^2}\right)\left(1 - \frac{r_s}{r_p}\right)^2\left[2 - \left(1 - \frac{r_s}{r_p}\right)^2\right]$$

In the Steric Hindrance Pore Model, the reflection coefficient is calculated from the pore size of the membrane and the diameter of the molecule. It is assumed that all the pores have the same size. Therefore, the uniform pore size should not be interpreted as a real value for the diameter of the pores. The calculated pore size corresponds with the pore size of an imaginary membrane with uniform pores, for which the retention of uncharged molecules is equal to retention with the real membrane. In reality, not every pore has the same cylindrical diameter; the model is an approximation of the membrane's structure.

The membrane is thus represented as a bundle of cylindrical pores through which molecule in solution can permeate. During the transport these molecules encounter steric hindrance and interactions with the pore wall. A molecule which is smaller than the diameter of the membrane is partially retained through these effects. A molecule with the same size as the pore diameter is completely retained. For the calculation of the reflection coefficient as a function of the effective molecular diameter, an estimation of log-normal distribution for the pore size can be used. In the log-normal model, no steric hindrance in the pores or hydrodynamic lag is taken into account, and the value of σ (the reflection coefficient) corresponds to the fraction of membrane pore that are smaller than the molecules in solution. The equation that calculates the reflection coefficient with a molar radius r* is:

$$\sigma(r) = \int_0^r \frac{1}{S_p\sqrt{2\pi}}\frac{1}{r}\exp\left(-\frac{\left[\ln(r)-\ln(\bar{r})\right]^2}{2S_p^2}\right)dr$$

This equation involves two variables, S_p and $\bar{r}$, where S_p is the standard deviation on the pore size distribution. This standard deviation is measure for the distribution of pore size. As the retention curve corresponds to an integrated log-normal distribution, a small "S_p" represents a large slope of the retention and the large "S_p" represents a small slope. $\bar{r}$ is the size of molecule that is 50% retention, i.e., an 'averaged' membrane pore size.

Although the molecular weight is not a direct measure of the dimensions of a molecule, it still reflects the molecular size, and it is a readily accessible parameter, whereas complicated calculations are necessary to obtain the effective diameter. However, the log-normal model can be adapted by taking the correlation between molecular weight and the diameter of the molecule into account.

This relation was already derived for the Stokes diameter and was determined for the effective diameter here. The equation for the correlation was found as d_S=A(MW)^B, where A=0.065 and B=0.438. This correlation is valid for molar masses up to7600 Da, which includes the range where nanofiltration typically operates; the correlation is similar to the equation found for the Stokes diameter. The equations of the log-normal model can be written as:

$$\sigma(MW) = \int_0^{MW} \frac{1}{S_{MW}\sqrt{2\pi}}\frac{1}{MW}\exp\left(-\frac{\left(\ln(MW)-\ln(\overline{MW})\right)^2}{2S_{MW}^{\ 2}}\right)dMW$$

In the equation, $\sigma(MW)$ is the reflection coefficient of a molecule to the membrane, S_{MW} is the standard deviation which is proportional parameter here, $\overline{MW}$ is the average molar mass where the retention is 90%. An example of a model fitting is shown in Figure 4 [33].

5. The Role of Membrane Fouling in Nanofiltration

Fouling is a paramount problem in membrane applications. Many studies have shown that membrane characteristics such as roughness, internal structure, hydrophobicity and zeta potential influence the volumetric flux and the solute rejection, as well as membrane fouling (i.e., the change in flux and potentially retention as a function of time, or flow). However, the fundamental relationship between membrane characteristics, performance and solution parameters is still not well understood. The resistance model is commonly used to describe phenomena of flux decline. For nanofiltration, the water flux is written as:

$$J = \frac{\Delta P}{\eta R_{tot}}$$

where ΔP is the pressure difference over the membrane used as the driving force; η is the viscosity; and R_{tot} is the total resistance. The flux decline that was found in the experiments should be explained by an increase in the total resistance against mass transport. A more detailed model to describe the water flux is the Hagen-Poiseuille equation:

$$J = \frac{\varepsilon r^2}{8\eta\tau}\frac{\Delta P}{\Delta x}$$

here, the membrane resistance depends on the porosity (ε), the tortuosity (τ), the pore radius (r), and the membrane thickness (Δx). The Hagen-Poiseuille equation is valid when pure water is applied to the membrane. When solutions of organic molecules in water are applied, the water flux will often be lower. Different mechanisms of flux decline can be distinguished [37,38]. Adsorption inside the pores or at the membrane surface narrows the pores. When the molecules have a similar size as the pores, permeation can lead to pore blocking, a phenomenon that can be enhanced or caused by adsorption. Pore blocking has been observed for ultrafiltration, where macromolecules are filtered. For the filtration of non-macromolecular components with nanofiltration, this phenomenon has not yet been described.

The total resistance is the sum of different individual resistances, i.e., $R_{tot} = R_p + R_a + R_m + R_g + R_{cp} + R_i + R_d$ (R_p, resistance due to pore blocking; R_a, resistance due to adsorption inside the pores; R_m, membrane resistance (intrinsic); R_g, resistance caused by the formation of a gel layer; R_{cp}, concentration polarization resistance; R_i, resistance caused by specific interactions; R_d, resistance from deposits on the membrane). In the ideal case, e.g., filtration of pure water, the membrane resistance (R_m) is the only resistance involved. This is an intrinsic membrane characteristic that corresponds to the resistance calculated from, for example, the Hagen-Poiseuille equation and does not change during filtration or by changing the feed solution. It reflects the minimal resistance of the system against mass transport and thus determines the maximal water flux at a given pressure. The other phenomena can only make pores narrower (or the membrane thicker), resulting in an increase of the total resistance or the addition of an extra resistance term to the intrinsic membrane resistance.

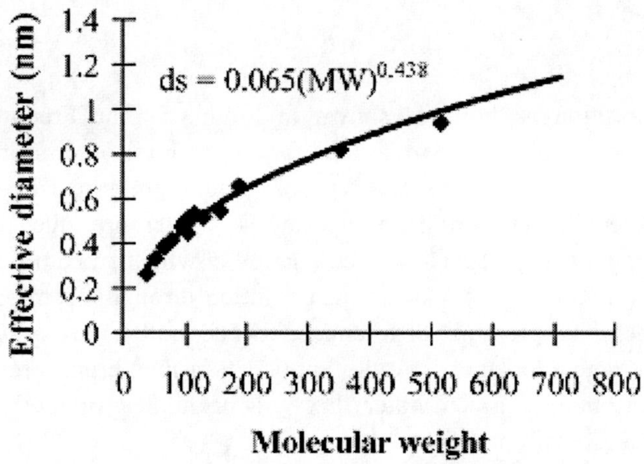

Figure 4. Correlation between molecular weight and effective diameter [36].

The gel layer resistance, the adsorption resistance, the pore blocking resistance, the deposition resistance, and the concentration polarization resistance depend strongly on the type of feed solution that is used. The gel layer resistance is often not considered, as the formation of a gel layer is related to macromolecules. Concentration polarization is not to be considered as a fouling effect, because it is reversible and only depends on the flow conditions. A high cross-flow velocity, for example, minimizes the effect of concentration polarization. The only factors that remain are the membrane resistance, pore blocking, and adsorption.

For uncharged organic compounds, adsorption is the process that is most likely to occur. Molecules can get attached to the membrane pores or to the membrane surface by adsorption or chemisorption. Inside the pores, they narrow the free pathway for the water flow, hence decreasing the net pore opening.

From the Hagen-Poiseuille equation, it can be seen that this should lead to a flux decline. When adsorption has a strong effect, it could even lead to pore blocking when the whole cross section of the pore is filled.

The Freundlich equation is typically employed to describe pore blocking and adsorption inside the membrane pores. The remaining flux decline can be explained by adsorption inside the membrane pores or at the membrane surface, possibly enhanced by pore blocking.

The Freundlich equation is:

$$q = K_f c^n$$

where c is the concentration of the component to be adsorbed at equilibrium and q is the amount of the component that is adsorbed on the material, divided by the amount of material. K_f and n are empirical constants. If it is assumed that adsorption and flux decline are proportional, q in the Freundlich equation can be replaced by the flux decline ΔJ :

$$\Delta J = K_f c^n$$

A typical adsorption isotherm is shown in Figure 5. The Freundlich isotherm is an approximation, and more complex expressions for adsorption isotherms can be found.

Closely related to flux decline is the effect of osmotic pressure. The retention of ions and small organic molecules in nanofiltration causes a pressure due to the concentration difference. This is a naturally occurring phenomenon in which an equilibrium of the chemical potential is attained by diluting a concentrated solution through a semipermeable membrane, causing a corresponding pressure difference between both sides of the membrane. This pressure has to be counterbalanced by the applied transmembrane pressure. Therefore, the pressure needed to obtain a given water flux will be higher, or the water flux at a given transmembrane pressure will be lower.

Thus, the osmotic pressure causes a lower flux, but this is due to a decrease of the driving force instead of an increase of the resistance against mass transport. This can be expressed by the phenomenologic equation for the water flux discussed above, in which the osmotic pressure is subtracted from the applied driving force.

The extent to which the osmotic pressure will play a role is determined by the retention of the components in the solution, their concentration, and their molar mass.

6. APPLICATIONS

Today, nanofiltration is widely applied in a range of environmental applications. The most dominant application is still is drinking water production.

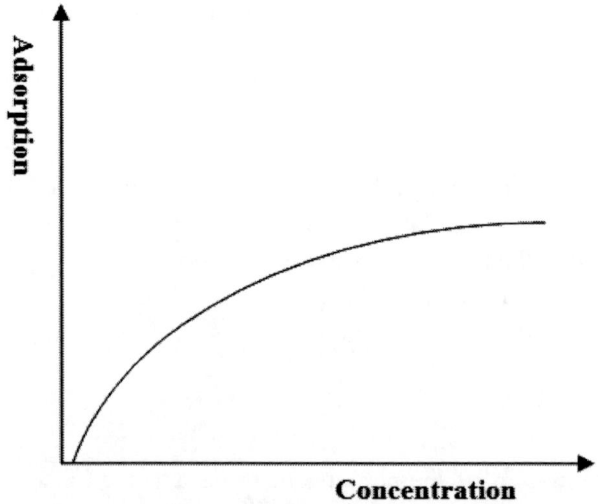

Figure 5. Typical adsorption isotherm of the Freundlich type.

Van der Bruggen and Vandecasteele [39] reviewed drinking water applications of nanofiltration, and described a large number of applications ranging from softening and removal of natural organic matter to specific applications such as uranium removal and partial removal of nitrates. Removal of micropollutants is still an important issue today, and new membranes and applications are reported. For example, Ghaemi et al. [40]studied the synthesis of new cellulose acetate membranes with the addition of a surfactant to obtain improved retentions for p-nitrophenol and 3,5-dinitrosalicylic acid in drinking water production. Plakas and Karabelas [41] describe the general challenge of pesticides in drinking water sources and removal by membrane technologies, with a focus on pilot and industrial-scale application of nanofiltration. The nanofiltration-based approach for pesticides remediation has been studied and applied for almost two decades [42], and is among the most reported applications on larger scale [43].

Nowadays, the attention has shifted from pesticides to other micropollutants, in particular endocrine disrupting compounds and pharmaceuticals but also chemical compounds such as plasticizers that can be found in drinking water sources, and personal care products. Bodzek et al. [44] reviewed the use of membrane processes for removal of a wide range of micropollutants, with attention for pharmaceutical active compounds and endocrine disrupting compounds, but also inorganic micropollutants such as nitrates, fluoride, boron, arsenic, chromium and heavy metals. Other applications for drinking water production considered on real scale are removal of color and pathogens [45], and treatment of brackish water [46], although it was found that the removal of salts was only moderate, between 44 and 66% for brackish groundwaters. Removal of arsenic is another important challenge for which nanofiltration has been studied extensively, now on the level of pilot plants. Saitua et al. [47] observed an arsenate rejection over 95% from naturally contaminated groundwater in a pilot plant.

Nanofiltration is also one of the prominent processes proposed for water recycling applications. Bellona et al. [48] compared ultralow-pressure reverse osmosis (ULPRO) and NF membranes for water reuse applications where lower operating pressures and a high permeate quality are desired, and they demonstrated that both ULPRO and NF membranes could meet potable water quality requirements in terms of total organic carbon, total nitrogen, and regulated and unregulated organic micropollutants. For water recycling applications, however, the risk of membrane fouling is higher and this limits the water yield because the fraction of the retentate has to be kept relatively high. To this end, Pervov and Andrianov [49] propose the elimination of the spacer mesh from the feed channels, which eliminates dead regions that provide scaling and fouling conditions and also reduces the risk of particle "trapping" and associated dramatic pressure increase. They claimed that this might lead to a nearly zero discharge of effluent, with a loss of no more than 1% [49]. Van Geluwe et al. [50] had the same objective, but rather consider the risk of scaling of sparingly soluble salts, such as $CaSO_4$ or $CaCO_3$, onto the membrane surface; they proposed to use electrodialysis as a method for scaling prevention, which results in a similarly high water yield.

Wastewater applications for nanofiltration have been studied extensively as well. Among many different kinds of wastewater, textile effluents emerge as the most studied waters; this is probably related to the fact that these contain very high concentrations of both organic and inorganic pollutants, and have been shown to be difficult to treat by any other method. Ellouze et al. [51] indicated some of the limitations; for example, high salt concentrations are difficult, and the volumetric reduction factor was limited to 3. Several other studies can be found on operational conditions of textile wastewater treatment by nanofiltration [52-55]. Vergili et al. [56] study the zero liquid discharge (ZLD) approach in treating textile dye bath wastewater via integrated membrane processes, and concluded that ZLD is feasible in processing textile dye bath waste and recycling treated wastewater.

Van der Bruggen et al. [57] considered this in a more visionary way, with nanofiltration used in treatment of textile effluents as part of an integrated separation system based on membranes. Several aspects of this idea come back in the study of Vergili et al. [56], which proves the feasilibity in principle.

Apart from textile wastewaters, a wide range of other applications has been suggested for nanofiltration. Urgun-Demirtas et al. [58] studied the use of nanofiltration for mercury removal in oil refinery's wastewater. Dairy wastewater is a typical application, which is still of interest today [59].

Many other industries can be found in studies, including cork processing [60], electroplating wastewaters [61], and the pulp and paper industry [62]. For a more extensive overview, the reader is referred to Schaefer et al. [63].

Conclusion

Nanofiltration is a versatile separation and purification process that was originally developed as a process for the production of potable water, but has emerged as a sustainable process for environmental remediation, in view of water recycling and process water production from unreliable water sources, including wastewater. It is scientifically well understood on the level of process modeling and estimation of fluxes (permeabilities) and retentions, although it is still difficult to develop reliable predictive models for the retention of specific compounds, particularly of ions.

Challenges in nanofiltration are related to further implementation issues, such as the occurrence of membrane fouling and the study of more advanced applications requiring fractionation or purification to very high purity such as in the removal of micropollutants in drinking water production. This also includes the development of reliable simulation tools for prediction of the performance of a wide range of nanofiltration membranes.

List of Symbols

J, J_V	Water flux (l/m^2h)
L_1, L_p	hydrodynamic permeability (water permeability) ($l/m^2h.bar$)
L_2, ω	osmotic permeability (solute permeability) ($mol/m^2h.bar$)
ΔP	pressure difference (bar)
J_d	diffusive flux (mol/m^2h)
J_s	solute flux (mol/m^2h)
$\Delta\pi$	osmotic pressure difference (bar)
σ	reflection coefficient (-)
$\overline{C_S}$	average concentration (mol/l)
η	viscosity (Pa.s)
R_{tot}	Total resistance for permeation ($1/m^3s$)

R	Retention/rejection (-)
P	Permeability (mol/m²h)
S_p	standard deviation on the pore size distribution (m)
K_f, n	empirical Langmuir constants (-)
ε	porosity (-)
τ	tortuosity (-)
r	pore radius (m)
Δx	membrane thickness (m)
MW	Molar mass (-)
$\overline{MW}$	Average molar mass where the retention is 90% (-)
r_p	Pore size (m)
r_s	Solute size (m)
$\bar{r}$	Average pore size (m)

REFERENCES

[1] Van der Bruggen, B.; Vandecasteele, C. Removal of pollutants from surface water and groundwater by nanofiltration: overview of possible applications in the drinking water industry. *Environ. Pollut.* 2003, 122 (3), 435-445.

[2] Van der Bruggen, B.; Vandecasteele, C.; Van Gestel, T.; Doyen, W.; Leysen, R. Pressure driven membrane processes in process and waste water treatment and in drinking water production. *Environ. Progr.* 2003, 22 (1), 46-56.

[3] Tanninen, J.; Nyström, M. Separation of ions in acidic conditions using NF. *Desalination*. 2002, 147, 295-299.

[4] Eriksson, P. Nanofiltration extends the range of membrane filtration. *Environm. Prog.* 1988, 7 (1), 58-62.

[5] Volkov, A.V.; Korneeva, G.A.; Tereshchenko, G.F. Organic solvent nanofiltration: prospects and application. *Russ. Chem. Rev.* 2010, 77 (11), 983-993.

[6] Van der Bruggen, B.; Everaert, K.; Wilms, D.; Vandecasteele, C. The use of nanofiltration for the removal of pesticides from ground water: an evaluation. *Water Sci. Technol.: Water Supply 2001,* 1(2), 99-106.

[7] Castelblanque J.; Salimbeni, F. NF and RO membranes for the recovery and reuse of water and concentrated metallic salts from waste water produced in the electroplating process. *Desalination 2004,* 167, 65-73.

[8] Van der Bruggen, B.; De Vreese, I.; Vandecasteele, C. Water Reclamation in the Textile Industry: Nanofiltration of Dye Baths for Wool Dyeing. *Ind. Eng. Chem. Res.* 2001, 40(18), 3973-3978.

[9] Braeken, L.; Van der Bruggen, B.; Vandecasteele, C. Regeneration of brewery waste water using nanofiltration. *Water Res.* 2004, 38 (13), 3075-3082.

[10] Platt, S.; Nyström, M. Recovery of brightness in paper mill water. *Desalination 2004,* 161, 123-136.

[11] Zhu, A.; Zhu, W.; Wu, Z.; Jing, Y. Recovery of clindamycin from fermentation wastewater with nanofiltration membranes. *Water Res.* 2003, 37, 3718-3732.

[12] Fähnrich, A.; Mavrov, V.; Chmiel, H. Membrane processes for water reuse in the food industry. *Desalination 1998*, 119, 213-216.

[13] Boussu, K.; De Baerdemaeker, J.; Dauwe, C.; Depla, D.; Aldea, S.; Vankelecom, I.F.J.; Vandecasteele, C.; Van der Bruggen, B. Physico-chemical characterization of nanofiltration membranes. *Chem. Phys. Chem.* 2007, 8 (3), 370-379.

[14] Johnson, D.J.; Al Malek, S.A.; Al-Rashdi, B.A.M.; Hilal, N. Atomic force microscopy of nanofiltration membranes: Effect of imaging mode and environment. *J. Membr. Sci.* 2012, 389, 486-498.

[15] Minnery, J.G.; Jacangelo, J.G.; Boden, L.I.; Vorhees, D.J.; Heiger-Bernays, B.G. Sensitivity Analysis of the Pressure-Based Direct Integrity Test for Membranes Used in Drinking Water Treatment. *Environ. Sci. Technol.* 2009, 43 (24), 9419-9424.

[16] Nakao, S.I. Determination of Pore Size Distribution 3. Filtration Membranes, *J. Membr. Sci.* 1994, 96, 131-165.

[17] Zhao, C.S.; Zhou, X.S.; Yue, Y.L. Determination of pore size and pore size distribution on the surface of hollow-fiber filtration membranes: a review of methods. *Desalination 2000*, 129 (2), 107-123.

[18] Kedem, O.; Katchalsky, A. Thermodynamic analysis of the permeability of biological membranes to nonelectrolytes. *Biochem. Biophys. Acta* 1958, 27, 229-246.

[19] Spiegler S.; Kedem, O. Thermodynamics of Hyperfiltration (Reverse Osmosis): Criteria for Efficient Membranes. *Desalination 1966*, 1, 311-326.

[20] Peppas, N.A.; Meadows, D.L. Macromolecular structure and solute diffusion in membranes: an overview of recent theories. *J. Membrane. Sci.* 1983, 16, 361-377.

[21] Robertson, B.C.; Zydney, A.L. A Stefan-Maxwell analysis of protein transport in porous membranes. *Sep. Sci. Technol.* 1988, 23, 1799-1811.

[22] Ferry, J.D. Statistical evaluation sieve constants in ultrafiltration. *J. Gen. Physiol.* 1936, 20, 95-104.

[23] Tsuru, T.; Urairi, M.; Nakao, S.; Kimura, S. Reverse osmosis of single and mixed electrolytes with charged membranes: Experiment and analysis. *J. Chem. Eng. Jap.* 1991, 24, 518–524.

[24] Bowen, W.R.; Mohammad, A.W. Diafiltration by nanofiltration: Prediction and optimization. *AIChE J.* 1998, 44 (8), 1799–1812.

[25] Bowen, W.R.; Doneva, T.A. Atomic force microscopy of nanofiltration membranes: surface morphology, pore size distribution and adhesion. *Desalination 2000*, 129, 163–172.

[26] Combe, C.; Guizard, C.; Aimar, P.; Sanchez, V. Experimental determination of four characteristics used to predict the retention of a ceramic nanofiltration membrane. *J. Membr. Sci.* 1997, 129, 147–160.

[27] Cooper, A.R.; van Derveer, D.S. Characterization of ultrafiltration membranes by polymer transport measurements. *Sep. Sci. Technol.* 1979, 14, 551–556.

[28] Mochizuki, S.; Zydney, A. L. Theoretical analysis of pore size distribution effects on membrane transport. *J. Membr. Sci.* 1993, 82, 211–227.

[29] Leypoldt, J.K. Determining pore size distributions of ultrafiltration membranes by solute sieving—mathematical limitations. *J. Membr. Sci.* 1987, 56, 339–354.

[30] Aimar, P.; Meireles, M.; Sanchez, V. A contribution to the translation of retention curves into pore size distributions for sieving membranes. *J. Membr. Sci.* 1990, 54, 321–338.

[31] Saksena, S.; Zydney, A.L. Pore size distribution effects on electrokinetic phenomena in semipermeable membranes. *J. Membr. Sci.* 1995, 105, 203–215.

[32] Van der Bruggen, B.; Schaep, J.; Vandecasteele, C.; Wilms, D. A Comparison of Models to Describe the Maximal Retention of Organic Molecules. *Separ. Sci. Technol.* 2000, 35 (2), 169-182.

[33] Van der Bruggen, B.; Vandecasteele, C. Modelling of the retention of uncharged molecules with nanofiltration. *Water Res.* 2002, 36 (5), 1360-1368.

[34] Kosutic, K.; Kunst, B. Removal of organics from aqueous solutions by commercial RO and NF membranes of characterized porosities. *Desalination 2002*, 142 (1), 47–56.

[35] Van der Bruggen, B.; Schaep, J.; Wilms, D.; Vandecasteele, C. Influence of molecular size, polarity and charge on the retention of organic molecules by nanofiltration. *J. Membr. Sci.* 1999, 156, 29-41.

[36] Wang, X.L.; Shang, W.J.; Wang, D.X.; Wu, L.; Tu, C.H. Characterization and applications of nanofiltration membranes: State of the art. *Desalination 2009*, 236 (1-3), 316-326.

[37] Mulder, M. Basic Principles of Membrane Technology (Second Edition), Kluwer Academic Publishers, Dordrecht, the Netherlands, 1996.

[38] Eykamp, W. In: Membrane Separations Technology, Principles and Applications; Noble, R. D., Stern, S. A., Eds.; Elsevier: Amsterdam, 1995, 1-43.

[39] Van der Bruggen, B.; Vandecasteele, C. Removal of pollutants from surface water and groundwater by nanofiltration: overview of possible applications in the drinking water industry. *Environ. Pollut.* 2003, 122 (3), 435-445.

[40] Ghaemi, N.; Madaeni, S.S.; Alizadeh, A.; Daraei, P.; Vatanpour, V.; Falsafi, M. Fabrication of cellulose acetate/sodium dodecyl sulfate nanofiltration membrane: Characterization and performance in rejection of pesticides. *Desalination 2012*, 290, 99-106.

[41] Plakas, K.V.; Karabelas, A.J. Removal of pesticides from water by NF and RU membranes - A review. *Desalination 2012*, 287, 255-265.

[42] Van der Bruggen, B.; Schaep, J.; Maes, W.; Wilms, D.; Vandecasteele, C. Nano-filtration as a treatment method for the removal of pesticides from ground waters. *Desalination 1998*, 117, 139-147.

[43] Ventresque, C.; Turner, G.: Bablon, G. Nanofiltration: from prototype to full scale. *J. AWWA* 1997, 89 (10), 65-76.

[44] Bodzek, M.; Konieczny, K.; Kwiecinska, A. Application of membrane processes in drinking water treatment-state of art. *Desal. Wat. Treat.* 2011, 35 (1-3), 164-184.

[45] Patterson, C.; Anderson, A.; Sinha, R.; Muhammad, N.; Pearson, D. Nanofiltration Membranes for Removal of Color and Pathogens in Small Public Drinking Water Sources. *J. Environ. Eng. – ASCE* 2012, 138 (1), 48-57.

[46] Galanakis, C.M.; Fountoulis, G.; Gekas, V. Nanofiltration of brackish groundwater by using a polypiperazine membrane. *Desalination 2012,* 286, 277-284.

[47] Saitua, H.; Gil, R.; Padilla, A.P. Experimental investigation on arsenic removal with a nanofiltration pilot plant from naturally contaminated groundwater. *Desalination 2011*, 274 (1-3), 1-6.

[48] Bellona, C.; Drewes, J.E.; Oelker, G.; Luna, J.; Filteau, G.; Amy, G. Comparing nanofiltration and reverse osmosis for drinking water augmentation. *J. AWWA* 2008, 100 (9), 102-+.

[49] Pervov, A.G.; Andrianov, A.P. Application of membranes to treat wastewater for its recycling and reuse: new considerations to reduce fouling and increase recovery up to 99 percent. *Desal. Wat. Treat.* 2011, 35 (1-3), 2-9.

[50] Van Geluwe, S.; Braeken, L.; Robberecht, T.; Jans, M.; Creemers, C.; Van der Bruggen, B. Evaluation of electrodialysis for scaling prevention of nanofiltration membranes at high water recoveries. *Resour. Conserv. Recycl.* 2011, 56 (1), 34-42.

[51] Ellouze, E.; Tahri, N.; Ben Amar, R. Enhancement of textile wastewater treatment process using Nanofiltration. *Desalination 2012,* 286, 16-23.

[52] Cheng, S.Y.; Oatley, D.L.; Williams, P.M.; Wright, C.J. Characterisation and application of a novel positively charged nanofiltration membrane for the treatment of textile industry wastewaters. *Water Res.* 2012, 46 (1), 33-42.

[53] Van der Bruggen, B.; Cornelis, G.; Vandecasteele, C.; I. Devreese, I. Fouling of nanofiltration and ultrafiltration membranes applied for waste water regeneration in the textile industry. *Desalination 2005,* 175, 111-119.

[54] Sojka-Ledakowicz, J.; Zylla, R.; Mrozinska, Z.; Pazdzior, K.; Klepacz-Smolka, A.; Ledakowicz, S. Application of membrane processes in closing of water cycle in a textile dye-house. *Desalination 2010*, 250 (2), 634-638.

[55] Ben Amar, N.; Kechaou, N.; Palmeri, J.; Deratani, A.; Sghaier, A. Comparison of tertiary treatment by nanofiltration and reverse osmosis for water reuse in denim textile industry. *J. Hazard. Mater.* 2009, 170 (1), 111-117.

[56] Vergili, I.; Kaya, Y.; Sen, U.; Gonder, Z.B.; Aydiner, C. Techno-economic analysis of textile dye bath wastewater treatment by integrated membrane processes under the zero liquid discharge approach. *Resour. Conserv. Recycl.* 2012, 58, 25-35.

[57] Van der Bruggen, B.; Curcio, E.; Drioli, E. Process intensification in the textile industry: the role of membrane technology. *J. Environ. Manage.* 2004, 73 (3), 267-274.

[58] Urgun-Demirtas, M.; Benda, P.L.; Gillenwater, P.S.; Negri, M.C.; Xiong, H.; Snyder, S.W. Achieving very low mercury levels in refinery wastewater by membrane filtration. *J. Hazard. Mater.* 2012, 215, 98-107.

[59] Luo, J.Q.; Ding, L.H.; Wan, Y.H.; Paullier, P.; Jaffrin, M.Y. Fouling behavior of dairy wastewater treatment by nanofiltration under shear-enhanced extreme hydraulic conditions. *Sep. Purif. Technol.* 2012, 88, 79-86.

[60] Bernardo, M.; Santos, A.; Catinho, P.; Minhalma, M. Biodegradation rate constants in different NF/UF fractions of cork processing wastewaters. *Desal. Wat. Treat.* 2011, 29 (1-3), 264-270.

[61] Luo, Q.; Huang, R.M.; Zhou, H.B.; Nie, L.Y. Application of ultra-filtration as pretreatment of nano-filtration for recycling electroplating wastewater. *Electroplating and Finishing 2011*, 30 (4), 47-49.

[62] Mänttäri, M.; Nyström, M. Membrane filtration for tertiary treatment of biologically treated effluents from the pulp and paper industry. *Water Sci. Technol.* 2007, 55 (6), 99-107.

[63] Schäfer, A.I.; Fane, A.G.; Waite, T.D. (Eds), Nanofiltration – principles and applications, Elsevier, Oxford, UK.

In: Membrane Processes for Sustainable Growth
Editors: Angelo Basile and Alfredo Cassano ISBN: 978-1-62618-446-6

Chapter 9

HIGH TEMPERATURE PERFORMANCE POLYMER ELECTROLYTE MEMBRANES

Suwabun Chirachanchai**, Autchara Pangon and Chatchai Jarumaneeroj
The Petroleum and Petrochemical College,
Chulalongkorn University, Bangkok, Thailand
Fuel Cell Research Group, Chulalongkorn University, Bangkok, Thailand

ABSTRACT

Proton exchange membrane fuel cell (PEMFC) and direct methanol fuel cell (DMFC) require high operating temperature to enhance proton transfer mechanism without fuel crossover, and bearing the heat generated during the operation. The present review focuses on the development of high temperature resistance polymer electrolyte membranes (PEM) which function in water retention ability, enhancing reaction kinetics, and reducing electrode CO poisoning.

Hydrous membranes which rely on the water clusters in the membrane to transfer proton can be stabilized to 80-100 °C with the conductivity 10^{-1}-10^{-2} S/cm when the polymers are sulfonated modified aromatics or organic-inorganic hybridized composites. Anhydrous membranes compromise the conductivity based on the proton movement under resonance structure of heterocycles, and hence can be applied in water free system. Several types of membranes with heterocyclic compounds, especially imidazoles and benzimidazoles, prepared by mixing with acids, incorporating with other thermally stable polymers, copolymerizing with other polymers (such as poly(4-vinylphosphonic aicd-*co*-4-vinylimidazole), poly(4,5-vinylimidazole-*co*-acrylic acid)), and conjugating at polymer side chains (such as benzimidazole grafted polydimethylsiloxane, multi-benzimidazole functionalized branched polyethylenimine), are reported, and in the successful cases, the conductivities are as high as 10^{-2}-10^{-5} S/cm at the temperature above 120 °C. A series of benzimidazole model compounds lead us to an understanding that proton transfer mechanism is under synergistic effects of hydrogen bond network and molecular mobility.

* Corresponding Author; e-mail: Suwabun.C@chula.ac.th.

Apart from proton conductivity, other important parameters of PEM are water uptake and swelling, gas or methanol permeability, ion exchange capacity, and single cell performance.

Keywords: Polymer electrolyte membrane (PEM), direct methanol fuel cell (DMFC), high operating temperature PEMFC, heterocycles, imidazoles, benzimidazoles

NOMENCLATURE

List of Acronyms

ABPBI	Poly(2,5-benzimidzole)
AL	Alginic acid
Atet	5-aminotetrazole
Atri	3-amino-1,2,4-triazole
bimiPEOAGE	Benzimidazole tethered on linear poly(ethylene glycol)
bimiPSAGEx	Benzimidazole-tethered poly(stryrene-g-allyl glycidyl ether)
bimiPVMSx	Benzimidazole-grafted polydimethylsiloxane
bimixPSU	Polysulfones tethered benzimidazole
BipPBI	Poly[2,20-(2-benzimidazole-p-phenylene)-5,50-bibenzimidazole]
BMI-BTSI	1-butyl-3-methylimidazolium-bis(trifluoromethylsulfonyl)imide
BuMeImBF4	1-butyl-3-methylimidazolium-tetrafluoroborate
DMAc	N,N'-dimethylacetamide
DMF	N,N'-dimethylformamide
DMFC	Direct methanol fuel cell
EMI-TFSI	1-ethyl-3-methylimidazolium-bis(trifluoromethanesulfonyl)imide
HBPBI	Hyperbranched polybenzimidazole
MD	Molecular dynamics
MDP	Monododecyl phosphate
MEA	Membrane electrode assembly
MeIm-BF4	N-methyl imidazolium tetrafluoroborate
MMT	Montmorillonite
MPA	Molybdophosphoric acid
MPEI	Multi-benzimidazole funtionalized branched polyethylenimine
NGDE	1,3-bis(2,3-epoxypropoxy)-2,2-dimethylpropane
NMP	N-methyl-2-pyrrolidone
ORR	Oxidation reduction reaction
PAA	Poly(acrylic acid)
PBI	Polybenzimidazole
PBIANI	Poly(benzimidazole-co-aniline)
PEGMEA	Poly(ethylene glycol) methyl ether acrylate
PEM	Polymer electrolyte membrane
PEMFC	Proton exchange membrane fuel cell
PFSA	Perfluorosulfonic acid
Poly(VPA-co-Vim)	Poly(vinylphosphonic acid-co-4-vinylimidazole)

PPA	Polyphosphoric acid
PTFE	Poly(tetrafluoroethylene)
PVA-SAA	Poly(vinyl alcohol)-sulfonic acid
PVDF	Polyvinylidine fluoride
PVPA	Poly(vinylphosphonic acid)
Py	Pyrazol
RH	Relative humidity
SA	Silicontungstic acid
SEBS	Sulfonated polystyrene ethylene butadiene polystyrene
SPBI	Sulfonated polybenzimidazole
SPEEK	Sulfonated poly(ether ether ketone)
SPEK	Sulfonated polyetherketone
SPI	Sulfonated polyimide
SPSU	Sulfonated polysulfone
TEA-TF	Trifluoromethanesulfonic of triethylammonium
TH	Thiophene
TP	Tridecyl phosphate
TPA	Tungstophosphoric acid
Tri	1H-1,2,4-triazole
UI	2-undecylimidazole

List of Symbols

A	Membrane surface area
A	Exposed area of membrane (eq. 4)
A_{dry}	Area before immersing in distilled water
A_{wet}	Area after immersing in distilled water
C_A	Methanol concentration in the compartments A
C_B	Methanol concentration in the compartments B
C_{MeOH}	Methanol concentration
C_{NaOH}	NaOH concentration
dP	Pressure difference between feed and permeate parts
ΔV_{NaOH}	Volume differences of NaOH solution
IEC	Ion exchange capacity
J_{MeOH}	Methanol flux
J_{tot}	Total flux
L	Membrane thickness
P	Gas permeablility coefficient (eq. 3)
P	Methanol permeability (eq. 4)
θ	Phase angle
R	Gas constant
T	Absolute temperature
t	Permeation time
V_B	Solution volume of compartment B
V_m	Molar volume of the gas at standard temperature and pressure

V_p	Permeate volume
W_d	Weight of dry membrane
W_{perm}	Weight of permeated methanol solution
W_w	Weight of immersed membrane
Z'	Real part of impedance
Z"	Imaginary part of impedance

1. INTRODUCTION

Among various types of fuel cells, polymer electrolyte membrane fuel cell or proton exchange membrane fuel cell (PEMFC), of which the polymer membrane functions as electrolyte and allows the proton transfer through the membrane to obtain electrical energy, is a promising system with attractiveness in terms of the light weight, the compactness, including the high power density and is considered to be suitable for portable electrical devices and automotives.

Electrochemical reactions in PEMFC take place simultaneously on both anode and cathode of the cell by applying hydrogen as a fuel and oxygen or air as an oxidant. At the anode, hydrogen is oxidized by the catalyst on the membrane surface to generate protons and electrons. The protons obtained move through the polymer electrolyte membrane and enter to the cathode whereas the electrons consequently flow through the electrical circuit with generating electricity. At the cathode, the oxygen is reduced by recombining with protons and electrons providing water and heat as by-products.The reactions are as following:

Anode: $$H_2 \rightarrow 2H^+ + 2e^- \quad (1)$$

Cathode: $${}^1/_2O_2 + 2H^+ + 2e^- \rightarrow H_2O \quad (2)$$

Overall: $$H_2 + {}^1/_2O_2 \rightarrow H_2O + \text{electricity} + \text{heat} \quad (3)$$

In addition, PEMFC is similar to direct methanol fuel cell (DMFC) that uses solid polymers as an electrolyte but DMFC employs methanol as the fuel instead of hydrogen. The basic electrochemical reactions of DMFC are:

Anode: $$CH_3OH + H_2O \rightarrow CO_2 + 6H^+ + 6e^- \quad (4)$$

Cathode: $$ {}^3/_2O_2 + 6H^+ + 6e^- \rightarrow 3H_2O \quad (5)$$

Overall: $$CH_3OH + {}^3/_2O_2 \rightarrow CO_2 + 2H_2O \quad (6)$$

Theoretically, the electrical potential (difference) between cathode and anode is equal to 1.229 V (at 25 °C and 1 atm) when pure hydrogen and pure oxygen are used as fuel and oxidant, respectively. [1] In practice, energy losses are always found in the fuel cell, as a consequence, the cell potential (difference) is not ideal and diminishes to be ~ 0.7-0.8 V for the optimal fuel cell performance. [2]

It should be noted that working temperature of PEMFC is depending on the polymer electrolyte membrane material used (typically lower than 100 °C). Although, PEMFC has been developed for decades, many limitations including high cost of the materials, inadequate water and heat management, durability of each component, intolerant to impurities of hydrogen such as CO, and slow kinetic reaction of the cathode are points to be solved. Operating at high temperature is one of the solutions and it brings the following performances.

i) *Enhancement of Reaction Kinetics:* Since oxidation reduction reaction (ORR) is a rate determining step of electrochemical reactions taking place in PEMFC, an enhancement of reaction kinetics, especially ORR, as a result of working temperature increment, improves the cell efficiency. [3]
ii) *Reduction of CO Poisoning on Catalyst Surface:* In general, platinum (Pt) is considered as a highly active catalyst for oxidation of hydrogen. The fact that hydrogen is normally produced by reforming carbon-containing resources such as methanol, ethanol and fossil fuel, etc. the hydrogen harvested is always contaminated with trace amount of CO. When CO is absorbed on Pt surfaces, the catalyst surface area available for hydrogen oxidation will be reduced resulting in a decrease of cell efficiency. Wilson et al. showed that the polarization curve of PEMFC with H_2 containing 5 ppm CO/air system was significantly decreased about two times at a current density of 800 mA cm^{-2} compared to that of PEMFC with pure H_2/air system. [4] It is known that the CO coverage on the catalyst surface is diminished at elevated temperature, in other words, high operating temperature allows PEMFC being CO toleration and performing with the reformed hydrogen. [5, 6]
iii) *Ease of Heat and Water Managements:* Forty to fifty percent of energy produced by PEMFC is lost as heat. [7] During PEMFC operation, an amount of heat will be generated. The use of low temperature PEM needs a cooling system to maintain PEM efficiency. In contrast, the use of high temperature PEM needs no extra process to exclude the heat and water generated.

2. Polymer Electrolyte Membrane

One of the most important parts in PEMFC systems is the polymer electrolyte membrane (PEM) which works as a proton exchanging part as well as a barrier for fuel and oxidant. There are several requirements of PEM needed for PEMFCs which are high proton conductivity, excellent electronic insulator, good mechanical and chemical stabilities in the fuel cell environment, low fuel crossover, and low cost. [8, 9]

Consequently, several polymer membranes have been developed for the use as electrolytes in PEMFC. Typically, based on the type of proton conductive species, PEMs are classified into two main ones: hydrous-based membranes (e.g. protons transport though water molecules) and anhydrous-based membranes (e.g. protons transport through other species such as heterocycles and oxo acids).

3. HYDROUS–BASED MEMBRANE

3.1. High Performance Hydrous-Based Membrane: Perfluorosulfonic Acid Polymer

At present, perfluorosulfonic acid (PFSA) polymer is the key material for polymer electrolytes. Nafion® developed by DuPont was firstly applied in 1966 and is still the most widely used as ion-exchange membrane in PEMFC prototypes. Apart from Nafion®, other PFSA membranes such as Flemion®, Aciplex®, and Dow® were developed (Figure 1). The PFSA chain consists mainly of a poly(tetrafluoroethylene) (PTFE) backbone, which statistically forms segments of several units in length, and a few perfluorinated vinyl polyether branches. The branches, functionating as a flexible pendant, combine the PTFE backbone which carries a terminal acidic group providing ion clusters for proton migration. As a consequence, proton can be transferred through the membrane. The high electronegativity or electron affinity of the fluorine atom, in α-position to the SO_3H group, makes the sulfonic acid to be a super strong acid. [10] This results in an excellent proton conductivity of PFSA membranes, e.g. $9x10^{-3}$-$1.2x10^{-1}$ S/cm at 80 °C in a relative humidity (RH) range of 34-100%. [8] In addition, the PFSA membranes, fully fluorinated polymers, exhibit exceptionally high chemical stability, for example, in 30% H_2O_2 at temperature up to 80 °C. Therefore this type of membrane is applicable in harsh conditions. [11, 12]

3.2. Development of New Hydrous-Based Membrane

Current research on hydrous-based membranes is related to (i) proton migration mechanism within the membranes and (ii) overcoming membrane limitations especially on addition of additives for water retention including for improvement of proton transfer efficiency.

3.2.1. Proton Migration Mechanism

As a result of phase separation between hydrophobic fluorocarbon backbone and water-surrounded hydrophilic end groups of sulfonic acid species, water cluster networks with narrow channels connected were formed, based on the studies on statistical mechanics, molecular dynamics, and microscopic phenomena. [14, 15] For hydrous membrane, proton conductivity critically depends on the degree of membrane hydration. [16, 17, 18, 19]

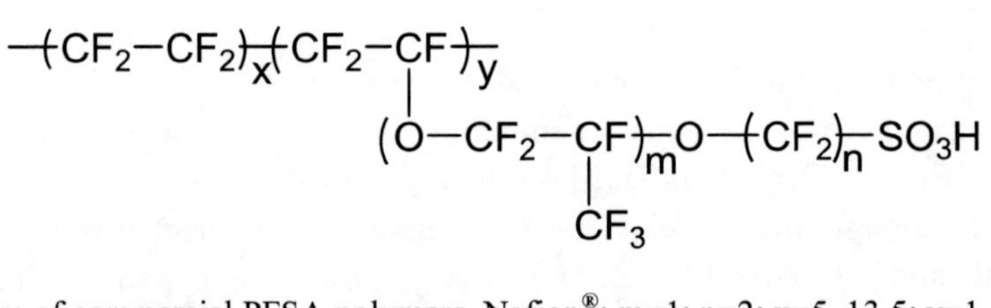

Figure 1. Structure of commercial PFSA polymers. Nafion®: m=1; n=2; x=5–13.5; y=1, Flemion®: m=0,1; n=1–5; x=5–13.5; y=1, Aciplex®: m=0; n=2–5; x=1.5–14; y=1, Dow®: m=0; n=2; x=3.6–10, y=1. [13]

The higher the amount of water inside the membrane, the higher the proton conductivity. This is due to the fact that water molecules act as proton conductive species with two principal mechanisms, i.e. vehicle and Grotthus mechanisms. In the case of vehicle mechanism, translational dynamics of hydronium ions (H_3O^+) with molecular diffusion of unprotonated vehicles (H_2O) allow the proton migration, whereas, for Grotthuss mechanism, the proton transfers though hydrogen bond network via molecular reorientation. [20]

3.2.2. Overcoming of Membrane Limitations

Although PFSA membranes are currently intensively used in PEMFC as they meet the requirements of high proton conductivity (9-120 mS/cm at 80 ºC in RH range of 34-100%), [8] outstanding chemical stability (in H_2O_2 at temperature up to 80 ºC), [12, 13] and longevity (60,000 h at 80 °C), [21] there are several major drawbacks as follows.

i) *High cost:* Nafion® membrane and related polymers are expensive, for example Nafion® 117 the price is of about US$ 800-2000 m^{-2}. [22]
ii) *Special cautions required in manufactural process*: Safety concerns arise from evolution of toxic intermediate and corrosive gases liberated at temperatures above 150 °C and the decomposed products during the manufacturing processes.
iii) *System complication*: As additional hydration is required, the systems become complicated and expensive. [23]
iv) *Working temperature limitation*: By increasing temperature, the proton conductivity becomes significant, however; membrane dehydration, water affinity reduction, mechanical strength decrease, and fuel crossover are observed at high temperature. [23, 24, 25, 26, 27]
v) *Fuel and ruthenium (Ru) anodic co-catalyst crossover*: PFSA membranes tend to allow fuel and Ru anodic crossover, for example; permeation of methanol through the Nafion® membrane was around 10^{-7}-10^{-6} cm^2/s in DMFC, [28, 29, 30] as a consequence; the DMFC performances dramatically drop. [31]

To overcome the drawbacks of the conventional PFSA hydrous-based polymer especially the Nafion® membrane, various approaches have been made, i.e. adding additives to retain water, enhancing proton conductivity and increasing PEMFC working temperature with auto-humidification to accelerate proton movement. [32] In addition, development of new membranes is also alternative choice.

3.2.3. PFSA Hydrous-Based Polymer Additives

Additives improve water retention, thermal stability, preventing fuel crossover, and proton transfer efficiency. Up to present, various types of additives for example acids, i.e. Brønsted acids, [33] polymer, i.e. PTFE, [34, 35] inorganic, i.e. metal oxide, [17, 25, 36, 37, 38, 39, 40, 41, 42, 43, 44, 45, 46] and clay including organic, i.e. heterocyclic compounds [47] have been investigated.

Savinell et al. reported that, by adding phosphoric acid, the performance of Nafion® 117 membrane was enhanced. [33] The reaction kinetics accompanied by high proton conductivity (<0.05 Ω^{-1} cm^{-1}) up to 200 °C as well as lower methanol crossover as compared to Nafion® 117 membrane were achieved. Moreover, Mex et al. [34] and Finsterwalder et al. [35] found that coating a thin layer of plasma polymerized tetrafluoroethylene with vinyl phosphoric acid and chlorosulfonic acid on Nafion® membranes reduced methanol crossover in DMFC.

Antonucci et al. demonstrated that the high water uptake of Nafion®-silica composite membrane allowed the use of DMFC at 145 °C with the power density of 240 mW cm^{-2}. [25] Dispersion of other inorganic materials such as silicon dioxide (SiO_2), [17, 36, 37, 38] titanium dioxide (TiO_2), [17, 38, 39, 40] zirconium dioxide (ZrO_2), [41, 42] and zirconium phosphate (ZrP), [43, 44, 45, 46] etc. in the PTFE membranes not only to retain water at elevated temperature but also to improve mechanical properties were investigated. A significant improvement in proton conductivity at elevated temperature was found in PFSA incorporating perfluorinated ionomers and heteropolyacids such as phosphotungstic acid, phosphomolybdenic acid, phosphotin acid, and cesium hydrogen salt of heteropolyacids. [48, 49, 50, 51, 52] Tazi and Savadago prepared the composite membrane of Nafion® 117 with the incorporation of silicotungstic acid (SA) or thiophene (TH) in order to enhance the ionic conductivity and power density. [47] The substantial improvement in ionic conductivity was observed in the case of Nafion® 117 membrane with SA as a result of an increase in water uptake for 33% as compared to the Nafion® 117 membrane (27% water uptake). The highest current density of 810 mA cm^{-2} at 600 mV was achieved from TH incorporated Nafion® 117 while that of Nafion® 117 was only 640 mA cm^{-2}. Moreover, an addition of silicon dioxide and phosphotungstic acid into Nafion® 115 membrane not only performed the same level for proton conductivity with plain Nafion® 115 at high temperature under 100% relative humidity but also showed an improvement of crystallinity and thermal stability. [53]

It is important to note that the addition of additives might induce phase separation, pores, roughness, etc. which results in defects of membranes. Gosalawit et al. demonstrated an approach to develop miscibility between silica and Nafion®. [54] The fact that the structure of Krytox is similar to that of Nafion®, a coupling reaction between silica and Krytox brought not only well-dispersed silica in the membrane without phase separation but also an enhancement of water retention and thermal stability. An addition of 5 wt.% Krytox-Si in Nafion® was found to maintain the proton conductivity at 10^{-3} S/cm even the temperature was as high as 130 °C. Yana et al. further confirmed an effective H_3O^+ diffusion through 5 wt.% Krytox-Si–Nafion® under high temperature range (60-100°C) by using molecular dynamics (MD) simulations. [55] The calculation also pointed out that the hydrolyzation of SO_3^- in Krytox-Si–Nafion® membrane was greater than that in Nafion®. Gosalawit et al. applied the concept of modifying silica with Krytox to the case of montmorillonite (MMT) and found that the organoclay obtained (Krytox-MMT) was in exfoliation. The Krytox-MMT-Nafion® composite membrane not only improved water uptake (>26%) and thermal resistance (>30 °C) of the system but also initiated a substantial reduction in methanol crossover (50%) as compared to Nafion®. [56]

3.2.4. Investigating of Alternative Membranes

For decades, many types of partially fluorinated and non-fluorinated membranes either aliphatic or aromatic polymers have been developed because they are less expensive, commercially available, and easy to modify with various functional groups. Moreover, some specific polymer structures, for example aromatic polymers are thermally stable resulting in the application at elevated temperature. [57] Proton transfer can be initiated by some specific functional groups, i.e. sulfonic acid, carboxylic acid, and phosphoric acid. The dissociation of these acid groups including electro-osmotic drag and diffusion of water induces proton movement though the hydrated membrane. The examples of acid-modified aliphatic polymers as potential hydrous PEM are poly(vinyl alcohol), [58, 59, 60] poly(vinylphosphoric acid), [61] polyacrylic acid, [62, 63] polyethylenimine, [64] chitosan, [62, 65] and fluorinated copolymers [66] (Figure 2).

It is important to note that earlier than aliphatic PEMs, since 1950s, the aromatic ones, i.e. cross-linked polystyrene-divinylbenzene-sulfonic acid and sulfonated phenol-formaldehyde (Figure 3) have received much attention. In addition, it was reported that fluorine-substitution prolonged the membrane shelf life and increased the chemical stability as seen in the case of polytrifluorostyrene sulfonic acid. [67]

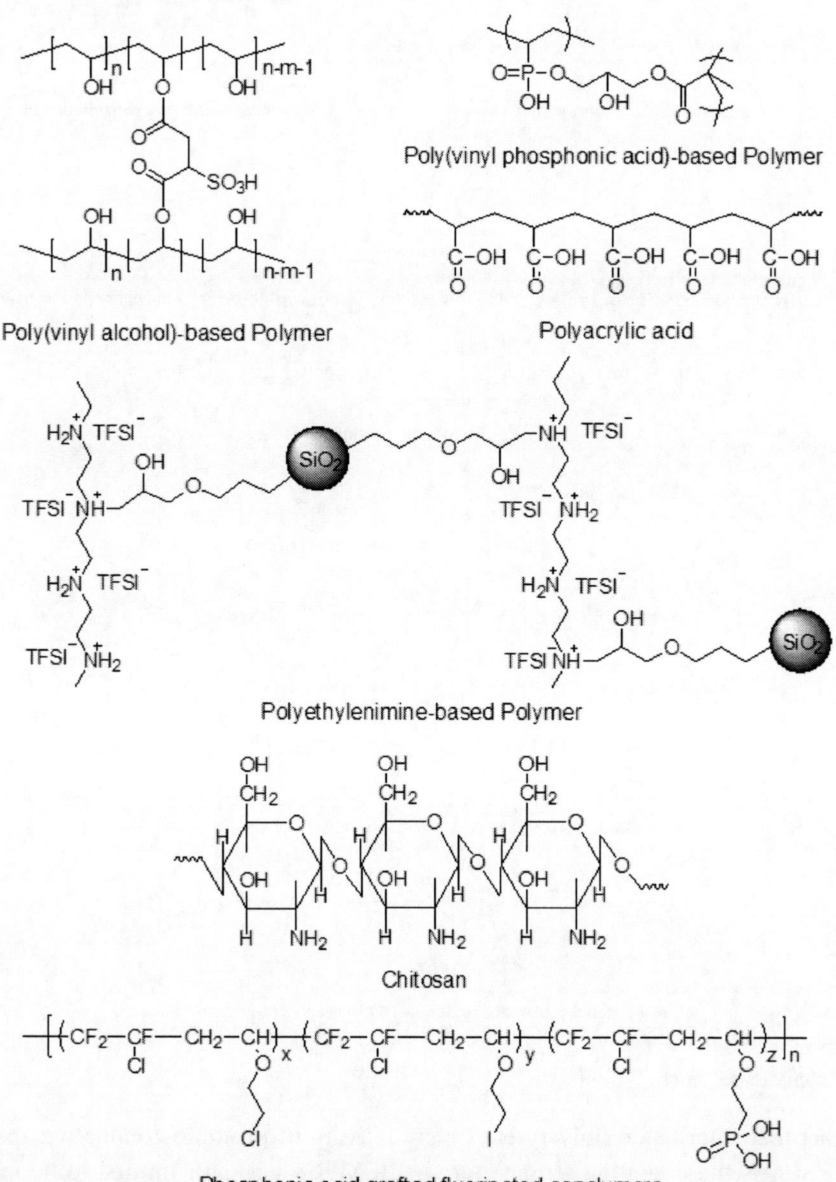

Figure 2. Aliphatic polymers.

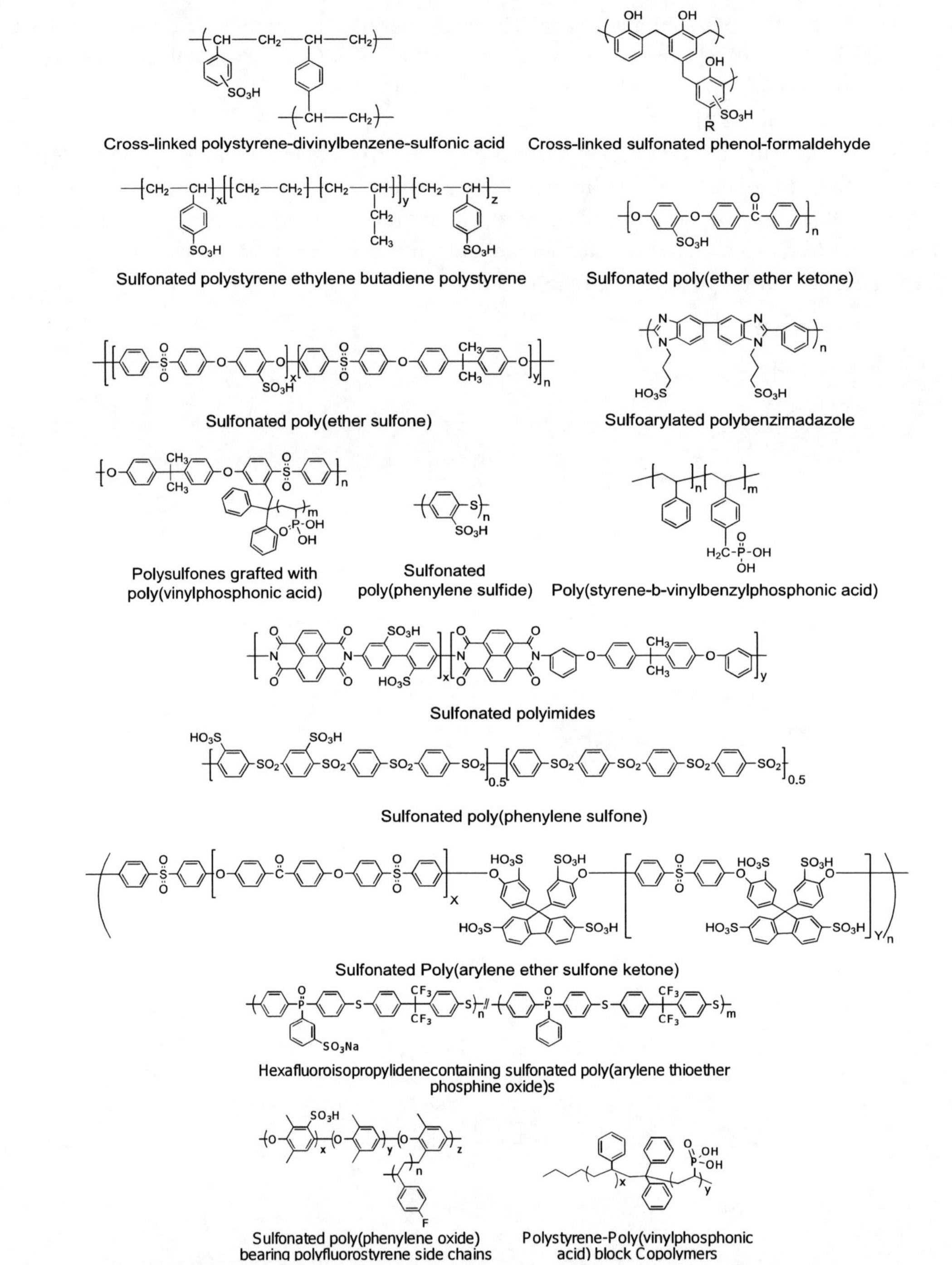

Figure 3. Aromatic polymers.

The fact that fluorinated polystyrenes were lacking of protonic conductive species, such as acids or water, the operating temperature of PEMFCs was still limited to be less than 80 °C. Sulfonated ethylenebutadiene polystyrene (SEBS) triblock copolymer (Figure 3) was also proposed as another type of polystyrene-based membrane. [68] This membrane with a water content of 18.6 wt.% revealed an excellent conductivity in the order of 10^{-2} S/cm at room temperature. Thereafter, Mokrini and Honeault blended this SEBS with polyvinylidine fluoride (PVDF) to find that SEBS portion offered good proton conductivity while the PVDF portion provided good dimensional stability and chemical resistance. [69]

Currently, acid-functionalized aromatic membranes, e.g. sulfonated poly(ether ether ketone), [70, 71, 72] sulfonated poly(ether sulfone), [73, 74, 75] polysulfones grafted with poly(vinylphosphonic acid), [76] sulfoarylated polybenzimadazole, [77] sulfonated polyimides, [78, 79, 80] sulfonated poly(phenylene sulfide), [81, 82, 83] sulfonated poly(phenylene sulfone), [84] sulfonated poly(arylene ether sulfone ketone), [85] poly(styrene-*b*-vinylbenzylphosphonic acid), [86] sulfonated poly(phosphine oxide), [87, 88] sulfonated poly(phenylene oxide), [89] and polystyrene-poly(vinylphosphonic acid) block copolymers [90] (Figure 3) have been widely investigated for PEM owning to their good thermal, mechanical, and chemical stabilities with affordable cost and wide availability.

Sulfonated poly(ether ether ketone) (SPEEK) is one of the most successful aromatic PEMs due to its good performance in conducting protons and reducing methanol crossover. SPEEK composite membranes embedded with heteropolyacids, i.e. tungstophosphoric acid (TPA) and molybdophosphoric acid (MPA) were reported about their significant thermal stability (T_d above 250 °C) and mechanical stability along with good proton conductivity either at room temperature (10^{-2} S/cm) or above 100 °C (10^{-1} S/cm). [70] It should be noted that the proton conductivity of SPEEK depends on the degree of sulfonation. [70, 91] Although the high degree of sulfonation is a main factor for PEM performance, in the case of DMFC, methanol crossover and mechanical instability under alcohol environment are the points to be concerned when the degree of sulfonation is increased. To overcome this, inorganic fillers such as silicate layers and silica particles were investigated. Ponce et al. prepared sulfonated polyether ketone (SPEK) with different heteropolyacids and inorganic frameworks of ZrO_2 and $RSiO_{3/2}$ for DMFCs and found that the presence of ZrO_2 decreased methanol and water permeability, and heteropolyacids bleeding. [92] Moreover, they showed a significant conductivity (86-110 mS/cm at 25 °C) for the membrane containing tungstophosphoric acid. Gaowen et al. showed that SPEEK composites containing exfoliated montmorillonite performed methanol crossover reduction as well as high proton conductivity (close to Nafion® 115 at 90 °C for 1.2×10^{-2} S/cm). [93] Besides, the composite membranes of sulfonic acid functionalized silica and SPEEK prepared by solvent casting were reported by Sambandam and Ramani. [94] These membranes showed the proton conductivity at 80 °C for 0.05 S/cm and 0.01 S/cm under 75% and 50% relative humidity, respectively. The hydrogen crossover through the membrane was studied by using linear sweep voltammetry based on the condition with oxidation of hydrogen at cathode determined, and at that time the hydrogen crossover current density for all MEAs was about 1-2 mA/cm^2.

Gosalawit et al. proposed the use of silane coupling agent, (3-aminopropyl) triethoxysilane, to enrich montmorillonite with acid species of 4-sulfophthalic acid. [95] The acid immobilized montmorillonite (1 wt.% and 3 wt.% contents) embedded in SPEEK showed an improvement of proton conductivity at 100 °C to be as high as 0.1 S/cm^2 as a consequence of water maintaining. The composite membranes also performed good water, methanol aqueous solution, and mechanical stabilities. The maximum power density in the DMFC system was observed at 21 mW/cm^2 in the case of SPEEK containing 1 wt.% sulfonated montmorillonite which was higher than that of SPEEK and Nafion® under acid treatment for 12 h and 24 h, respectively.

4. ANHYDROUS–BASED MEMBRANES

As mentioned above, the increase of operating temperature to more than 120 °C leads to a significant improvement in CO tolerant of electrode catalysts, especially platinum, resulting in applicability of hydrogen gas obtained from reforming process. [96] Proton movement acceleration at higher temperature also provides a higher proton conductivity. However, evaporation of water inside the membrane limits working temperature to be lower than 100°C. This leads to a limitation of hydrous-based membrane, therefore many researchers focus on alternative polymers for PEMFC which functions in water-free or anhydrous conditions. [97]

Heterocycles are potential compounds as proton carriers due to (i) amphoteric behavior which can either donate or accept protons and (ii) high thermal stability leading to a development of heterocycles to use as proton conductive species instead of water in anhydrous PEM. Kreuer et al. investigated the use of heterocycles, i.e. imidazole and pyrazole as acidic proton transfer species. [98] Imidazole is one of the simplest heteroaromatic compounds that have been successively studied as proton conductive species due to its melting point and boiling point at 90 °C and 256 °C, respectively. Proton transfer mechanism of the neat imidazole was reported by Münch et al. [99] The mechanism involved with a diffusion of excess protons in the form of proton transfer and local cooperative reorientation was proposed. Consequently, investigations of imidazole based PEMFC, e.g. mixing imidazole derivatives with other acids, incorporating into polymer matrices, and conjugating with other polymers proved that imidazoles effectively function in proton conductivity as summarized in Table 1. The fact that imidazole and other derivatives for anhydrous PEM mostly require protonic sources to initiate proton transfer, adding strong oxo acids (CF_3SO_3H, H_3PO_4) or functionalizing acid groups onto polymers (carboxylic acid, phosphonic acid, and sulfonic acid) was also reported. In the absence of water, oxo acids perform self-dissociation [100] to allow proton hopping through (i) structural diffusion and (ii) breaking and forming of hydrogen bond among acidic species.

For type I, imidazole derivatives mixed with acids were studied to understand how proton transfer occurs through heterocyclic structure. Kreuer et al. showed that imidazole acts as proton donor-acceptor in the proton transfer process as observed its conductivity increased with temperature up to 120 °C (~2 Ω^{-1} cm^{-1} K). [98] When acids, like sulfanilic acids were added, protonated imidazoles were formed to allow an enhancement of proton conductivity. Schuster et al. functionalized imidazoles onto ethylene oxide oligomers and demonstrated proton transfer of immobilized heterocycles along hydrogen bond pattern as a result of local dynamics of aggregated structures. [101] Proton conductive pathway of self-assembled acid-base composite was investigated by Yamada et al. through a case study of acidic surfactant monododecyl phosphate (MDP) and basic surfactant 2-undecylimidzole (UI) system. [102] The ordered molecular arrangement of UI-MDP resulted in lamellar packing structure with two-dimensions of proton conducting pathway.

In the case of type II, as reviewed in Table 1, imidazole derivatives were developed as anhydrous membranes by incorporated in the polymer matrices, for example, sulfonated polyetherketone (SPEK), Nafion®, poly(acrylic acid) (PAA), poly(vinylphosphonic acid) (PVPA), poly(vinyl alcohol)–sulfosuccinic acid (PVA-SAA), alginic acid (AL), sulfonated poly(ether ether ketone) (SPEEK), and sulfonated polyimide (SPI).

Table 1. Examples of imidazole-based compounds for fuel cell membranes

Imidazole-based compounds	Proton Conductivity	References
Type I (Mixing imidazole derivatives with other acid species)		
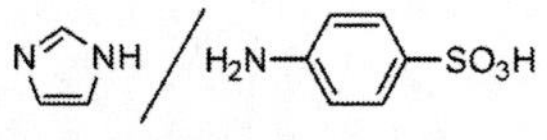 Imidazole/Sulfanilic acid	10^{-2} S/cm at 120 °C	98
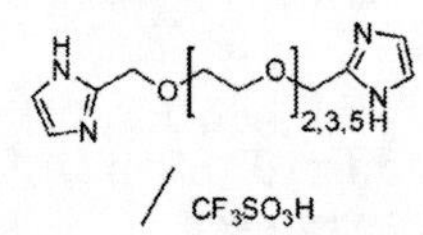 Imidazole-terminated ethyleneoxide oligomer/ Trifluoromethanesulfonic	$5x10^{-3}$ S/cm at 120 °C	101
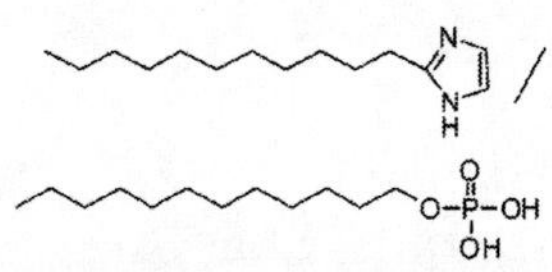 Monododecyl phosphate (MDP)/ 2-undecylimidzole (UI)	10^{-3} S/cm at 150 °C	102
Type II (Incorporating into polymer matrices)		
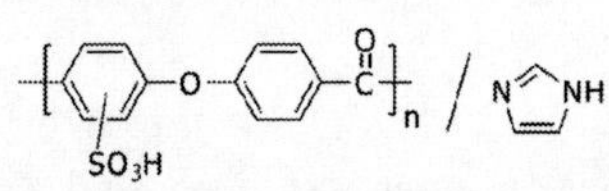 Sulfonated polyetherketone (SPEK)/Imidazole	~10^{-2} S/cm at 200 °C	98
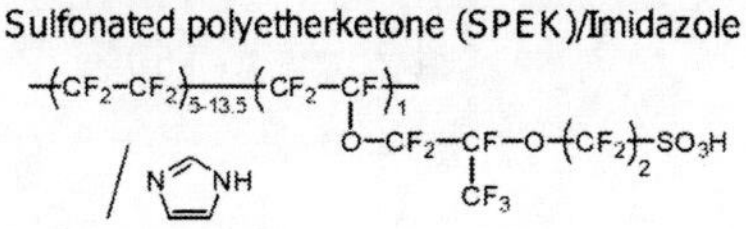 Nafion/Imidazole	~10^{-1} S/cm at 170 °C	96
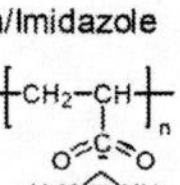 Poly(acrylic acid) (PAA)/Imidazole	10^{-3} S/cm at 120 °C	103
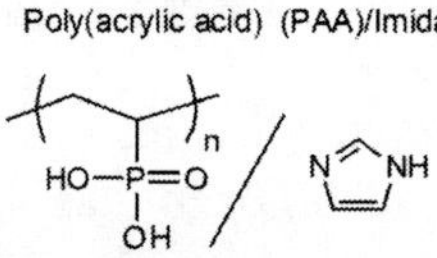 Poly(vinylphosphonic acid) (PVPA)/Imidazole	$7x10^{-3}$ S/cm at 150 °C	104
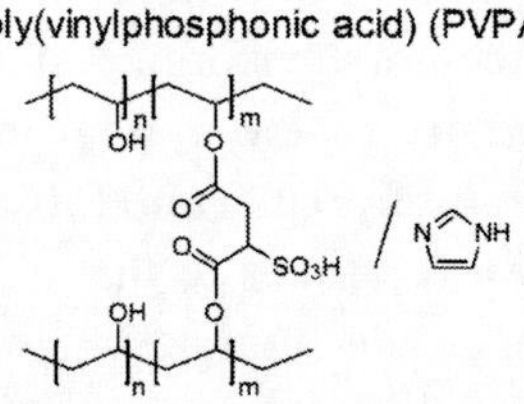 Poly(vinyl alcohol)–sulfosuccinic acid (PVA-SAA) /Imidazole	$1.4x10^{-3}$ S/cm at 120 °C	105
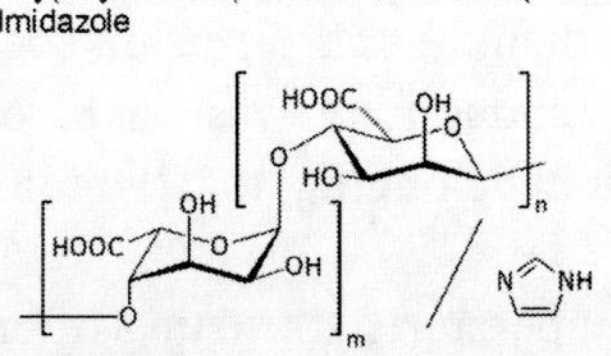 Alginic acid (AL)/Imidazole	$2x10^{-3}$ S/cm at 120 °C	106

Table 1. (Continued)

Imidazole-based compounds	Proton Conductivity	References
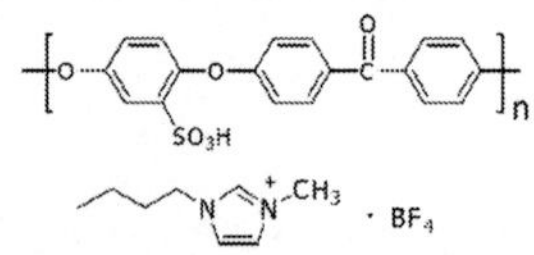 Sulfonated poly(ether ether ketone) (SPEEK)/1-butyl-3-methylimidazolium-tetrafluoroborate (BuMeImBF4)	1.04×10^{-2} S/cm at 170 °C	107
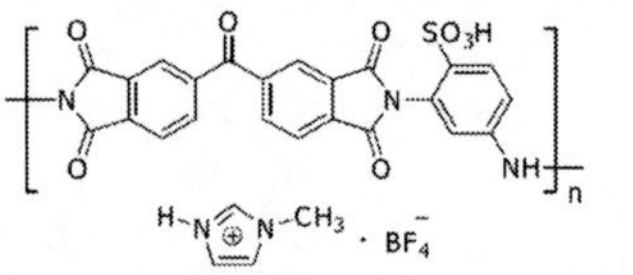 H_2SO_4 doped sulfonated polyimide (sPI)/*N*-methyl imidazolium tetrafluoroborate	5.6×10^{-2} S/cm at 180 °C	108
Type III (Conjugating with polymer chains)		
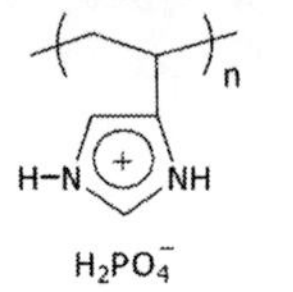 Poly(4-vinylimidazole)/phosphoric acid	1×10^{-4} S/cm at 25 °C	109
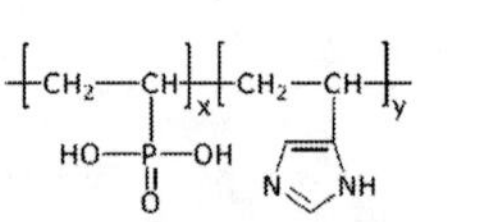 Poly(vinylphosphonic acid-*co*-4-vinylimidazole) (poly(VPA-co-4-VIm))	10^{-6}-10^{-12} S/cm at 25 °C to 160 °C	110
Poly(vinylphosphonic acid-*co*-4-vinylimidazole) (poly(VPA-co-4-VIm)) 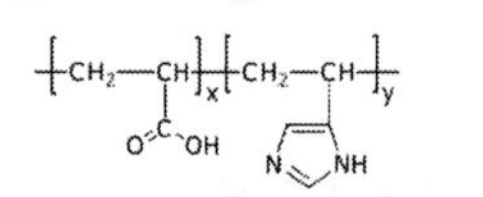 4(5)-Vinylimidazole-*co*-acrylic acid copolymer	10^{-2} S/cm at 120 °C	111

The blend membranes showed the maximum proton conductivities in the range of 10^{-1}-10^{-3} S/cm at high temperature (>120 °C). The properties of imidazole as solvent for acidic proton transfer in polymer matrix of sulfonated polyetherketone were studied. [98] It was found that imidazole, acting as proton donor and acceptor, performed similar behavior to water in proton transfer process. The proton conductivity of the system increased with an increase of imidazole concentration and temperature. Up to 150 °C, sulfonated polyetherketone with the mole number of imidazole for 6.7 times to that of SO_3H showed a higher conductivity than the pure sulfonated polyetherketone saturated with water. It should be noted that, at that time, diffusion coefficient evaluated by the conductivity using Nernst-Einstein relationship was found to be much higher than the self diffusion coefficient observed by ^{1}H-PFG-NMR. This indicated that the ionic conductivity was based on intermolecular proton transfer reactions rather than self diffusion belonging to all protons of imidazole including acid protons.

Recently, ionic liquids, liquid salts at room temperature, consisting of organic cation and inorganic anion, as thermally stable charge carrier to initiate proton conductivity in high temperature PEM under anhydrous condition have received much more attention. Imidazolium salts (*N*-methyl imidazolium tetrafluoroborate (MeIm-BF4), [108] 1-butyl-3-methylimidazolium-tetrafluoroborate (BuMeImBF4), [107] 1-butyl-3-methylimidazolium-bis(trifluoromethylsulfonyl)imide (BMI-BTSI), [112] 1-ethyl-3-methylimidazolium-bis(trifluoromethanesulfonyl)imide (EMI-TFSI), [113] and 1-propyl-3-methylimidazolium dihydrogen phosphate [114]) and ammonium salts (trifluoroacetic propylamine [115] and trifluoromethanesulfonate of triethylammonium (TEA-TF) [116]) are examples of ionic liquids used in PEM. Noto et al. proposed the use of TEA-TF in Nafion® to improve mechanical and thermal properties including water uptake. [116] Composite SPEEK membranes containing BuMeImBF4, were found to show the highest ionic conductivity (1.04×10^{-2} S/cm at 170 °C under anhydrous conditions) when using *N,N'*-dimethylformamide (DMF) for casting the membranes, compared to other solvents, i.e. *N,N'*-dimethylacetamide (DMAc), and *N*-methyl-2-pyrrolidone (NMP). [107] At that time, DMF might induce the formation of inter-microchannel connected ionic clusters to facilitate proton transfer. Deligöz et al. developed ionic liquid composite membrane of sulfonated polyimide (SPI)/MeIm-BF4 and found the improvement of conductivity reaching to 5.59×10^{-2} S/cm at 180 °C with H_2SO_4 doping and excellent thermo-mechanical properties (storage module 0.91 GPa at 300 °C) including conductive stability. [108]

For type III, Bozkurt and Meyer showed the doping of poly(4-vinylimidazole) with H_3PO_4 increased the proton conductivity for 10^9 times at ~300 K as compared to the non-doping ones with high thermal stability (150 °C). [109] However, the fact that the leakage of small molecules during membrane operation led to catalyst deactivation or cell efficiency reduction, modifications of polymer chains containing heterocycles as proton solvent and acid functional groups as proton source were the points to be considered. Bozkurt et al. also showed that the copolymer of poly(vinylphosphonic acid-*co*-4-vinylimidazole) (poly(4-VIm-*co*-VPA)) performed an increase of proton conductivity not only at high temperature but also when increasing the 4-vinylimidazole contents. [110] At that time, the proton transfer of the system was suspected to be based on a structural diffusion mechanism. Anhydrous membranes of 4(5)-vinylimidazole-*co*-acrylic acid copolymer with various contents of imidazole and acrylic acid were investigated by Jithunsa et al. [111, 117] A significant increase in proton conductivity was observed at temperature above 80 °C up to 10^{-2} S/cm in the case of the copolymer containing high imidazole content (66%). The studies of the structure changes related to temperature revealed that, in the case of copolymers with imidazole content more than 60%, the maintaining of hydrogen bond between carboxylic acids and imidazoles led to an enhancement of proton conductivity. Therefore, the conductivity was as high as 10^{-2} S/cm at the temperature up to 120 °C under anhydrous condition. They also showed that, in the case of copolymer with 4% imidazole content, a decrease of proton conductivity at high temperature was a consequence of anhydride formation due to dehydration of carboxylic acids.

Up to present, not only imidazole but also other *N*-containing heterocycles, i.e. benzimidazole, pyrazole, triazole, and tetrazole have been applied as proton carrier in PEMFCs. Benzimidazole-based PEMs have received much attention due to their various preparation methods and their stiff structure allowing high thermal stability. Table 2 summarizes developments of benzimidazoles in terms of their structures, i.e. (i) main chain structures, (ii) side chain structures, and (iii) free molecules including their proton conductivities.

Table 2. Examples of benzimidazole-based membranes

Benzimidazole-based compounds	Proton Conductivity
Type IV: Main chain structures	
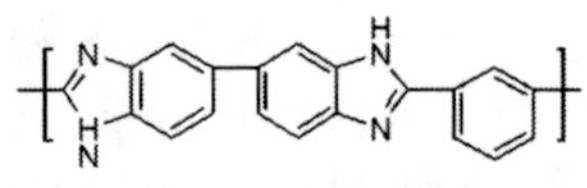 Polybenzimidazole, (PBI)	
338 mol% H_3PO_4/PBI	10^{-4} S/cm at 130 °C [118]
3.25 mol H_2SO_4/PBI	10^{-4} S/cm at 30 °C [119]
3.05 mol H_3PO_4/PBI	$7x10^{-6}$ S/cm at 30 °C [119]
20-40 mol H_3PO_4/PBI	0.26 S/cm at 200 °C [120]
1.8 Tridecyl Phosphate/PBI	~10^{-4} S/cm at 140 °C [121]
39 mol H_3PO_4/sulfonated PBI	0.326 S/cm at 180 °C [122]
193% H_3PO_4/epoxy-based cross-linked PBI	0.017 S/cm at 200 °C [123]
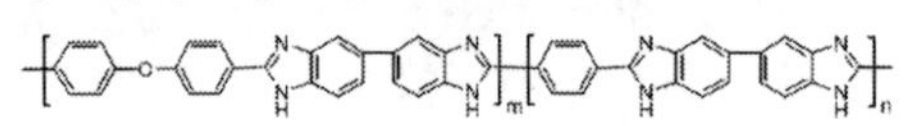 Poly(aryl ether benzimidazole) copolymers	
7.5 mol H_3PO_4/poly(aryl ether benzimidazole) copolymers	$9.8x10^{-2}$ S/cm at 180 °C [124]
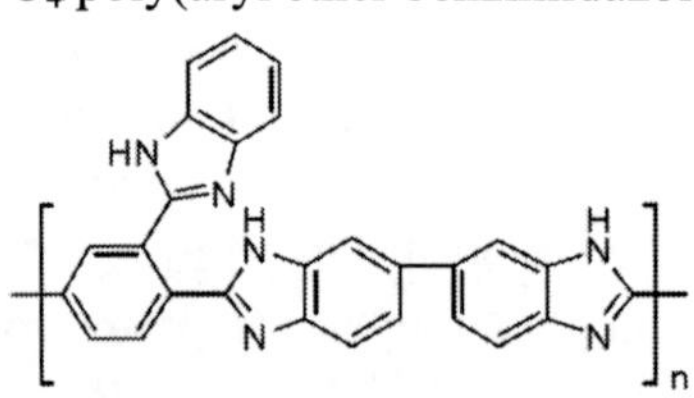 Poly[2,20-(2-benzimidazole-*p*-phenylene)-5,50-bibenzimidazole], BI*p*PBI	
>81 wt.% H_3PO_4/BI*p*PBI	0.16 S/cm at 180 °C [125]
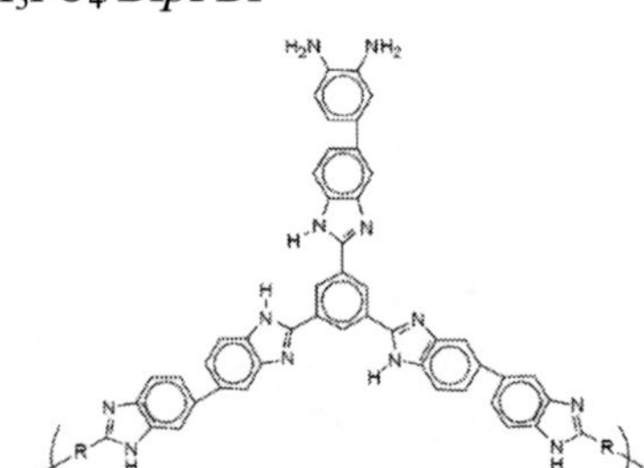 Hyperbranched polybenzimidazoles Phosphoric acid-doped hyperbranched polybenzimidazole	$6.4x10^{-2}$ S/cm at 170 °C [126]
Phosphoric acid-doped hyperbranched polybenzimidazole	
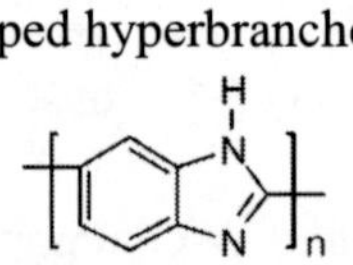 Poly(2,5-benzimidazole), ABPBI	
60 wt.% H_3PO_4/ABPBI	$6.0x10^{-2}$ S/cm at 110 °C [127]
2.7 H_3PO_4/ABPBI	$2.5x10^{-2}$ S/cm at 180 °C [128]
4.6 H_3PO_4/Sulfonated ABPBI	$3.5x10^{-2}$ S/cm at 185 °C [129]
84 wt.% H_3PO_4/Cross-linked ABPBI	0.12 S/cm at 150 °C [130]

Benzimidazole-based compounds	Proton Conductivity
Type V: Side chain structures	
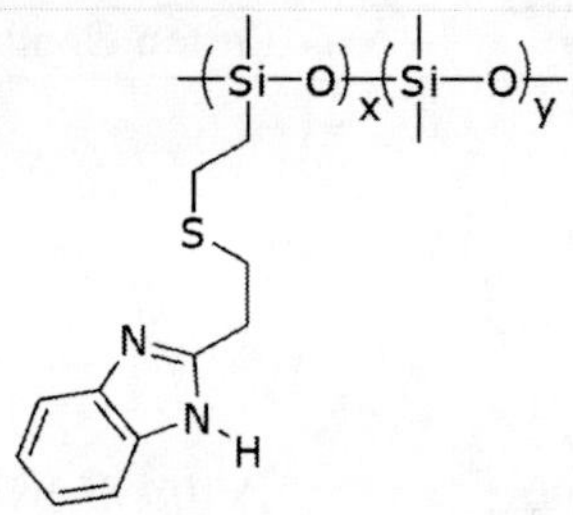 Benzimidazole-grafted polydimethylsiloxane, bimiPVMS	$7x10^{-6}$ S/cm at 160 °C [131]
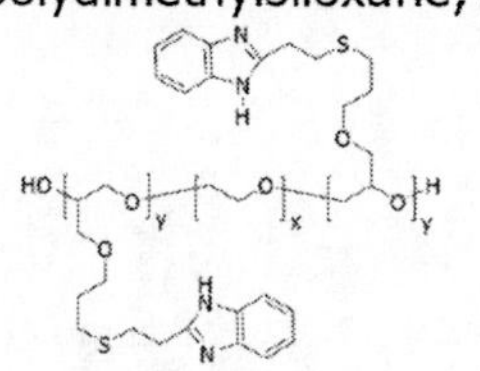 Benzimidazole tethered on linear poly(ethylene glycol), bimiPEOAGE	$80x10^{-6}$ S/cm at 160 °C [132]
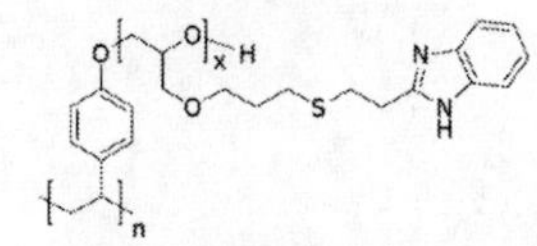 Benzimidazole tethered poly(styrene-g-allyl glycidyl ether)	$6.6x10^{-6}$ S/cm at 160 °C [133]
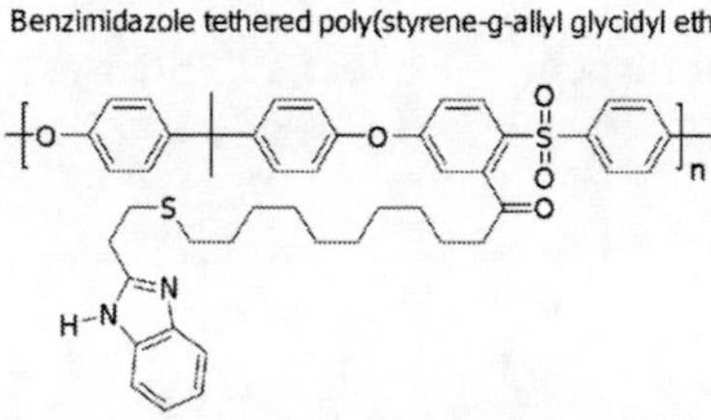 Polysulfones tethered with benzimidazole, $bimi_x$PSU	$34x10^{-9}$ S/cm at 180 ºC [134]
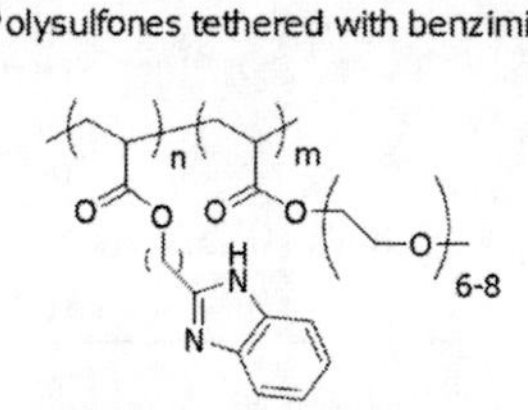 Benzimidazole functional acrylate polymer and poly (ethylene glycol) methyl ether acrylate (PEGMEA) copolymer, B5A-PEGMEA	10^{-5} S/cm at 200 °C [135]
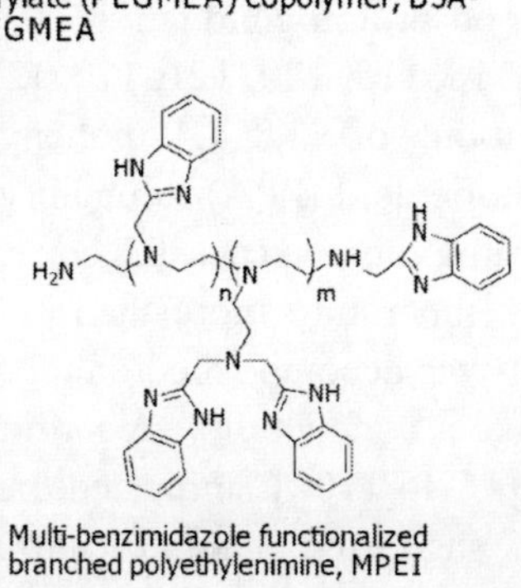 Multi-benzimidazole functionalized branched polyethylenimine, MPEI	10^{-5} S/cm at190 °C [136]

Table 2. (Continued)

Benzimidazole-based compounds	Proton Conductivity
Type VI: Free molecules	
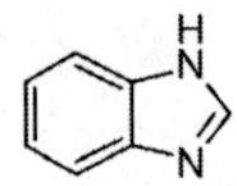	
Benzimidazole	
Benzimidazole/44 mol% Monododeccyl phosphate	1x10^{-2} S/cm at 120 °C [137]
Benzimidazole/SPEEK	7.6x10^{-4} S/cm at 120 ºC [138]
Benzimidazole/Polyphosphoric acid	4.6x10^{-2} S/cm at 170 ºC [139]
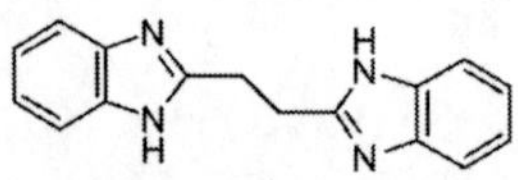	
Ethylene-1,2-di-2-benzimidazole	
Ethylene-1,2-di-2-benzimidazole/SPEEK	8.0x10^{-8} S/cm at 120 ºC [138]
Ethylene-1,2-di-2-benzimidazole/Polyphosphoric acid	7.8x10^{-2} S/cm at 170 ºC [139]
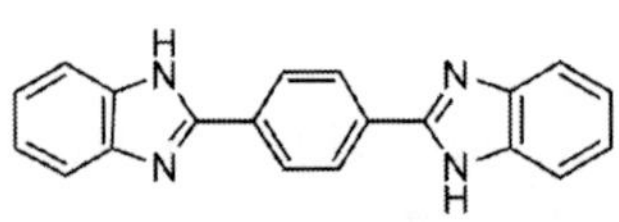	
1,4-Bis(2-benzimidazolyl)benzene	
1,4-Bis(2-benzimidazolyl)benzene/Polyphosphoric acid	9.2x10^{-3} S/cm at 170 ºC [139]
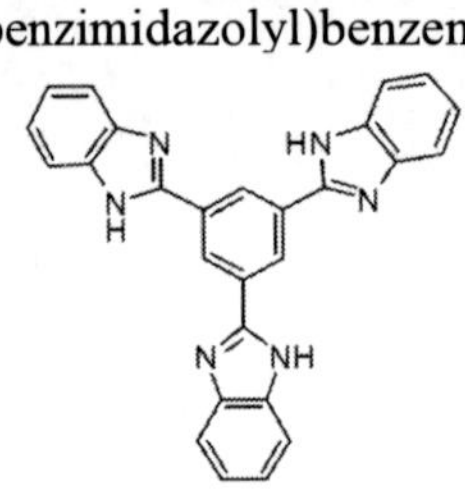	
1,3,5-Tri(2-benzimidazolyl)benzene	
1,3,5-tri(2-benzimidazolyl)benzene/SPEEK	8.1x10^{-} S/cm at 120 ºC [138]
1,3,5-tri(2-benzimidazolyl)benzene/Polyphosphoric acid	7.8x10^{-2} S/cm at 170 ºC [139]

For type IV, polybenzimidazole (PBI) is a commercial available membrane frequently used in high temperature or anhydrous PEMFC since it illustrates high thermal, chemical, oxidative, and mechanical stabilities as well as good proton conductivity (~10^{-1} S/cm under acid treatment even at temperature up to 200 ºC). [118, 119, 120, 121, 122, 123, 140]

PBI was synthesized from polycondensation of 3,3',4,4'-biphenyltetramine and terephthalic acid or isophthalic acid in polyphosphoric acid (PPA) to obtain *p*PBI and *m*PBI, respectively. [140] Phosphoric acid-doped PBI is the most successful one as it showed the proton conductivity as high as 10^{-1} S/cm at the temperature more than 100 ºC. When the polymer was mixed with acid, protonation of PBI was accomplished under strong hydrogen bond formation between PBI and acid. Its proton conductivity according to Arrhenius behavior under the temperature in the range of 25 ºC to 130 ºC indicated the proton transfer under Grotthuss mechanism. [140] Jiang et al. suggested a significant role of tridecyl phosphate (TP) in increasing proton conductivity with less migration as compared to that of

H_3PO_4 in PBI. For example, it was found that the blending content of TP and H_3PO_4 for 1.8 mol per PBI repeating unit brought the proton conductivity for 10^{-4} S/cm and 10^{-5} S/cm at 140 °C, respectively. [121] Although PBI, especially acid-doped one, showed a good performance in proton conduction at high temperature, some drawbacks were reported, e.g. the difficulty of the preparation of high molecular weight PBI, the polymer solubility, the membrane durability, the limitation of acid doping amount, and the retention of acid during operation, etc. Modified PBIs such as sulfonated PBI (SPBI), [122] epoxy-based cross-linked PBI, [123] poly(aryl ether benzimidazole) copolymers, [124] poly[2,20-(2-benzimidazole-*p*-phenylene)-5,50-bibenzimidazole] (BI*p*PBI), [125] and hyperbranched polybenzimidazoles (HBPBI) [126] were reported as alternative materials.

Mader et al. succeeded in preparing high molecular weight SPBI (M_v>1.0 dL g^{-1}) based on the PPA process (PPA: as a polycondensation agent and as a solvent for polymerization and membrane casting). [122] The membrane contained acid loading as high as 30-35 mol PA/PBI with robust mechanical properties (tensile stress for 0.804 MPa and strain for 69.64%) leading to an excellent proton conductivity more than 0.1 S/cm at temperature above 100 °C. The performance of the membranes was evaluated by fuel cell testing without external humidification. PA-doped SPBI showed a promising PEM efficiency for 0.62-0.68 V (at 0.2 A cm^{-2}, 160 °C) and 0.69-0.76 V (at 0.2 A cm^{-2}, 160 °C) under hydrogen/air system and hydrogen/oxygen system, respectively, with long-term stability for ~2900 h and a degradation rate of -0.030 mV h^{-1}. Most reports showed that PBI with higher acid doping gave rise to higher proton conductivity; in contrast, its chemical and mechanical properties became significantly reduced at high acid doping. [140, 141] To improve mechanical properties of PBI with high acid level, epoxy-based cross-linked PBI was proposed. [123] PBI with 5 wt.% crosslinker, 1,3-bis(2,3-epoxypropoxy)-2,2-dimethylpropane (NGDE) under acid uptake for 193%, not only exhibited good proton conductivity (0.017 S/cm at 200 °C) with excellent chemical and oxidative stabilities but also improved its mechanical and dimension stabilities. The introduction of aryl ether linkages into rigid structure of PBI significantly improved solubility of the polymer in polymerization solvent (PPA) and *N,N'*-dimethylacetamide/LiCl. [124] Kim et al. developed BI*p*PBI membranes, PBI containing bulky group of benzimidazole, showing an enhancement of acid uptake and acid retention capacity as a consequence of acid-base interaction. Moreover, as compared to *m*PBI, the BI*p*PBI was found to be more soluble in aprotic polar solvents such as DMAc, DMSO, and NMP at concentrations up to 1 wt.% at 100 °C without LiCl salt. [125] Xu et al. studied the use of hyperbranched polybenzimidazoles (HBPBI) membrane for high temperature PEM. They found that HBPBI with terephthaldehyde as a crosslinking agent not only showed high proton conductivity (0.064 S/cm at 170 °C and 0% RH) but also maintained high mechanical strength (tensile strength: 7.0 MPa and elongation at break: 44%) even at high PPA doping (436 g/100 g polymer) due to its insolubility in PPA which is superior to the common linear PBI one. [126]

Poly(2,5-benzimidazole) (ABPBI) is another promising benzimidazole-based polymer which can be easily synthesized by condensation of a single monomer of 3,4-diaminobenzoic acids with various methods such as polymerization in polyphosphoric acid or methanesulfonic acid/P_2O_5 medium. [128, 141, 142] The progress of the reaction significantly enhanced viscosity, tensile strength, and elongation of the materials obtained. [127] ABPBI is normally doped with acids, for example phosphoric acid and sulfuric acid to obtain potential membranes used in PEMFC. Proton conductivities of this type of membrane dramatically depended on acid doping level. For example, an enhancement of proton conductivity from $2.6x10^{-2}$ S/cm to $6.0x10^{-2}$ S/cm (110 °C) was found when varying the acid concentration from 35 wt.% to 60 wt.%, respectively. Asensio et al. demonstrated a method to enhance proton conductivity of ABPBI membranes by sulfonating the membrane in sulfuric acid followed by heat treatment at 450 °C for 3-5 min in the air. [129] The sulfonated ABPBI was impregnated in phosphoric acid and showed the sulfonation bringing in high PA absorption and as a result significant proton conductivity ($3.5x10^{-2}$ S/cm at 185 °C under dry state). Cross-linked structures of poly(2,5-benzimidazole) showed an improvement of chemical stability and an increase of PA absorption compared to ABPBI and *m*PBI membranes. The higher the content of PA, the higher the proton conductivity of the cross-linked ABPBI (0.12 S/cm at 150 °C with PA doping >84wt.%). Although ABPBI could be prepared simply and showed a good performance in transferring proton, poor solubility in organic solvent limited the membrane preparation.

Persson et al. developed various types of benzimidazole tethered polymer chains (Type V) via thiol-ene coupling reaction, i.e. benzimidazole-grafted polydimethylsiloxane (bimiPVMS), [131] benzimidazole tethered on linear poly(ethylene glycol) (bimiPEOAGE), [132] benzimidazole-tethered poly(styrene-*g*-allyl glycidyl ether), (bimiPSAGE*x*), [133] and polysulfones tethered benzimidazole (bimi*x*PSU). [134] Considering bimiPVMS with various degrees of benzimidazole functionalities, it was found that the glass transition temperature (T_g) increased as a consequence of inter- and intramolecular hydrogen bonds among benzimidazole units. [131] Proton transfer efficiency of the bimiPVMS membranes was found to be promoted when the polymers showed high segmental mobility (low T_g) at low temperature (60 °C) and high benzimidazole concentration at high temperature (140 °C). However, thermal stability of the bimiPVMS membrane was limited at 165 °C. The study of bimiPEOAGE triblock copolymer membranes also showed that both segmental mobility and benzimidazole concentration were the main factors controlling proton conductivity with an improvement of the thermal stability as high as 210 °C. [131] Persson et al. also reported that bimiPSAGE7.0 with lower T_g (T_g = 50 °C) showed a higher proton conductivity than bimiPSAGE2.4 (T_g = 57 °C) especially at low temperature. [131] For bimi*x*PSU systems, it was found that their proton conductivities increased with an increase of benzimidazole substitution. However, as a result of low concentration of benzimidazole units, the percolation of hydrogen bond network was not high enough to allow an efficient proton transfer, resulting in only modest proton conductivity of $\sim 10^{-9}$ S/cm at 180 °C. [134] Woudenberg et al. also studied the effect of glass transition temperature on proton conductivity of benzimidazole functional acrylates and poly(ethylene glycol) methyl ether acrylate copolymer (B5A-PEGMEA) and suggested that lowering the T_g from 62 °C to 2 °C resulted in an increase of proton conductivity from 0.0092 to 0.011 mS/cm at 200 °C. [135] Pangon et al. developed benzimidazoles branched (MPEI) with varied benzimidazole substitutions for 19.7-90.5% and found that proton conductivity was enhanced through a higher molecular mobility of the polymer chain under hydrogen bond network. They found that MPEI with a benzimidazole content for 19.7% allowed preferable proton conductivity as high as 10^{-4} S/cm at 190 °C. Therefore, they concluded that this level of benzimidazole substitution brought a good balance between hydrogen bond and chain mobility to favor the proton movement. [136]

In the case of type VI, Totsatitpaisan et al. studied a series of benzimidazole model compounds with varied numbers of benzimidazole units, i.e. mono-, di-, and tri-functional benzimidazole model compounds to investigate how the structures develop their hydrogen bond systems to favor the proton transfer. [138, 139, 143] Single crystal analysis revealed that hydrogen bond systems of the model compounds were developed from lamella hydrogen bond network in the case of mono- and di-functional benzimidazoles to columnar hydrogen bond network in the case of tri-functional benzimidazole. Proton conductivities of the model compounds were evaluated by doping them with PPA to find that the proton conductivity was the highest for tri-functional type ($7.8x10^{-2}$ S/cm at 170 °C). By blending this tri-functional compound (15 phr) to the SPEEK, the conductivity became as high as $8.1x10^{-4}$ S/cm at 120°C or 1.3 times higher than the pure one.

Other nitrogen-containing heterocycles for example pyrazole, triazole, and tetrazole derivatives were also investigated as polymer membranes used in high temperature PEMFC (Figure 4).

Pyrazole (Py) was studied as a solvent for proton transfer by intercalating into sulfonated polyetherketone (SPEK) membrane. [98] The performance in transferring protons of SPEK mixed with Py in a 10 times mole ratio was comparable with Nafion® under hydrated state. Pure 1*H*-1,2,4-triazole performed self-dissociation in order to generate proton charge carriers allowing a proton movement similar to the case of imidazole. They showed proton conductivity as high as $1.5x10^{-4}$ S/cm at 115 °C and $1.2x10^{-3}$ S/cm at melting point (120°C). [144] Mixing triazoles with large acid molecules like 4-dodecylbenzenesulfonic acid increased the proton charge carrier concentration and suppressed acid species motion resulting in an increment of proton conductivity ($2.0x10^{-2}$ S/cm at 120 °C with 9 mol% acid). Triazoles showed the potential use as PEM at temperature above 100 °C as observed the proton conductivities of the triazole blended sulfonated polysulfones (SPSU) and H_3PO_4 doped triazoles grafted onto hybrid inorganic-organic polymer network membranes (Figure 5) at around 10^{-3} S/cm (140 °C) under anhydrous system. Anhydrous Nafion-based membranes containing various types of azole compounds, i.e. 1*H*-1,2,4-triazole (Tri), 3-amino-1,2,4-triazole (ATri), and 5-aminotetrazole (ATet) were studied by Sen et al. [145] In the Nafion-Tri and Nafion-ATri systems, hydrogen bond networks were formed as a result of proton dissociation of sulfonic acid and partial protonation of azole. They found that the introduction of azole units, especially ATri, into hydrophilic channels of Nafion® led to significant proton conductivity at 10^{-3} S/cm (at 180 °C under anhydrous conditions) and lower methanol permeability of 2 times compared to the pristine Nafion®112.

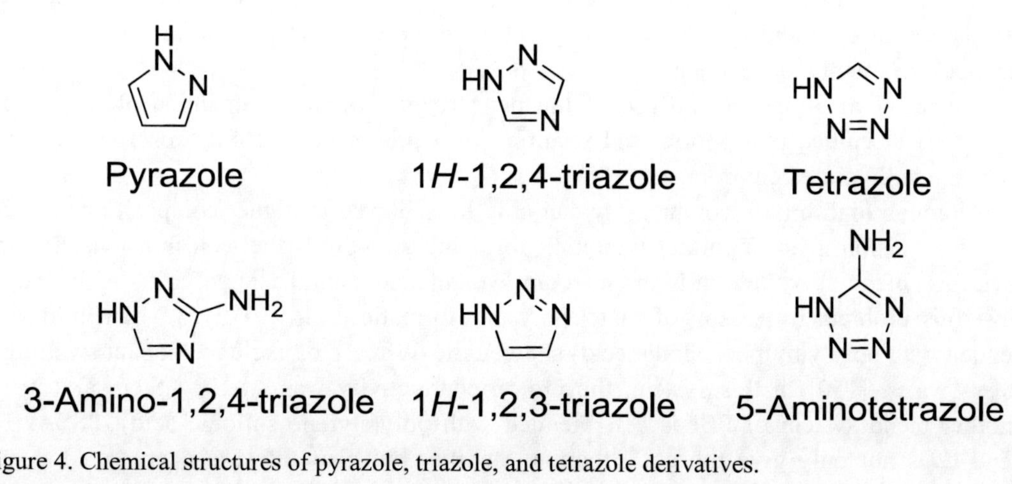

Figure 4. Chemical structures of pyrazole, triazole, and tetrazole derivatives.

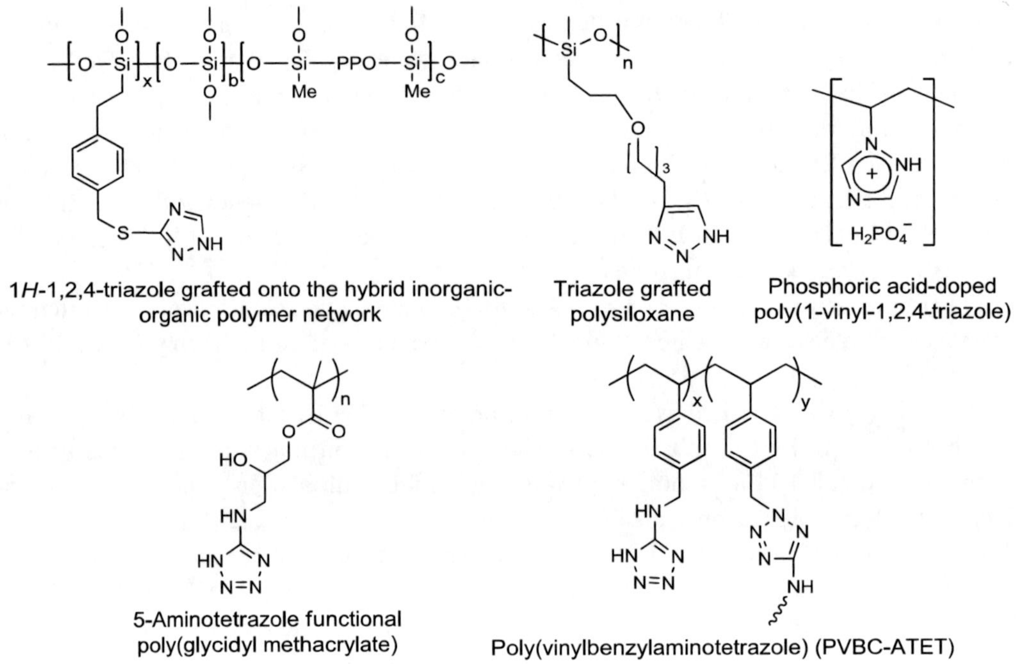

Figure 5. Triazole and tetrazole based PEMs.

Granados-Focil et al. studied 1*H*-1,2,3-triazole grafted polysiloxanes with variable alkyl chain spacer length as a water free PEM and concluded that the proton conductivity of the system was mainly controlled by the glass transition temperature of the polymer matrix and the volume fraction of proton carriers. [146] They also found that the pKa of the conductive heterocycles insignificantly influenced on their performance in transferring protons. Nonlinear temperature dependence proton conductivity of the polysiloxane-based polymers suggested proton transfer under mobility of the polymer matrix. Çelik et al. developed homopolymer membranes of poly(1-vinyl-1,2,4-triazole) via free radical polymerization, and studied their proton transfer efficiency at high temperature. [147]

It should be noted that heterocycles, in some cases, were applied in hydrous system to find that the conductivities were in the range of 10^{-3} S/cm. In aqueous, the high degree of dissociation of oxo acids, i.e. phosphoric acid, sulfuric acid, and perchloric acid in solution enhances the membrane efficiency.

Bhadra et al. proposed the use of the net structure of poly(benzimidazole-*co*-aniline) (Figure 6) to embed phosphoric acid resulting in a proton conductivity for 167 mS/cm at 120°C under hydrated environment. [148]

Although high proton conductivity can only be achieved at higher compositions of acid, dopant exclusion is an important limitation for using anhydrous fuel cell in a long lifetime. Therefore, blending system of high molecular weight acidic and basic polymers is the way to solve this problem by reason of their 'polysalt' formation. [149] Poly(2,5-benzimidazole) blended with poly(vinylphosphonic acid) showed the decrease of excluded dopant swelling in excess water. [150] In this system, the proton conductivity was 1.8×10^{-6} S/cm at 150 °C. Another blend system of PBI is PBI blended with poly(styrene sulfonic acid), PSSA. The blend does not only give the good thermal stability but also shows a decrease of dopant exclusion due to the complexation of PBI and PSSA.

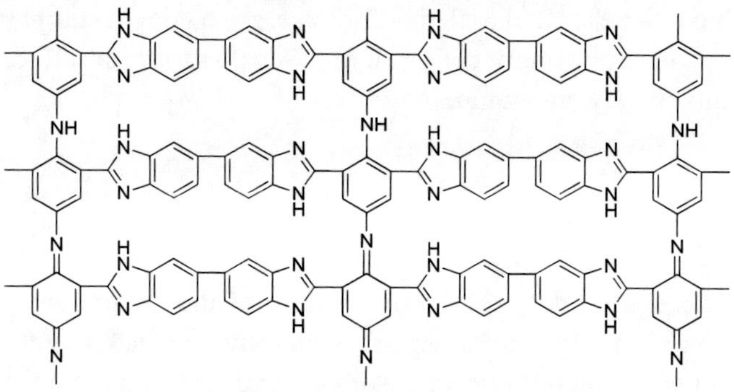

Poly(benzimidazole-*co*-aniline), PBIANI

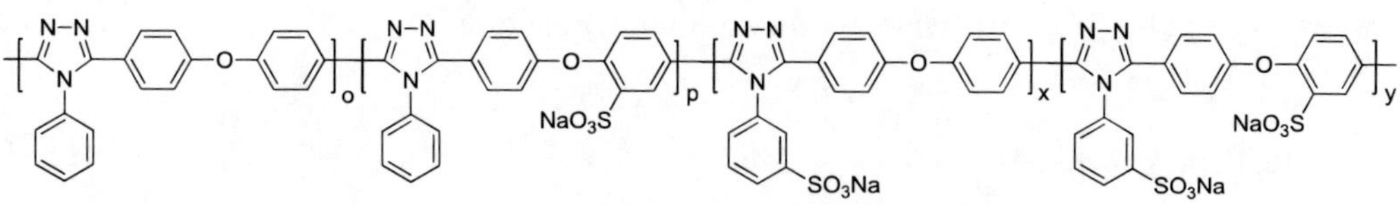

Sulfonated polytriazole polymer (S-DPE-PT)

Figure 6. Heterocyclic-based PEMs operating under relatively low humidity or high temperature.

The proton conductivity of this blended membrane was around 10^{-3} S/cm at high temperature for PBI:PSSA blend with 1:4 molar ratio. However, the acid-base membrane is not the only way to improve dopant exclusion and the thermal stability simultaneously. [151] Kerres and coworkers have reported several acid-base blends of sulfonated polysulfones with various types of *N*-containing basic polymers such as polybenzimidazole and poly(4-pyridine). [21, 152] Poly(acrylic-*co*-4,5-vinylimidazole) was synthesized as single chain acid-base polymer however this membrane is fragile and not practical to use in anhydrous fuel cell system. The blend of poly(acrylic-*co*-4,5-vinylimidazole) with SPEEK can give better thermal stability and mechanical properties for forming as a membrane. The proton conductivity of this membrane is 2.4×10^{-2} S/cm at high temperature (approximately 130 °C under anhydrous state). [153]

Pangon and coworker proposed the concept of proton donor-acceptor system of PAA containing multi-benzimidazole units to facilitate the proton transfer with a balance of hydrogen bond network and chain mobility. They successfully synthesized the modified polymers with various degrees of substitution of branching tribenzimidazole. The proton conductivity of the membrane with tribenzimidazole content for 2.8 mol% was around 10^{-5} S/cm at high temperature. [154]

5. Evaluations of PEM

Apart from proton conductivity, PEM performances are also related to other physicochemical properties. For hydrous system, membrane water uptake, swelling ratio, and ion exchange capacity are points to be considered. The use of membrane in anhydrous system under high temperature needs information related to thermal properties, i.e. glass transition temperature (T_g) and degradation temperature (T_d). Gas permeability and methanol crossover of the membranes have to be well clarified before use. Also, single cell performance is a basic guideline to further develop stack cells. Herein, evaluations of all different parameters including experimental details are summarized.

5.1. Water Uptake

Water uptake is based on the molecular structure of the membrane, i.e. ion cluster size. [155] The water uptake in the membrane is generally evaluated with the procedures as follows: (i) immersing the membrane in distilled water at room temperature for 48 h and weighed to obtain W_w; (ii) drying the wet membrane under vacuum at 80 °C for 24 h to obtain dry membrane (W_d); and (iii) calculating the water uptake from the weight difference between wet and dry membranes using the following equation:

$$\text{Water Uptake} = \left(\frac{W_w - W_d}{W_w}\right) \times 100\% \qquad ; W = \text{weight} \qquad \text{(eq. 1)}$$

In addition, the water uptake is also evaluated by thermogravimetric analysis. The water content in the membrane is considered as the percent weight loss of the hydrated membrane in the temperature range of 60-150 °C which is the water evaporation temperature.

5.2. Swelling Ratio

Membranes are cut with area size of 10 mm × 50 mm and weighed before measurement. [72] After that, the membranes were immersed in distilled water at room temperature for 24 h. The weight and dimensional change of films were measured. Swelling ratio is calculated based on the area (A) of the membrane before and after immersing in distilled water.

$$\text{Swelling Ratio} = \left(\frac{A_{wet} - A_{dry}}{A_{dry}}\right) \times 100 \; ; A = \text{Area} \qquad \text{(eq. 2)}$$

It should be noted that when a swelling ratio of the membrane is high, the membrane resistance will be high resulting in poorer proton conductivity.

5.3. Gas Permeabilty

Gas permeation is an important parameter of the membranes used in PEMFCs since the system needs membrane without gas leakage ideally.

From Figure 7, the membrane under test is placed in between the filter and sealing ring. Feed side is supplied by gas or vapor while permeate side is a vacuum. The gas permeability is measured by the pressure changed at the permeate side.

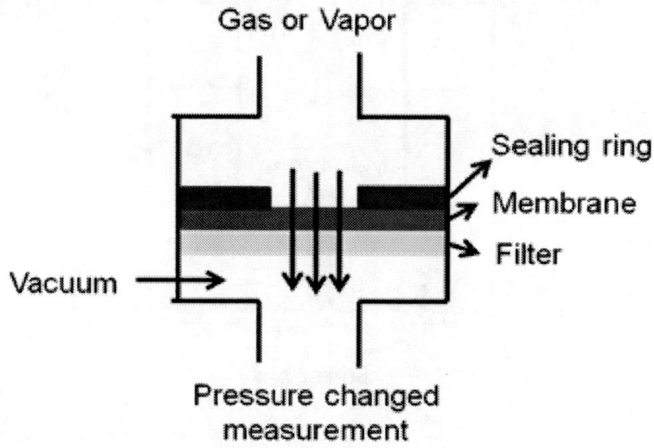

Figure 7. Schematic draw for a gas permeability measurement.

The gas permeation is expressed by the gas permeability coefficient, P, and can be calculated as follow: [156]

$$P = \frac{L \times V_p \times V_m}{A \times T \times \Delta P \times R}\left(\frac{dP}{dt}\right) \quad \text{(eq. 3)}$$

where L is the membrane thickness (cm), V_p is the permeate volume (cm^3), V_m is the molar volume of the gas at standard temperature and pressure, A is the membrane surface area (cm^2), T is the absolute temperature (K), dP is the pressure difference between feed and permeate parts (cmHg), and dP/dt is the changing rate of permeate pressure (cmHg/s).

5.4. Methanol Permeability

In the case of the membranes for DMFC systems, the ability to be a barrier not only for gas permeation and electron transfer but also for methanol crossover has to be considered. Methanol permeability and excessive swelling in alcohol environment cause an over potential at the cathode and the decrease in mechanical strength, respectively. [71] In general, methanol permeability through the membrane is determined by either diffusion or pervaporation system. From Figure 8, the diffusion system composes of two compartments with the test membrane in between.

Compartment A is filled with methanol solution while compartment B contains water. The sample solutions from both compartments are taken at 1-1.5 h interval of time. The methanol concentration in each solution is analyzed by gas chromatography (GC). The methanol permeability, P (cm^2/s), is calculated as the following. [157]

$$P = \frac{V_B \times L \times \frac{dC_{B(t)}}{dt}}{C_A \times A} \quad \text{(eq. 4)}$$

where C_A and C_B are the methanol concentrations in the compartments A and B (%(v/v)), respectively, t is the permeation time (s), A is the exposed area of the membrane (cm^2), L is the thickness of the membrane (cm), and V_B is the solution volume of the compartment B.

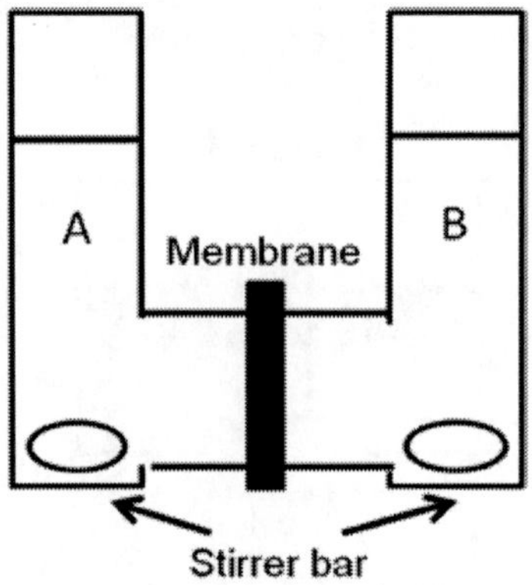

Figure 8. Diffusion system.

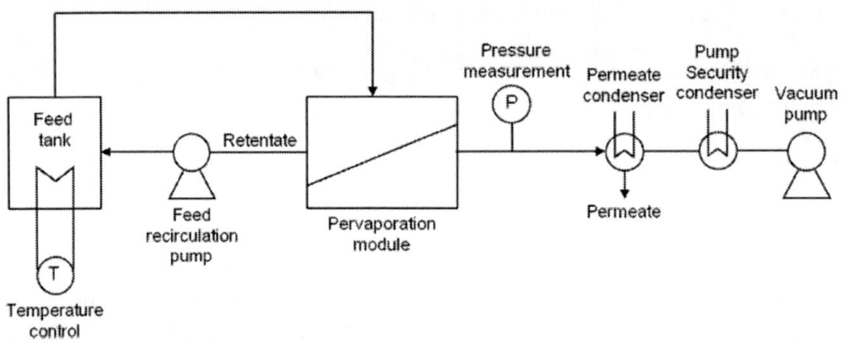

Figure 9. Pervaporation system.

In the case of the pervaporation system (Figure 9), the membrane is placed in a feed chamber with circulated methanol, at the same time, the permeate side is evacuated. The permeated vapor is collected via liquid nitrogen every 1-1.5 h. The vapor is weighed and analyzed by gas chromatography. The methanol permeability is estimated by the following equations based on the total (J_{tot}) and methanol (J_{MeOH}) fluxes.

$$J_{tot} = \frac{W_{perm} \times L}{t \times A} \qquad \text{(eq. 5)}$$

$$J_{MeOH} = \frac{J_{tot} \times C_{MeOH}}{100} \qquad \text{(eq. 6)}$$

where W_{perm} is the weight of permeated methanol solution, L and A are thickness and surface area of the membrane, respectively, t is permeation time, and C_{MeOH} is the methanol concentration.

5.5. Ion Exchange Capacity

Ion exchange capacity (IEC) of the membrane is an important factor to indicate the efficiency of protons transferring through the membrane. In general, the membrane is immersed in 1 M NaCl for 48 h at ambient temperature. By titration of NaCl solution with

0.01 M NaOH solution using phenolphthalein as indicator, IEC can be varied depending on the number proton exchangeable groups. The volume of NaOH solution titrated is expressed in term of milli-equivalent of the exchanged proton per dry polymer weight (meq_{H+}/g). [158]

$$IEC\,(mmol/g) = \left(\frac{\Delta V_{NaOH} \times C_{NaOH}}{Dry\,sample\,weight}\right) \times 100 \qquad \text{(eq. 7)}$$

5.6. Proton Conductivity

The complex impedance spectroscopy is used in order to evaluate the electrical property of the membrane by applying alternating current with various frequencies. A typical impedance response (Z′–Z″ plot) is obtained (see Figure 10). The impedance (Z) is defined as the ratio of the voltage to the current at a given frequency which has a complex quantity of real (Z′) and imaginary (Z″) parts with a phase angle (θ) as shown in the equation 8-10.

$$Z' = |Z| \cos\theta \qquad \text{(eq. 8)}$$

$$Z'' = |Z| \sin\theta \qquad \text{(eq. 9)}$$

$$\theta = \tan^{-1}(Z'/Z'') \qquad \text{(eq. 10)}$$

Ohmic loss is a key driving performance of fuel cell. There are three potential causes of ohmic voltage loss: (a) resistance to ion migration in the electrolyte, (b) resistance to electron transport in the cell components, and (c) contact resistances. Only small amounts of ohmic resistance (on the order of milliohms) have a drastically effect on overall efficiency. This leads to smaller current densities in fuel cells. However, the internal resistance of the cell during operation consists of a complex function of many parameters, e.g. temperature, hydration, current densities, etc. [159]

Generally, the proton conductivity (σ) at various temperatures is calculated by $\sigma = L/RA$, where L and A are the thickness and surface area of the membrane, respectively. The ohmic resistance of the membrane, R, is obtained from the extrapolation of x-axis intercept.

In order to develop novel PEM material, complex impedance measurements of commercial Nafion® have been demonstrated frequently in many research papers. Theoretically, a dry polymer membrane shows dielectric relaxation, so-called insulator, especially at low temperatures. The dielectric relaxation is the reorientation of electrical dipoles where the centers are fixed in position. It is fascinating that even dry SPEEK has significantly high ion mobility. In fact, it is extremely difficult to remove all traces of water from the material. Figure 10 is an example of the plot obtained from the authors' laboratory illustrating a complex impedance diagram of SPEEK which frequencies increase from right to left. The intercept of arc is shifted from high to low resistance as water content increases significantly. The calculated conductivity can be easily replotted in terms of proton conductivity related to temperature. Figure 11 is the proton conductivity of SPEEK membrane prepared in the authors' laboratory which shows the conductivity for 0.38 mS/cm at 70°C.

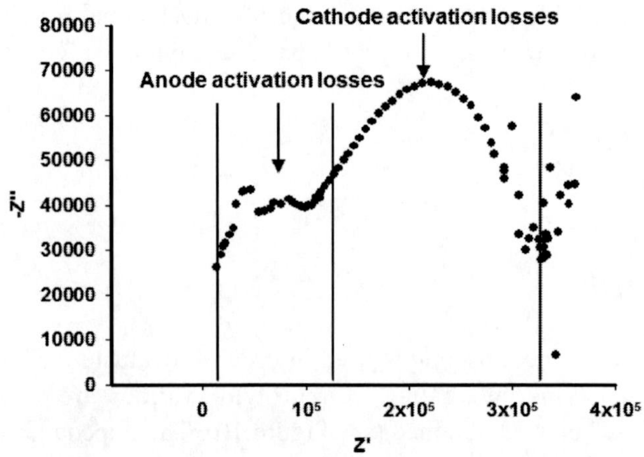

Figure 10. Example of Nyquist (impedance) plot of a hypothetical PEM fuel cell stack and two losses in the system.

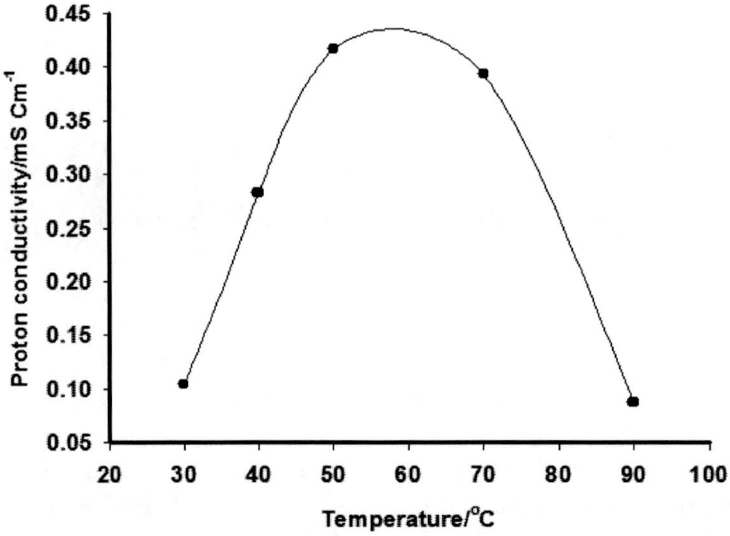

Figure 11. Proton conductivity as a function of temperature of an in-house prepared SPEEK.

5.7. Single Cell Performance

As fuel cells are electrochemical devices that convert the chemical energy of redox reactions directly into electric energy, the performance of the fuel cells based on the generated electricity and voltage is evaluated by the single cell test either with humidified hydrogen gas or aqueous methanol solution for supplying the anode.

Membrane electrode assembly (MEA) (Figure 12) is prepared by compressing the membrane in between two electrodes with temperature and pressure (80 °C and 2 tons, respectively). In the case of hydrogen-based PEMFC, Pt black loaded on carbon cloth are used for both electrodes while for those of DMFC, Pt-Ru alloy and Pt black are used for anode and cathode, respectively. Thereafter, the MEA is put together with the bipolar plates and gas diffusion layers (GDLs) to obtain the single fuel cell. Figure 13 shows a cylindrical cell developed in the author's lab.

The current and voltage generated from the single cell are measured and revealed as the plot of voltage and power density versus the current density named as polarization curve. Theoretically, the standard potential (E°) of a single fuel cell using hydrogen and oxygen as fuel and oxidant is 1.229 V. The actual potential is continually decreased from its equilibrium because of irreversible losses. The losses, which are called polarization, over-potential or over-voltage, originate primarily from three sources: (i) activation polarization; (ii) ohmic polarization; and (iii) concentration polarization. These losses result in reduction of cell voltage (E) which is less than the ideal potential, E° ($E = E° - E_{losses}$).

Figure 12. (a) Membrane electrode assembly (MEA) and (b) a single cell and instruments for fuel cell performance evaluation.

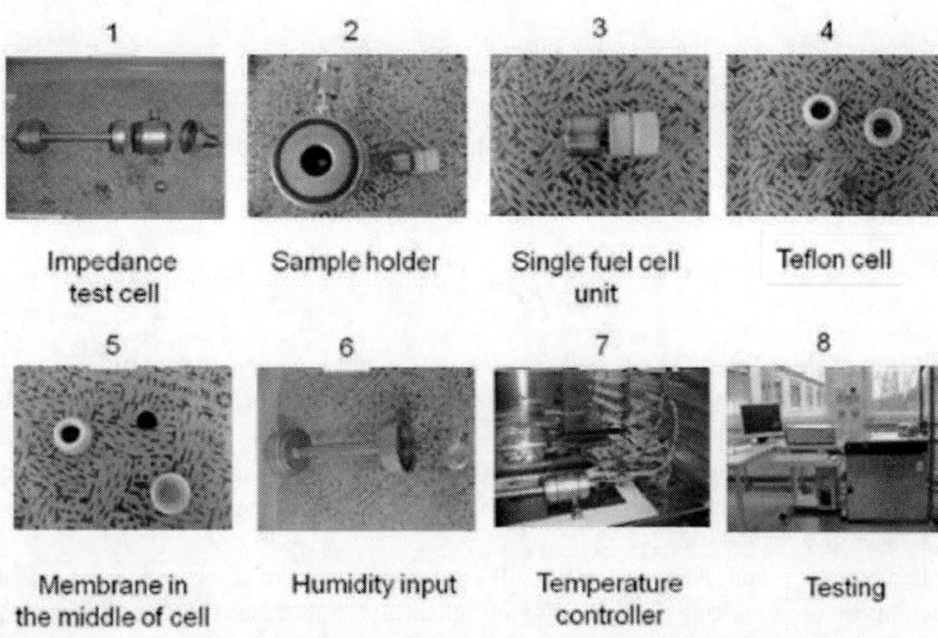

Figure 13. Cylindrical cell for proton conductivity measurement under anhydrous system.

CONCLUSION

At present, perfluorosulfonated membranes are well accepted as the most reliable PEM. The fact that the membranes perform under water saturation, the best performance is limited at below moisture evaporation condition. The high operating temperature accelerates proton transfer as well as cleaning up the CO contamination (which comes with H_2) on the Pt catalyst, therefore, the high temperature resistance polymer electrolyte membranes are needed. In the current years, for hydrous membranes, the uses of organic and inorganic additives, and the composites with inorganic compounds were found to improve the thermal stability (to 80-100 °C) as well as the lowering of fuel crossover. In those cases, the problem related to phase separation between polymer matrices and additives is the point to be further developed. For anhydrous membranes, heterocycles, especially imidazoles and benzimidazoles are the molecules that allow proton transfer via their resonance structures. Several types of copolymers and polymer blends between sulfonated polymers, e.g. SPEEK, SPEK, SPSU, SPI, etc. and heterocyclic compounds were found to be good approaches to obtain membranes with conductivity in the range of 10^{-5}-10^{-2} S/cm at the operating temperature as high as 120-180 °C. Model case studies on the variation of benzimidazole structures lead us to an understanding of proton transfer related to hydrogen bond network and molecular mobility of the heterocyclic-based proton conductive species.

Although heterocyclic polymers are alternative materials for anhydrous PEM membranes, high efficient proton conductivity is still on the expectation. On this viewpoint, a depth understanding of the primary structure such as resonance structure, hydrogen bonds, chain mobility, and the secondary structure such as chain packing, crystalline-amorphous phase, including microdomains and morphologies should give a guideline for polymer designs and syntheses in the future.

ACKNOWLEDGMENTS

The authors' works were supported by the Higher Education Research Promotion and National Research University Project of Thailand, Office of the Higher Education Commission (EN276B).

REFERENCES

[1] Williams, M. C. 2002. *Fuel Cell Handbook*, 7th ed., U.S. department of Energy, Office of Fossil Energy, National Energy Technology Laboratory, Morgantown West Virginia Press.

[2] Costamagna, P., and Srinivasan, S. 2001. Quantum jumps in the PEMFC science and technology from the 1960s to the year 2000 Part I. Fundamental scientific aspects. *J. Power Sources,* 102: 242–252.

[3] Jiang, J. H., and Kucernak, A. 2004. Investigations of fuel cell reactions at the composite microelectrode/solid polymer electrolyte interface. I. Hydrogen oxidation at the nanostructured Pt/Nafion® membrane interface. *J. Electroanal. Chem.,* 567: 123–137.

[4] Wilson, M. S., Derouin, C., Valerio, J., and Gottesfeld, S. 1993. *Electrocatalysis Issues in Polymer Electrolyte Fuel Cells. Proc. 28th IECEC*, 1.1203-1.1208.

[5] Yang, C., Costamagna, P., Srinivasan, S., Benziger, J., and Bocarsly, A. B. 2001. Approaches and technical challenges to high temperature operation of proton exchange membrane fuel cells. *J. Power Sources*, 103: 1–9.

[6] Kwon, K., Yoo, D. Y., and Park, J. O., 2008. Experimental factors that influence carbon monoxide tolerance of high-temperature proton-exchange membrane fuel cells. *J. Power Sources*, 185: 202–206.

[7] Frank, G. 2003. Proceedings of the 2nd European PEFC Forum, June 30-July 4, Lucerne, Switzerland: 749.

[8] Alberti, G., Casciola, M., Massinelli, L., and Bauer, B. 2001. Polymeric proton conducting membranes for medium temperature fuel cells (110-160 ºC). *J. Memb. Sci.*, 185: 73-81.

[9] Devanathan, R. 2008. Recent developments in proton exchange membranes for fuel cells. *Energy Environ. Sci.*, 1: 101-119.

[10] Kreuer, K. D. 2001. On the development of proton conducting polymer membranes for hydrogen and methanol fuel cells. *J. Membr. Sci.*, 185: 29–39.

[11] Curtin, D. E., Lousenberg, R. D., Henry, T. J., Tangeman, P. C., and Tisack, M. E. 2004. Advanced materials for improved PEMFC performance and life. *J. Power Sources*,131: 41-48.

[12] Kinumoto, T., Inaba, M., Nakayama, Y., Ogata, K., Umebayashi, R., Tasaka, A., Iriyama, Y., Abe, T., and Ogumi, Z. 2006. Durability of perfluorinated ionomer membrane against hydrogen peroxide. *J. Power Sources*, *158*: 1222-1228.

[13] Heitner-Wirguim, C. 1996. Recent advances in perfluorinated ionomer membranes: structure properties and applications. *J. Membr. Sci.,* 120: 1-33.

[14] Gierke, T. D., Hsu, W. Y., Eisenberg, A., and Yeager, H.L. 1982. Perfluorinated ionomer membranes. *Paper presented at ACS Symposium Series No*. 180, American Chemical Society, Washington, DC, USA.

[15] Eikerling, M., Kornyshev, A. A., and Stimming, U. 1997. Electrophysical properties of polymer electrolyte membranes: A random network model. *J. Phys. Chem. B,* 101: 10807-10820.

[16] Alberti, G., Boccali,L., Casciola, M., Massinelli, L., and Montoneri, E. 1996. Protonic conductivity of layered zirconium phosphonates containing –SO3H groups. III. Preparation and characterization of γ-zirconium sulfoaryl phosphonates. *Solid State lonics*, 84: 97-104.

[17] Shao, Z.-G., Xu, H., Li, M., and Hsing, I.-M. 2006. Hybrid Nafion–inorganic oxides membrane doped with heteropolyacids for high temperature operation of proton exchange membrane fuel cell. *Solid State Ionics*, 177: 779–785.

[18] Kreuer, K. D., Schuster, M., Obliers, B., Diat, O., Traub, U., Fuchs, A., Klock, U., Paddison, S. J., and Maier, J.2008. Short-side-chain proton conducting perfluorosulfonic acid ionomers: Why they perform better in PEM fuel cells. *J. Power Sources*, 178: 499–509.

[19] Perona,J., Mani, A., Zhao, X., Edwards, D., Adachi, M., Soboleva, T., Shi, Z., Xie, Z., Navessin, T., and Holdcroft, S. 2010. Properties of Nafion® NR-211 membranes for PEMFCs. *J. Membr. Sci.*, 356: 44–51.

[20] Kreuer, K. D., Weppner, W., and Rabenau, A. 1982. Vehicle mechanism, a new model for the interpretation of the conductivity of fast proton conductors. *Angew. Chem. Int. Ed. Engl.*, 21: 208-209.

[21] Kerres, J. A. 2001. Development of ionomer membranes for fuel cells. *J. Membr. Sci.*, 185:3-27.

[22] Zaidi, S. M. J., and Matsuura T. 2009. Research trends in polymer electrolyte membranes for PEMFC. *Polymer Membranes for Fuel Cells*, 7-25.

[23] Smitha, B., Sridhar, S., and Khan, A. A. 2005. Solid polymer electrolyte membranes for fuel cell applications-a review. *J. Membr. Sci.*, 259: 10-26.

[24] Surampudi, S., Narayanan, S. R., Vamos, E., Frank, H., Halpert, G., LaConti, A., Kosek, J., Surya Prakash, G. K., and Olah, G. A. 1994. Advances in direct oxidation methanol fuel cells, *J. Power Sources*, 47: 377–385.

[25] Antonucci, P. L., Arico, A. S., Creti, P., Ramunni, E., and Antonucci, V. 1999. Investigation of a direct methanol fuel cell based on a composite Nafion-silica electrolyte for high temperature operation. *Solid State Ionics*, 125: 431–437.

[26] Rikukawa, M., and Sanui, K. 2000. Proton-conducting polymer electrolyte membranes based on hydrocarbon polymers. *Prog. Polym. Sci.*, 25: 1463-1502.

[27] Hietala, S., Maunu, S. L., and Sundholm, F. 2000. Sorption and diffusion of methanol and water in PVDF-*g*-PSSA and Nafion 117 polymer electrolyte membranes, *J. Polym. Sci., B: Polym. Phys.* 38: 3277-3284.

[28] Cho, K.-Y., Eom, J.-Y., Jung, H.-Y., Choi, N.-S., Lee, Y. M., Park, J.-K., Choi J.-H., Park, K.-W., and Sung, Y.-E. 2004. Characteristics of PVdF copolymer/Nafion blend membrane for direct methanol fuel cell (DMFC). *Electrochim. Acta*, 50: 583–588.

[29] Deligöz, H., Yılmaztürk, S., Karaca, T., Özdemir, H., Koç, S. N., Öksüzömer, F., Durmuş, A., and Gürkaynak, M. A. 2009. Self-assembled polyelectrolyte multilayered films on Nafion with lowered methanol cross-over for DMFC applications. *J. Membr. Sci.* 326: 643–649.

[30] Ladewig, B. P., Knott, R. B., Martin, D. J., Costa, J. D., and Lu, G. Q. 2007. Nafion-MPMDMS nanocomposite membranes with low methanol permeability. *Electrochem. Commun.*, 9: 781–786.

[31] Neburchilov, V., Martin, J., Wang, H., and Zhang, J. 2007. A review of polymer electrolyte membranes for direct methanol fuel cells. *J. Power Sources*, 169: 221–238.

[32] Hogarth, W. H. J., and Benzigera, J. B. 2006. Dynamics of autohumidifed PEM fuel cell operation, *J. Electrochem. Soc.*, 153: A2139–A2146.

[33] Savinell, R., Yeager, E., Tryk, D., Landau, U., Wainwright, J., Weng, D., Lux, K., Litt, M., and Rogers, C. 1994. A Polymer electrolyte for operation at temperatures up to 200 °C. *J. Electrochem. Soc.* 141: L46-L48.

[34] Mex, L. and Muller, J. 1999. Plasma-polymerized electrolyte membrane for miniaturized DMFC. *Membrane Technology*, 115: 5-9.

[35] Finsterwalder, F. and Hambitzer, G. 2001. Proton conductive thin films prepared by plasma polymerization. *J. Membr. Sci.*, 185: 105-124.

[36] Adjemian, K. T. Srinivasan, S. Benziger, J., and Bocarsly, A. B. 2002. Investigation of PEMFC operation above 100 °C employing perfluorosulfonic acid silicon oxide composite membranes. *J. Power Sources*, 109: 356–364.

[37] Ke, C.-C., Li, X.-J., Shen, Q., Qua, S.-G., Shao, Z.-G., and Yi, B.-L. 2011. Investigation on sulfuric acid sulfonation of in-situ sol-gel derived Nafion/SiO2 composite membrane. *Int. J. Hydrogen Energy*, 36: 3606-3613.

[38] Adjemian, K. T., Dominey, K., Krishnan, L., Hitoshi, O., Majsztrik, P. Zhang, T., Mann, J. Kirby, B. Gatto, L., Velo-Simpson, M., Leahy, J., Srinivasan, S., Benziger, J. B. Bocarsly, A. B. 2006, Function and Characterization of Metal Oxide-Nafion Composite Membranes for Elevated-Temperature H2/O2 PEM Fuel Cells. *Chem. Mater.*, 18:2238-2248.

[39] Jian-hua, T., Peng-fei, G. Zhi-yuan, Z., Wen-hui, L., and Zhong-qiang, S., 2008. Preparation and performance evaluation of a Nafion-TiO2 composite membrane for PEMFCs. *Int. J. Hydrogen Energy*, 33: 5686-5690.

[40] Zhengbang, W., Haolin, T., and Mu, P. 2011. Self-assembly of durable Nafion/TiO2 nanowire electrolyte membranes for elevated-temperature PEM fuel cells. *J. Membr. Sci.*, 369: 250–257.

[41] Saccà, A. Gatto, I., Carbone, A., Pedicini, R., and Passalacqua E.2006. ZrO2–Nafion Composite membranes for polymer electrolyte fuel cells (PEFCs) at intermediate temperature. *J. Power Sources*, 163: 47–51.

[42] D'Epifanio, A., Navarra, M. A., Weise, F. C. Mecheri, B., Farrington, J., Licoccia, S., and Greenbaum, S. 2010. Composite Nafion/sulfated zirconia membranes: effect of the filler surface properties on proton transport characteristics. *Chem. Mater.*, 22: 813–821.

[43] Costamagna, P., Yang, C., Bocarsly, A. B., and Srinivasan, S. 2002. Nafion®115 /zirconium phosphate composite membranes for operation of PEMFCs above 100 °C. *Electrochim. Acta.*, 47: 1023–1033.

[44] Bauer, F., and Willert-Porada, M. 2005. Characterisation of zirconium and titanium phosphates and direct methanol fuel cell (DMFC) performance of functionally graded Nafion® composite membranes prepared out of them. *J. Power Sources*, 145: 101–107.

[45] Chena, L.-C., Yu, T. L., Lin, H.-L., and Yeh, S.-H. 2008. Nafion/PTFE and zirconium phosphate modified Nafion/PTFE composite membranes for direct methanol fuel cells. *J. Membr. Sci.* 307: 10–20.

[46] Arbizzani, C., Donnadio, A., Pica, M., Sganappa, M., Varzi, A. Casciola, M., and Mastragostino, M. 2010. Methanol permeability and performance of Nafion-zirconium phosphate composite membranes in active and passive direct methanol fuel cells. *J. Power Sources*, 195: 7751–7756.

[47] Tazi, B., and Savadago, O. 2000. Parameters of PEM fuel-cells based on new membranes fabricated from Nafion®, silicotungstic acid and thiophene. *Electrochim. Acta*, 45: 4329–4339.

[48] Bahar, B., Hobson, A. R., Kolde, J. A., and Zukerbrod, D. 1996. U.S. Patent.5:547-551.

[49] Staiti, P., Aricò, A. S., Baglio, V., Lufrano, F., Passalacqua, E., and Antonucci, V. 2001. Hybrid Nafion–silica membranes doped with heteropolyacids for application in direct methanol fuel cells. *Solid State Ionics*, 145: 101–107.

[50] Xu, W., Lu, T., Liu, C., and Xing, W. 2005. Low methanol permeable composite Nafion/silica/PWA membranes for low temperature direct methanol fuel cells. *Electrochimi. Acta*, 50: 3280–3285.

[51] Ramani, V., Kunz, H. R., and Fenton, J. M. 2006. Metal dioxide supported heteropolyacid/Nafion® composite membranes for elevated temperature/low relative humidity PEFC operation. *J. Membr. Sci.*, 279: 506–512.

[52] Amirinejad, M., Madaeni, S, S., Rafiee, E., Amirinejad, S. 2011. Cesium hydrogen salt of heteropolyacids/Nafion nanocomposite membranes for proton exchange membrane fuel cells. *J. Membr. Sci.*, 377: 89– 98.

[53] Shao, Z.-G., Joghee, P., and Hsing, I.-M. 2004. Preparation and characterization of hybrid Nafion®-silica membrane doped with phosphotungstic acid for high temperature operation of proton exchange membrane fuel cells. *J. Membr. Sci.*, 229: 43-51.

[54] Gosalawit, R., Chirachanchai, S., Manuspiya, H., and Traversa,E. 2006. Krytox-Silica–Nafion® composite membrane: A hybrid system for maintaining proton conductivity in a wide range of operating temperatures. *Catal. Today*, 118: 259–265.

[55] Yana, J., Nimmanpipug, P., Chirachanchai, S., Gosalawit, R., Dokmaisrijan, S., Vannarat, S., Vilaithong, T., Lee, V. S., 2010. Molecular dynamics simulations of Krytox-Silica-Nafion composite for high temperature fuel cell electrolyte membranes. *Polymer*, 51: 4632-4638.

[56] Gosalawit, R., Chirachanchai, S., Shishatskiy, S., Nunes, S. P. 2007. Krytox-montmorillonite-Nafion® nanocomposite membrane for effective methanol crossover reduction in DMFCs. *Solid State Ionics*, 178: 1627–1635.

[57] Peighambardoust, S. J., Rowshanzamir, S., and Amjadi, M. 2001. Review of the proton exchange membranes for fuel cell applications. *Int. J. Hydrogen Energy*, 35,9349-9384.

[58] Rhim, J.-W., Park, H. B., Lee, C.-S., Jun, J.-H., Kim, D. S., and Lee, Y. M. 2004. Crosslinked poly (vinyl alcohol) membranes containing sulfonic acid group: proton and methanol transport through membranes. *J. Membr. Sci.*, 238: 143–151.

[59] Anis, A., Banthia A.K., and Bandyopadhyay, S. 2008. Synthesis and characterization of polyvinyl alcohol copolymer/phosphomolybdic acid-based crosslinked composite polymer electrolyte membranes. *J. Power Sources*, 179: 69–80.

[60] Tripathi, B. P., and Shahi, V. K. 2009. 3-[[3-(Triethoxysilyl) propyl] amino]propane-1-sulfonic acid-poly(vinyl alcohol) cross-linked zwitterionic polymer electrolyte membranes for direct methanol fuel cell applications. *ACS Appl. Mater. Interfaces*,1: 1002–1012.

[61] Aslan, A., Çelik, S. Ü., and Bozkurt, A. 2009. Proton-conducting properties of the bembranes based on poly(vinyl phosphonic acid) grafted poly(glycidyl methacrylate). *Solid State Ionics*, 180: 1240–1245.

[62] Smitha, B., Sridhar, S., and Khan, A. A. 2004. Polyelectrolyte complexes of chitosan and poly(acrylic acid) as proton exchange membranes for fuel cells. *Macromolecules,* 37: 2233-2239.

[63] Kim, D. S., Park, H. B., Rhim, J. W., and Lee, Y. M. 2005. Proton conductivity and methanol transport behavior of cross-linked PVA/PAA/silica hybrid membranes. *Solid State Ionics*, 176: 117–126.

[64] Yang, Z., Coutinho, D. H., Yang, D.-J., Balkus Jr, K. J., and Ferraris, J. P. 2008. Proton-conducting membranes based on HTFSI-doped PEI/SiO2 nanocomposites. *J. Membr. Sci.*, 313: 91–96.

[65] Osifo, P. O., and Masala, A. 2010. Characterization of direct methanol fuel cell (DMFC) applications with H2SO4 modified chitosan membrane. *J. Power Sources*, 195: 4915–4922.

[66] Tayouo, R., David, G., Améduri, B., Rozière, J. and Roualdès, S. New fluorinated polymers bearing pendant phosphonic acid groups. Proton conducting membranes for fuel cell. *Macromolecules*, 43: 5269-5276.

[67] Steck, A. E., Stone, C., Savadogo, O. and Roberge, P. R. 1997. New materials for Fuel Cell and Modern Battery Systems. *Proceedings of a Symposium, Ecole Polytechnique, Montreal*, PQ (Canada).

[68] Edmondson, C. A., Fontanella, J. J., Chung, S. H., Greenbaum, S.G., and Wnek, G. E. 2001. Complex impedance studies of S-SEBS block polymer proton-conducting membranes. *Electrochim. Acta*, 46: 1623–1628.

[69] Mokrini, A., and Huneault, M. A. 2006. Proton exchange membranes based of PVDF/SEBS blends. *J. Power Sources*, 154: 51-58.

[70] Zaidi, S. M. J., and Mikhailenko, S. D., Robertson, G. P., Guiver, M. D., and Kaliaguine, S. 2000. Proton conducting composite membranes from polyether ether ketone and heteropolyacids for fuel cell applications. *J. Membr. Sci.*, 173: 17–34.

[71] Li, L., Zhang, J., and Wang, Y. 2003. Sulfonated poly(ether ether ketone) membranes for direct methanol fuel cell. *J. Membr. Sci.,* 226: 159–167.

[72] Xing, P., Robertson, G. P., Guiver, M. D., Mikhailenko, S. D., Wang, K., Kaliaguine, S. 2004. Synthesis and characterization of sulfonated poly(ether ether ketone) for proton exchange membranes. *J. Membr. Sci.,* 229: 95–106.

[73] Wang, F., Hickner, M., Kim, Y. S., Zawodzinski, T. A., and McGrath, J. E. 2002. Direct polymerization of sulfonated poly(arylene ether sulfone) random (statistical) copolymers: candidates for new proton exchange membranes. *J. Membr. Sci.*, 197: 231–242.

[74] Kim, H.-J., Krishnan, N. N., Lee, S.-Y., Hwang, S. Y., Kim, D., Jeong, K. J., Lee, J. K. Cho, E. A. Lee, J., Han, J., Ha, H. Y., and Lim, T.-H. 2006. Sulfonated poly(ether sulfone) for universal polymer electrolyte fuel cell operations. *J. Power Sources*, 160: 353–358.

[75] Dai, H., Guan, R., Li, C., and Liu, J. 2007. Development and characterization of sulfonated poly(ether sulfone) for proton exchange membrane materials. *Solid State Ionics*, 178: 339–345.

[76] Parvole, J., and Jannasch, P. 2008. Polysulfones Grafted with poly(vinylphosphonic acid) for highly proton conducting fuel cell membranes in the hydrated and nominally dry state. *Macromolecules*, 41: 3893-3903.

[77] Kawahara, M., Rikukawa, M., Sanui, K., and Ogata, N. 2000. Synthesis and proton conductivity of sulfopropylated poly(benzimidazole) films. *Solid State Ionics*, 136-137, 1193–1196.

[78] Genies, C., Mercier, R., Sillion, B., Cornet, N., Gebel, G., and Pineri, M. 2001. Soluble sulfonated naphthalenic polyimides as materials for proton exchange membranes. *Polymer*, 42: 359–373.

[79] Fang, J., Guo, X., Harada, S., Watari, T., Tanaka, K., Kita, H., and Okamoto, K.-I. 2002. Novel sulfonated polyimides as polyelectrolytes for fuel cell application. 1. synthesis, proton conductivity, and water stability of polyimides from 4,4'-diaminodiphenyl Ether-2,2'-disulfonic Acid. *Macromolecules*, 35: 9022-9028.

[80] Qiu, Z., Wu, S., Li, Z., Zhang, S., Xing, W., and Liu, C. 2006. Sulfonated poly(arylene-co-naphthalimide)s synthesized by copolymerization of primarily sulfonated monomer and fluorinated naphthalimide dichlorides as novel polymers for proton exchange membranes. *Macromolecules*, 39: 6425-6432.

[81] Miyatake, K., Iyotani, H., Yamamoto, K., and Tsuchida, E. 1996. Synthesis of poly(phenylene sulfide sulfonic acid) via poly(sulfonium cation) as a thermostable proton-conducting polymer. *Macromolecules*, 29: 6969-6971.

[82] Schauer, J., and BrožováL. 2005. Heterogeneous ion-exchange membranes based on sulfonated poly(1,4-phenylene sulfide) and linear polyethylene: preparation, oxidation stability, methanol permeability and electrochemical properties. *J. Membr. Sci.,* 250: 151–157.

[83] Schauer, J., Kůdela, V., Richau, K., and Mohr, R. 2006. Heterogeneous ion-exchange membranes based on sulfonated poly(1,4-phenylene sulfide). *Desalination*, 198: 256–264.

[84] Dai, H., Zhang, H. M., Zhong, H. X., Li, X. F., Xiao, S. H., and Mai, Z. S. 2010. High performance composite membranes with enhanced dimensional stability for use in PEMFC. *Int. J. Hydrogen Energy*, 35: 4209-4214.

[85] Bae, B., Miyatake, K., and Watanabe, M. 2010. Sulfonated poly(arylene ether sulfone ketone) multiblock copolymers with highly sulfonated block. Synthesis and properties. *Macromolecules*, 43, 2684–2691.

[86] Cho, C. G., Kim, S. H., Park, Y. C., Kim, H., and Park, J.-W. 2008. Fuel cell membranes based on blends of PPO with poly (styrene-b-vinylbenzylphosphonic acid) copolymers. *J. Membr. Sci.*, 308: 96-106.

[87] Fu, L., Xiao, G., and Yan, D. 2010. Synthesis and characterization of sulfonated poly(arylene ether ketone/ketone phosphine oxide)s as proton exchange membranes. *J. Membr. Sci.*, 362: 509-516.

[88] Gui, L., Zhang, C., Kang, S., Tan, N., Xiao, G., and Yan, D. 2010. Synthesis and properties of hexafluoroisopropylidene-containing sulfonated poly(arylene thioether phosphine oxide)s for proton exchange membranes. *Int.J. Hydrogen Energy*, 35, 2436-2445.

[89] Ingratta, M., Jutemar, E. P., and Jannasch, P. 2011. Synthesis, nanostructures and properties of sulfonated poly(phenylene oxide) bearing polyfluorostyrene side chains as proton conducting membranes. *Macromolecules*, 44: 2074-2083.

[90] Perrin, R., Elomaa, M., and Jannasch, P. 2009. Nanostructured proton conducting polystyrene-poly(vinylphosphonic acid) block copolymers prepared via sequential anionic polymerizations. *Macromolecules*, 42: 5146-5154.

[91] Gosalawit, R., Chirachanchai, S., Basile, A., and Iulianelli, A. 2009. Thermo and electrochemical characterization of sulfonated PEEK-WC membranes and Krytox-Si-Nafion® composite membranes. *Desalination*, 235, 293-305.

[92] Ponce, M. L., Prado, L., Ruffmann, B., Richau, K., Mohr, R., and Nunes, S. P. 2003. Reduction of methanol permeability in polyetherketone-heteropolyacid membranes. *J. Membr. Sci.*, 217: 5-15.

[93] Gaowen, Z., and Zhento, Z. 2005. Organic/inorganic composite membranes for application in DMFC. *J. Membr. Sci.*, 261: 107-113.

[94] Sambandam, S., and Ramani, V. 2007. SPEEK/functionalized silica composite membranes for polymer electrolyte fuel cells. *J. Power Sources*, 170: 259-267.

[95] Gosalawit, R., Chirachanchai, S., Shishatskiy, S., and Nunes, S. P. 2008. Sulfonated montmorillonite/sulfonated poly(ether ether ketone) (SMMT/SPEEK) nanocomposite membrane for direct methanol fuel cells (DMFCs). *J. Membr. Sci.*, 323, 337-46.

[96] Yang, C., Costamagna, P., Srinivasan, S., Benziger, J., and Bocarsly, A. B. 2001. On the development of proton conducting polymer membranes for hydrogen and methanol fuel cell. *J. Power Sources*, 103: 1-15.

[97] Beuschern, U., Cleghorn, S. J. C., and Johnson, W. B. 2005. Challenges for PEM fuel cell membranes. *Int. J. Energy Res.*, 29, 1103-1112.

[98] Kreuer, K. D., Fuchs, A., Ise, M., Spaeth, M., and Maier, J. 1998. Imidazole and pyrazole-based proton conducting polymers and liquids. *Electrochim. Acta*, 43: 1281-1288.

[99] Munch, W., Kreuer, K. D., Silvestri, W., Maier, J., and Seifert. G. 2001. The diffusion mechanism of an excess proton in imidazole molecule chains: first results of an ab initio molecular dynamics study. *Solid State Ionics*, 145: 437-443.

[100] Munson, R. A., 1964.Self-Dissociative equilibria in molten phosphoric acid.*J. Phys. Chem.*, 68: 3374-3377.

[101] Schuster, M., Meyer, W. H., Wegner, G., Herz, H. G., Ise, M., Schuster, M., Kreuer, K. D., and Maier, J. 2001. Proton mobility in oligomer-bound proton solvents: imidazole immobilization via flexible spacers. *Solid State Ionics*, 145: 85-92.

[102] Yamada, M., and Honma, I. 2004. Anhydrous protonic conductivity of a self-assembled acid-base composite material. *J. Phys. Chem. B*, 108: 5522-5526.

[103] Bozkurt, A., Meyer, W. H., and Wegner, G. 2003. PAA/imidazol-based proton conducting polymer electrolytes. *J. Power Sources*, 123:126-131.

[104] Yamada, M., and Honma, I. 2005. Anhydrous proton conducting polymer electrolytes based on poly(vinylphosphonic acid)-heterocycle composite material. *Polymer*, 46: 2986-2992.

[105] Boroglu, M. S., Celik, S. U., Bozkurt, A., and Boz, I. 2011. The synthesis and characterization of anhydrous proton conducting membranes based on sulfonated poly(vinyl alcohol) and imidazole. *J. Membr. Sci.*, 375: 157–164.

[106] Yamada, M., and Honma, I. 2004. Alginic acid–imidazole composite material as anhydrous proton conducting membrane. Polymer, 45: 8349–8354.

[107] Li, W., Zhang, F., Yi, S., Huang, C., Zhang, H., and Pan, M. 2012. Effects of casting solvent on microstructrue and ionic conductivity of anhydrous sulfonated poly(ether ether ketone)-inoic liquid composite membranes. *Int.J. Hydrogen Energy*, 37: 748-754.

[108] Deligöz, H. and Yımazoğlu, M. 2011. Development of a new highly conductive and thermomechanically stable complex membrane based on sulfonated polyimide/ionic liquid for high temperature anhydrous fuel cells. *J. Power Sources*, 196: 3496-3502.

[109] Bozkurt, A., and Meyer, W. H. 2001. Proton conducting blends of poly(4-vinylimidazole) with phosphoric acid. *Solid State Ionics*, 138: 259-265.

[110] Bozkurt, A., Meyer, W. H., Gutmann, J., and Wegner, G. 2003. Proton conducting copolymers on the basis of vinylphosphonic acid and 4-vinylimidazole. *Solid State Ionics*, 164: 169-176.

[111] Jithunsa, M., Tashiro, K., Nunes, S. P., and Chirachanchai, S. 2008. Preparation of 4(5)-vinylimidazole-co-acrylic acid copolymer and thermal performances related to applicability as PEM fuel cells. *Polym. Degrad. Stab.*, 93: 1389-1395.

[112] Subianto, S., Mistry, M. K., Choudhury, N. R., Dutta, N. K., and Knott, R. 2009. Composite polymer electrolyte containing ionic liquid and functionalized polyhedral oligomeric silsesquioxanes for anhydrous PEM applications. *ACS Applied Materials and Interfaces*, 1: 1173-1182.

[113] Lakshminarayana, G., and Nogami, M. 2010. Inorganic-organic hybrid membranes with anhydrous proton conduction prepared from tetramethoxysilane/methyl-trimethoxysilane/trimethylphosphate and 1-ethyl-3-methylimidazolium-bis (trifluoro-methane sulfonyl) imide for H2/O2 fuel cells. *Electrochim. Acta*, 55: 1160-1168.

[114] Ye, H., Huang, J., Xu, J. J., Kodiweera, N. K. A. C., Jayakody, J. R. P., and Greenbaum, S. G. 2008. New membranes based on ionic liquids for PEM fuel cells at elevated temperatures. *J. Power Sources*, 178: 651-660.

[115] Che, Q., Sun, B., and He, R. 2008. Preparation and characterization of new anhydrous, conducting membranes based on composites of ionic liquid trifluoroacetic propylamine and polymers of sulfonated poly (ether ether) ketone or polyvinylidenefluoride. *Electrochim. Acta*, 53: 4428-4434.

[116] Di Noto, V., Negro, E., Sanchez, J.-Y., and Iojoiu, C.2010. Structure-relaxation interplay of a new nanostructured membrane based on tetraethylammonium trifluoro methanesulfonate ionic liquid and neutralized Nafion 117 for high-temperature fuel cells. *J. Am. Chem. Soc.*, 132, 2183-2195.

[117] Jithunsa, M., Tashiro, K., and Chirachanchai, S. 2009. Investigation of structural changes related to temperature: An understanding of H-bond based proton transfer in 4(5)-vinylimidazole and acrylic acid copolymer membrane. *Solid State Ionics*, 180: 132-140.

[118] Wainright, J. S., Wang, J., Weng, D., Savinell, R. F., and Litt, M. 1995. Acid-doped polybenzimidazoles: A new polymer electrolyte. *J. Electrochem. Soc.* 142: L121–L123.

[119] Bouchet, R., and Siebert, E. 1999. Proton conduction in acid doped polybenzimidazole. *Solid State Ionics*, 118: 287-299.

[120] Xiao, L. X., Zhang, H. F., Scanlon, E., Ramanathan, L. S., Choe, E. W., Rogers, D., Apple, T., and Benicewicz, B. C. 2005. High-temperature polybenzimidazole fuel cell membranes via a sol-gel process. *Chem. Mater.*, 17: 5328-5333.

[121] Jiang, F. J., Pu, H. T., Meyer, W. H., Guan, Y. S., and Wan, D. C. 2008. A new anhydrous proton conductor based on polybenzimidazole and tridecyl phosphate. *Electrochim. Acta*, 53: 4495-4499.

[122] Mader, J. A., and Benicewicz, B. C. 2010. Sulfonated polybenzimidazoles for high temperature PEM fuel cells. *Macromolecules*, 43: 6706-6715.

[123] Wang, S., Zhang, G., Han, M., Li, H., Zhang, Y., Ni, J., Ma, W., Li, M., Wang, J., Liu, Z., Zhang, L., and Na, H. 2011. Novel epoxy-based cross-linked polybenzimidazole for high temperature proton exchange membrane fuel cells. *Int.J. Hydrogen Energy.*, 36: 8412-8421.

[124] Kim, T.-H., Kim, S.-K., Lim, T.-W., and Lee, J.-C. 2008. Synthesis and properties of poly(aryl ether benzimidazole) copolymers for high-temperature fuel cell membranes. *J. Membr. Sci.*, 323: 362-370.

[125] Kim, S.-K., Kim, T.-H., Jung, J.-W., and Lee, J.-C. 2009. Polybenzimidazole containing benzimidazole side groups for high-temperature fuel cell applications. *Polymer*, 50: 3495–3502.

[126] Xu, H., Chen, K., Guo, X., Fang, J., and Yin, J. 2007. Synthesis of hyperbranched polybenzimidazoles and their membrane formation. *J. Membr. Sci.*, 288: 255–260.

[127] Kim, H.-J., Cho, S. Y., An, S. J., Eun, Y. C., Kim, J.-Y., Yoon, H.-K., Kweon, H.-J., and Yew, K. H. 2004. Synthesis of poly(2,5-benzimidazole) for use as a fuel-cell membrane. *Macromol. Rapid Commun.*, 25: 894–897.

[128] Asensio, J. A., Borrós, S., and Gómez-Romero, P. 2004. Proton-conducting membranes based on poly(2,5-benzimidazole) (ABPBI) and phosphoric acid prepared by direct acid casting. *J. Membr. Sci.*, 241: 89-93.

[129] Asensio, J. A., Borrós, S., and Gómez-Romero, P. 2004. Sulfonated poly(2,5-benzimidazole) (SABPBI) impregnated with phosphoric acid as proton conducting membranes for polymer electrolyte fuel cells. *Electrochim. Acta*, 49: 4461-4466.

[130] Kim, S.-K., Kim, T.-H., Ko, T., and Lee, J.-C. 2011. Cross-linked poly(2,5-benzimidazole) consisting of wholly aromatic groups for high-temperature PEM fuel cell applications. *J. Membr. Sci.*, 373: 80-88.

[131] Persson, J. C., and Jannasch, P. 2005. Intrinsically proton-conducting benzimidazole units tethered to polysiloxanes. *Macromolecules*, 38: 3283-3289.

[132] Persson, J. C., and Jannasch, P. 2006. Block copolymers containing intrinsically proton-conducting blocks tethered with benzimidazole units. *Chem. Mater.*, 18: 3096-3102.

[133] Persson, J. C., and Jannasch, P. 2006. Intrinsically proton-conducting comb-like copolymers with benzimidazole tethered to the side chains. *Solid State Ionics*, 177: 653–658.

[134] Persson, J. C., Josefsson, K., and Jannasch, P. 2006. Polysulfones tethered with benzimidazole. *Polymer*, 47: 991-998.

[135] Woudenberg, R. C., Yavuzcetin, O., Tuominen, M. T., and Coughlin, E. B. 2007. Intrinsically proton conducting polymers and copolymers containing benzimidazole moieties: Glass transition effects. *Solid State Ionics*, 178: 1135-1141.

[136] Pangon, A., Tashiro, K., and Chirachanchai, S. 2011. Polyethylenimine containing benzimidazole branching: A model system providing a balance of hydrogen bond network or chain mobility enhances proton conductivity. *J. Phys. Chem. B*, 115: 11359-11367.

[137] Kim, J.-D., and Honma, I. 2005. Anhydrous solid state proton conductor based on benzimidazole/monododecyl phosphate molecular hybrids. *Solid State Ionics*, 176: 979-984.

[138] Totsatitpaisan, P., Tashiro, K., and Chirachanchai, S. 2008. Investigating the Proton Transferring Route in a Heteroaromatic Compound Part I: A Trial to Develop Di- and Trifunctional Benzimidazole Model Compounds Inducing the Molecular Packing Structure with a Hydrogen Bond Network. *J. Phys. Chem. A*, 112: 10348-10358.

[139] Pangon, A., Totsatitpaisan, P., Eiamlamai, P., Hasegawa, K., Yamasaki, M., Tashiro, K., and Chirachanchai, S. 2011. Systematic studies on benzimidazole derivatives: Molecular structures and their hydrogen bond networks formation toward proton transfer efficiency. *J. Power Sources*, 196: 6144-6152.

[140] Lee, J. W., Lee, D. Y., Kim, H.-J., Nam, S. Y., Choi, J. J., Kim, J.-Y., Jang, J. H., Cho, E., Kim, S.-K., Hong, S.-A., and Lim, T.-H. 2010. Synthesis and characterization of acid-doped polybenzimidazole membranes by sol-gel and post-membrane casting method. *J. Membr. Sci.*, 357: 130-133.

[141] Matar, S., Higier, A., Liu, H. T. 2010. The effects of excess, phosphoric acid in a polybenzimidazole-based high temperature proton exchange membrane fuel cell. *J. Power Sources*, 195:181-184.

[142] Asensio, J. A., Borrós, S., Gómez-Romero, P., 2002. Proton-conducting polymers based on benzimidazoles and sulfonated benzimidazoles. *J. Polym. Sci., A: Polym. Chem.*, 40: 3703-3710.

[143] Totsatitpaisan, P., Nunes, S. P., Tashiro, K. and Chirachanchai, S. 2009. Investigation of the role of benzimidazole-based model compounds on thermal stability and anhydrous proton conductivity of sulfonated poly(ether ether ketone). *Solid State Ionics*, 180: 738-745.

[144] Li, S., Zhou, Z., Zhang, Y., Liu, M. and Li, W. 2005. 1H-1,2,4-Triazole: An Effective Solvent for Proton-Conducting Electrolytes. *Chem. Mater.*, 17: 5884-5886.

[145] Sen, U., Çelik,S.Ü ., Ata, A. and Bozkurt, A. 2008. Anhydrous proton conducting membranes for PEM fuel cells based on Nafion/Azole composites. *Int. J. Hydrogen Energy*, 33: 2808-2815.

[146] Granados-Focil, S., Woudenberg, R. C., Yavuzcetin, O., Tuominen, M. T. and Coughlin, E. B. 2007. Water-free proton-conducting polysiloxanes: A study on the effect of heterocycle structure. *Macromolecules*, 40: 8708-8713.

[147] Celik, S. U., Aslan, A., Bozkurt, A. 2008. Phosphoric acid-doped poly(1-vinyl-1,2,4-triazole) as water-free proton conducting polymer electrolytes. *Solid State Ionics*, 179: 683-688.

[148] Bhadra, S., Kim, N. H., and Lee, J. H. 2010. A new self-cross-linked, net-structured, proton conducting polymer membrane for high temperature proton exchange membrane fuel cells. *J. Membr. Sci.*, 349: 304-311.

[149] Jang, W., Sundar, S., Choi. S., Shul Y-G., and Han, H. 2006. Acid-base polyimide blends for the application as electrolyte membranes for fuel cells. *J. Membrane Sci.,* 280: 321-329.

[150] Acar, O., Sen, U., Bozkurt, A., and Ata, A. 2009. Proton conducting membranes based on poly(2,5-benzimidazole) (ABPBI)-poly (vinylphosphonic acid) blends for fuel cell. *Int. J. Hydrogen Energy,* 34: 2724-2730.

[151] Kosmala, B., Schauer, J. 2002. Ion-exchange membranes prepared by blending sulfonated poly(2,6-dimethyl-1,4-phenylene oxide) with polybenzimidazole, *J. Appl. Polym. Sci.*, 85 1118.

[152] Kerres, J., Ullrich, A., Meier, F., and Haring, T. 1999. Synthesis and characterization of novel acid-base polymer blends for application in membrane fuel cells. *Solid State Ionics,* 125: 243-249.

[153] Jithunsa, M., Tashiro, K., Nunes, S. P., Chirachanchai, S. 2011. Poly(acrylic acid-co-4-vinylimidazole)/Sulfonated poly(ether ether ketone) blend membranes: A role of polymer chain with proton acceptor and donor for enhancing proton transfer in anhydrous system. *Int. J. Hydrogen Energy*, 36: 10384-10391.

[154] Pangon, A., Chirachanchai, S. 2012. Poly(acrylic acid) containing multi-benzimidazole units: A simple approach to obtain polymer with proton donor-acceptor system. *Polymer*, 53: 3878-3884.

[155] Kreuer, K.D., 2001. On the development of proton conducting polymer membranes for hydrogen and methanol fuel cells. *J. Membrane Sci.,* 185: 29-39.

[156] Jansen, J. C., Machhione, M., and Drioli, E. 2007. On the unusual solvent retention and the effect on the gas transport in perfluorinated Hyflon® AD membrane. *J. Membrane Sci.,* 287: 132-137.

[157] Jiang, R., Kunz, H. R., and Fenton, J. M. 2006. Composite silica/Nafion® membranes prepared by tetraethylorthosilicate sol-gel reaction and solution casting for direct methanol fuel cells. *J. Membrane Sci.,* 272: 116-124.

[158] Basile, A., Patuzo, L., Lulianelli, A., Gatto, I., Passalacqua, E. 2006. Sulfonated PEEK-WC membrane for proton exchange membrane fuel cell: Effect of the increasing level of sulfonation on electrochemical performance. *J. Membrane Sci.,* 281: 377-385.

[159] Cooper, K. R., and Smith, M. 2006. Electrical test methods for on-line fuel cell ohmic resistance measurement. *J. Power Sources,* 160: 1088-1095.

In: Membrane Processes for Sustainable Growth
Editors: Angelo Basile and Alfredo Cassano
ISBN: 978-1-62618-446-6

Chapter 10

ADVANCES IN REVERSE OSMOSIS TECHNOLOGY FOR THE PROCESSING OF FRUIT JUICES

Navin K. Rastogi **[*]** ***and K. S. M. S. Raghavarao***
Department of Food Engineering,
Central Food Technological Research Institute,
Council of Scientific and Industrial Research, Mysore, India

ABSTRACT

Fruit juices are concentrated to improve economy in storage, transportation, distribution and commercial operations as well as conservation due to water activity reduction.

Traditional process of concentration involving high temperatures promotes significant sensory and nutritional changes of the fruit juice quality. On the other hand, reverse osmosis is a membrane process that can be used to pre-concentrate fruit juices avoiding heating leading to minimum thermal damage, besides requiring lower capital and operating costs.

This chapter addresses the significant progress in the area of reverse osmosis for the concentration of various fruit juices such as apple, pear, grapefruit, kiwi, pineapple, water melon, passion fruit, grape, black currant, citrus and other juices. It also provides an overview of recent advances and developments in the use of reverse osmosis technology with regard to fruit juice processing and considers their impact on product quality. The application of reverse osmosis and its possible integration with other processes have also been discussed.

Keywords: Reverse osmosis; fruit juice; concentration polarization; integrated processes

[*] Corresponding author; e-mail: nkrastogi@cftri.res.in.

NOMENCLATURE

Symbols

$\frac{dC}{dx}$ = Solute concentration gradient
ΔP = Transmembrane pressure
$\Delta \pi$ = Difference in the osmotic pressures of feed and permeate
A_w = Water permeability coefficient
C = Solute concentration retained on the membrane at any time
C_B = Boundary layer solute concentration
C_G = Gel concentration
C_P = Solute concentration on permeate side
D = Solute diffusion coefficient for solute transport through solvent
x = Thickness of the boundary layer
J_w = Water flux
K = Mass transfer coefficient

Acronyms

RO = Reverse osmosis
OMD = Osmotic membrane distillation

1. INTRODUCTION

Liquid foods such as fruit juices are important sources of bioactive compounds including vitamins, phenolic compounds, anthocyanins and carotenoids but the use of thermal processing technologies and subsequent storage conditions may cause significant changes in these compounds. These products are generally concentrated to achieve longer shelf-life by employing thermal evaporation under vacuum in order to reduce the storage and transportation costs. Such concentration process involving thermal treatment results in the loss of fresh juice flavour and colour besides the development of cooked taste. Increasing consumer demand for good quality fruit juices with natural flavour and taste, free from additives and preservatives, stimulated the need for the development of innovative approaches for food juice concentration. It triggered food industry to develop alternative approaches to produce fruit juice concentrate with minimum changes in nutritional, physicochemical or organoleptic characteristics. The techniques include freeze concentration, sublimation concentration, aroma recovery followed by concentration, careful process control as well as blending techniques. However, use of these methods did not significantly improved the quality of the product (Chen et al., 1993; Koseoglu et al., 1990).

Membrane processing involving microfiltration, ultrafiltration and reverse osmosis (RO) has been found to be the most promising alternative, which is extensively used for the clarification and the concentration for liquid foods. It has been widely applied in the dairy and beverage industries since the development of asymmetric membranes by Loeb and Sourirajan in the early 1960s (D'souza and Mawson, 2005). The advent of such membranes having high retention of low-molecular-weight organic compounds and good physical and chemical stability has enabled RO to be used more widely on a commercial scale. Membrane processes have been used in the food industry because these are more efficient (shorter processing time) and economical (less operation cost) as compared to thermal processing (Yazdanshenas et al., 2010). The most commonly used membrane processes are microfiltration, ultrafiltration and RO, which correspond to pressure-driven processes capable of separating particles in the approximate size ranges of 0.1–10 μm, 0.001–0.10 μm and less than 0.001 μm, respectively (Madaeni et al., 2006; Katasonova and Fedotov, 2009). Various separation processes based on sizes of the particles to be separated are presented in Figure 1.

RO is gaining worldwide acceptance in food processing especially for the concentration of fruit juices. It is a chemical engineering unit operation with a wide range of known and potential commercial applications. It is a separation technique, which operates at room temperature and can be used to concentrate or purify liquids without a phase change (Alvarez et al., 2002; Jiao et al., 2004).

Because less heat is used in RO, the thermal damage to the products is practically eliminated leading to the attainment of high-quality products having higher retention of nutritional, aroma and flavour compounds. Besides, being energy efficient and simple in operation (Alvarez et al., 2000), energy savings of 1/30 to 1/10 can be obtained when compared with single or triple effect evaporators (Girard and Fukumoto, 2000).

The RO process requires high operating pressures (ranging from 10 to 200 bar) in order to overcome the osmotic pressure of the juice (Matta et al., 2004). The major components in fruit juices contributing to osmotic pressure are sugars (hexoses and dissaccharides) and organic acids. The liquid foods can be concentrated up to a maximum concentration of 25–30 °Brix by RO due to high osmotic pressure limitation, whereas, thermal evaporation can concentrate the liquid food up to 45–65°Brix.

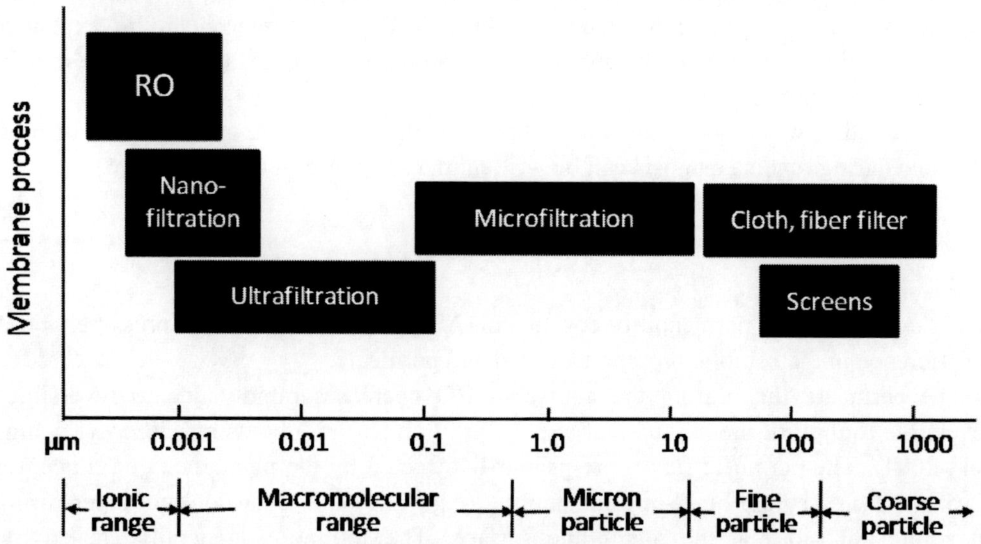

Figure 1. Various separation processes based on size of the particles.

Therefore, RO can be considered as a first-stage process that can be combined with other processes such as thermal evaporation, osmotic evaporation or freeze concentration (Girard and Fukumoto, 2000).

Fruit juices contain high solids and pectin content and form very viscous streams when directly submitted to concentration by RO resulting in lower permeate flux (Vaillant et al., 2001). Integration of RO with other pre-processes such as ultrafiltration, microfiltration, centrifugation, or treatment with pectic enzymes can reduce viscosity, which in turn, increases the transmembrane flux. Therefore, the integrated membrane process is a feasible option to achieve high quality product with lower energy consumption.

This chapter describes the basic concepts of RO and highlights its major applications in fruit juice processing.

2. Transmembrane Flux and Concentration Polarization

Osmosis is the basis for RO. When a semi-permeable membrane separates a dilute solution from a concentrated solution, the solvent (water) crosses from the dilute solution side to the concentrated solution side of the membrane to equalize concentrations on both the sides and this process is known as osmosis. The flow of solvent can be prevented by applying an opposing hydrostatic pressure to the concentrated solution. The magnitude of the pressure required to completely impede the flow of water is defined as the osmotic pressure. In case of RO, a hydraulic pressure greater than the osmotic pressure is applied for water to move from high solute to low solute concentration (Figure 2). It necessitates high-pressure requirements. Sugars, mainly hexoses and dissaccharides as well as organic acids are the osmo-active substances, which contribute to the osmotic pressure of fruit juices. It may vary from very low values of about 200 psi for fresh juice to more than 1500 psi for concentrated fruit juice. For instance, the osmotic pressure of orange juice for 10.5, 11.0, 21.5, 31.5 and 42°Brix was reported to be 210, 230, 430, 850 and 1370 psi, respectively (Merson and Morgan, 1968). The dominant mechanism governing transport through RO membranes can be explained as solution-diffusion, which involves preferential dissolution of solvent (water) in the membrane and its transport through diffusion.

The equation which gives the water flux through a membrane as a function of pressure difference during reverse osmosis can be written in a simple form (Mulder, 1996).

$$J_w = A_w(\Delta P - \Delta\pi) \quad (1)$$

where A_w is the water permeability coefficient, ΔP is the transmembrane pressure, and $\Delta\pi$ is the difference in the osmotic pressures of feed and permeate.

The permeate flux during the course of RO decreases rapidly due to reversible and irreversible fouling of membrane that significantly affect the process efficiency (Kwang-Sup et al., 2004). The permeate fluxes are primarily affected by the phenomena of concentration polarization (i.e., solute build-up) and fouling (e.g., microbial adhesion, gel layer formation and solute adhesion) at the membrane surface. The fouling is generally caused by the deposition of colloidal particles, inorganic and organic compounds, and microbes on the surface of the membrane (Chien-Hwa et al., 2010).

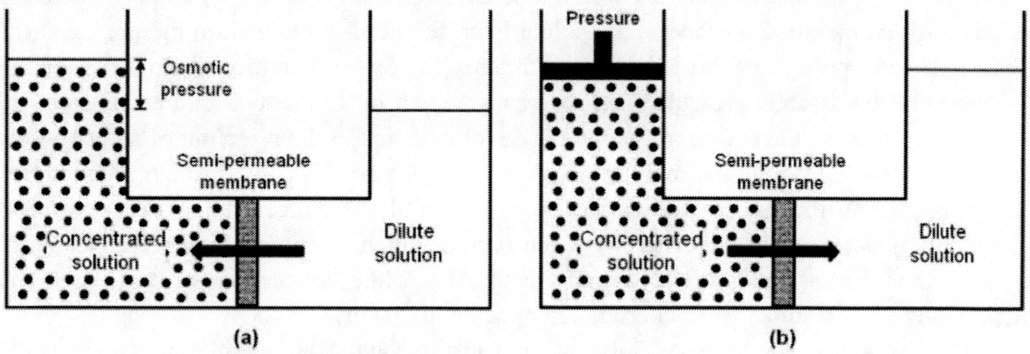

Figure 2. (a) Osmosis, (b) Reverse osmosis.

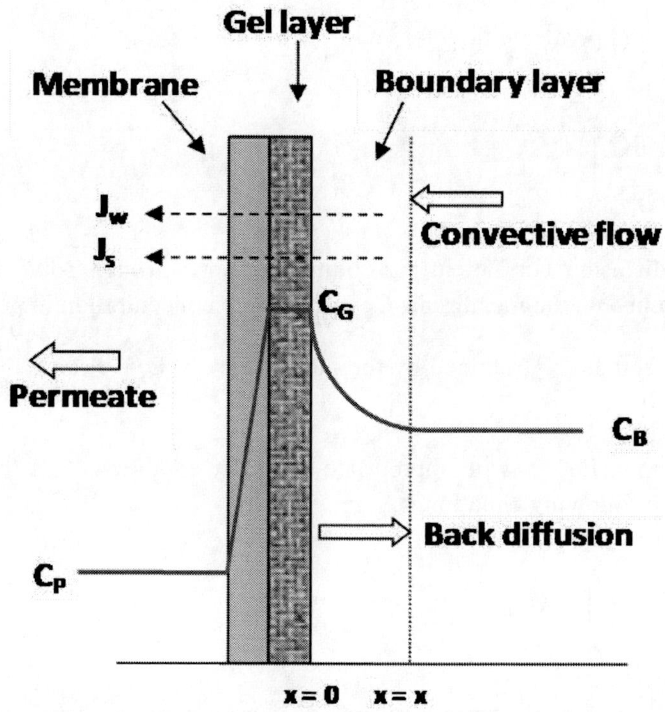

Figure 3. Formation of concentration gradient near the surface of membrane.

It is considered as a group of physical, chemical, and biological effects leading to irreversible loss of membrane permeability (Sablani et al., 2001; Chien-Hwa et al., 2010). In order to control membrane fouling during juice clarification, several flux enhancement methods have been proposed (Vladisavljevic et al., 2003). On the other hand, concentration polarization is mainly limited to the build up of retained solute i.e. accumulation of matter on the surface of the membrane. It is considered to be reversible and can be controlled in a membrane module by means of velocity adjustment, pulsation (back flush), ultrasound, or an electric field.

Considering a situation where a solution consisting of solvent and solute is subjected to a pressure driven membrane process, the solute is preferentially retained on membrane surface whereas solvent permeates through the membrane. As solvent is removed from the feed, the difference in the osmotic pressure ($\Delta\pi$) increases, which in turn results in a decrease in flux. At the same time, a concentration gradient is developed due to the rejection of solute with the passage of solvent through the membrane (Figure 3). The solvent flux through the membrane increases with an increase in the applied pressure until the concentration of solute on the membrane surface reaches a critical concentration, which is referred as gel concentration (C_G). Further increase pressure does not change the solute concentration at the membrane surface, however, results in a thicker and compacted gel layer.

At equilibrium, the rate of solvent transport through the membrane (flux) can be calculated on the basis of the convective transport of solutes to the membrane surface ($J_w.C$) by the solvent which is just equal to the sum of diffusive back transport of solute

$\left[-D\cdot\left(\frac{dC}{dx}\right)\right]$ and permeate flow ($J_w.C_p$), i.e.

$$J_w.C = \left(-D\frac{dC}{dx}\right) + J_w.C_p \tag{2}$$

where D is the diffusion coefficient for solute transport through solvent, C is the solute concentration retained on the membrane, C_p is the solute concentration in permeate (in case of RO, the solute is completely retained by the membrane i.e. $C_p = 0$) and $\left(\frac{dC}{dx}\right)$ is the solute concentration gradient.

The integration of Eq. (2) with appropriate boundary conditions ($x = 0$, $C = C_G$; $x = x$, $C = C_B$) results in the following equation.

$$J_w = \frac{D}{x}\cdot\ln\left(\frac{C_G}{C_B}\right) = k\cdot\ln\left(\frac{C_G}{C_B}\right) \tag{3}$$

where x is the thickness of the boundary layer, k is mass transfer coefficient, C_B and C_G are the bulk and gel layer solute concentrations, and $\frac{C_G}{C_B}$ is the concentration polarization modulus. The equation (3) shows that at this juncture the flux through the membrane is independent of transmembrane pressure drop or permeability and is dependent only on the solute characteristics (D and C_G,) and the boundary layer thickness (x). Concentration polarization is a reversible fluid dynamic phenomenon. If C_B reaches C_G, solutes will start to precipitate or deposit on the membrane forming a gel layer, which leads to the permanent loss of flux known as fouling (Echavarria et al., 2011).

In the fruit juice concentration process, membrane fouling may be caused by pectin, tannins, proteins, starch, hemicelluloses and cellulose (Santon et al., 2008). To enhance membrane performance, fruit juices are usually treated before filtration with enzymes aimed at hydrolyzing mainly soluble polysaccharides responsible for high viscosity. The mixture of enzymes consisting of pectinlyase, polygalacturonase and pectinesterase are generally used for the sedimentation and clarification of the juice (Echavarria et al., 2011).

3. Concentration of Fruit Juices Using Reverse Osmosis

The concentration of fruit juice by RO started with the pioneering work by Merson and Morgan (1968), who demonstrated that the oil-soluble aroma components of orange juice were retained using cellulose acetate membrane, whereas loss of water-soluble aroma components was noticed. Later, Peri et al. (1973) showed that the RO resulted in higher retention of sugars, acids, potassium, calcium, magnesium, and phosphorus. The use of polyamide membranes was found to have greater retention of flavors (~80%) and higher fluxes as compared to cellulose acetate membranes (Chua et al., 1987; Sheu and Wiley, 1983). Further, many publications worldwide have been emanated on the use of RO for the concentration of a variety of fruit juices such as apple (Sheu and Wiley, 1983; Alvarez et al., 2002), pear (Lee and Lee, 1999), grapefruit (Braddock et al., 1988), kiwi (Palmieri et al., 1990), pineapple (Bowden and Isaacs, 1989), water melon (Das Gupta and Jayaraman, 1996; Gomes et al., 2011), passion fruit (Vaillant et al., 2001), grape (Rektor et al., 2007; Santana et al., 2011), black currant (Kozák et al., 2008; Bánvölgyi et al., 2009; Pap et al., 2010), tomato (Pepper et al., 1985; Yildiz et al., 1993; Bottino et al., 2002) and citrus juices (Braddock et al., 1988; Cassano et al., 2003; Jesus et al., 2007a, b; Garcia-Castello et al., 2011). The main focus in these research works was the evaluation of different types of membranes and modules as well as the effect of operating conditions (such as pressure, flow rate and temperature) on the permeate flux and retention of inherent juice compounds. The following sections comprehensively summarize the significant progress made in the area of RO for the concentration of various fruit juices.

3.1. Apple Juice

A comparison of cellulose acetate and polyamide membranes for the concentration of apple juice indicated that the flux was higher for the polyamide membrane than that of cellulose acetate membrane (Hunter et al., 1987). The retention of total odor-active volatiles was also reported to be higher for polyamide membrane (45%) as compared to cellulose acetate membrane (23%). Later, Chou et al. (1991) evaluated the permeation of apple juice volatiles from the two thin film composite membranes made up of polyether-urea and polyamide. It was shown that apple juice concentrated using the polyamide membrane had more intense flavour than juice concentrated using the polyether-urea membrane.

Alvarez et al. (1997) predicted transmembrane flux by adopting the solution-diffusion model combined with the film model while concentrating apple juice by RO using polyamide tubular membranes. The increase in transmembrane pressure not only increased the transmembrane flux, but also the maximum achievable concentration. Pressure was found to be the most important variable controlling the process (Figure 4a). The permeate flux was found to increase linearly with temperature is due to a reduction of viscosity and increasing of diffusion coefficient of molecules (Figure 4b). Physico-chemical analyses of the concentrated juice showed that sugar and malic acid retention were 100 and 98%, respectively. These data indicated the good performance of polyamide membranes leading to maintenance of a constant sugar/acid ratio, which is an important variable to assure the final quality.

An integrated membrane process for producing apple juice and apple juice aroma concentrates (as shown in Figure 5) consisted of several steps such as juice extraction, clarification (enzyme treatment), pre-concentration (up to 25°Brix using RO), aroma recovery (using pervaporation) and final concentration (up to 72°Brix by evaporation). The product was more clear than apple juice produced by conventional methods.

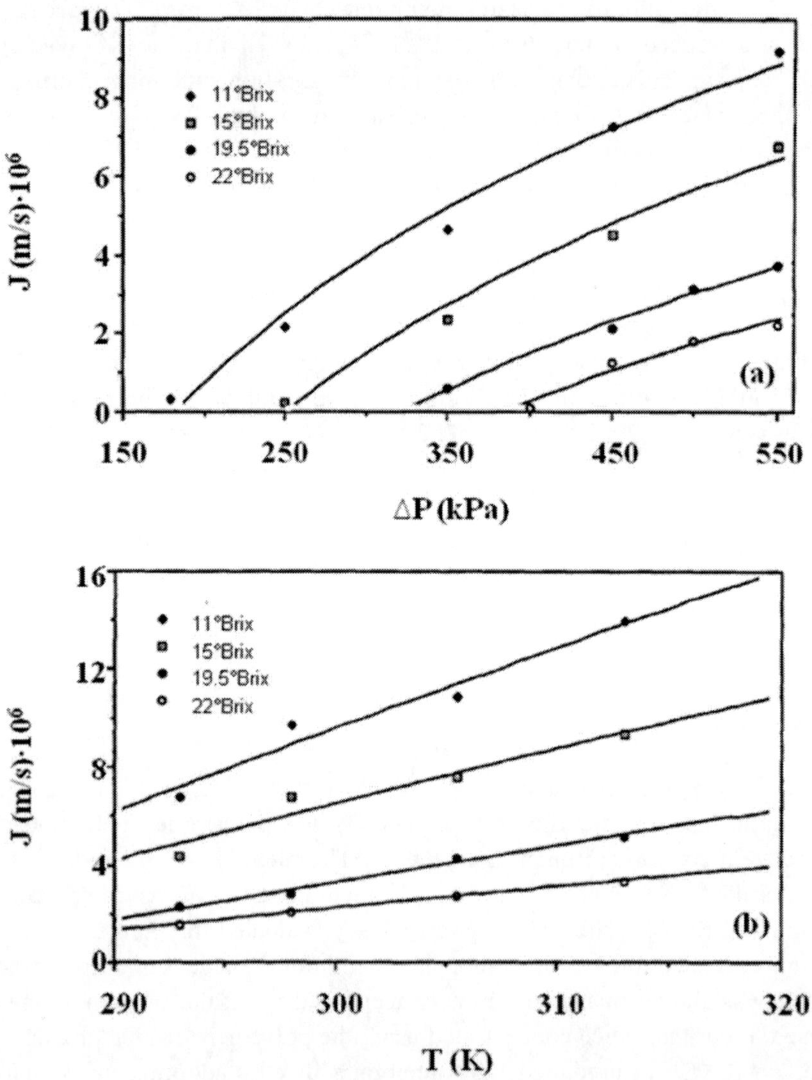

Figure 4. Influence of (a) transmembrane pressure and (b) temperature on permeate flux at constant concentration (Alvarez et al., 1997).

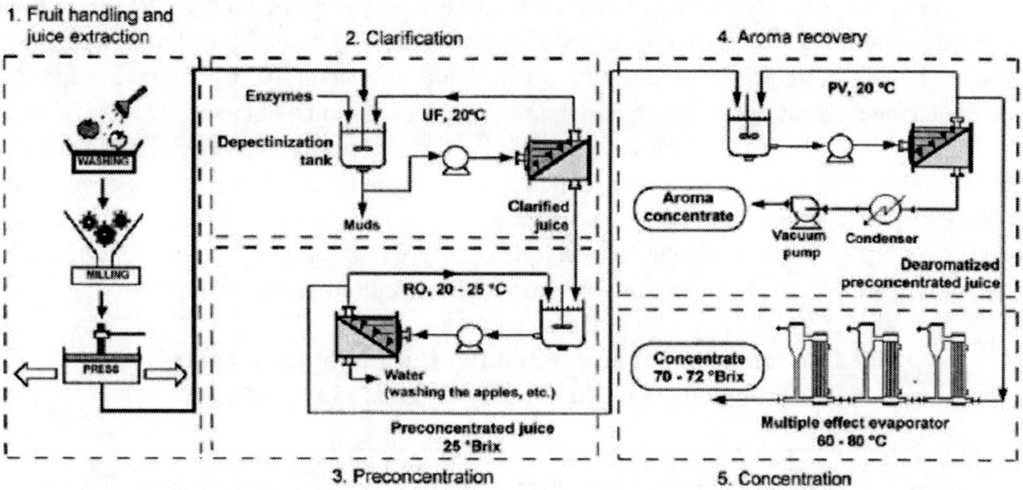

Figure 5. Integrated membrane process for the production of apple juice concentrate and apple juice aroma (Alvarez et al., 2000).

The integrated membrane process was found to be more economical than the conventional one (Alvarez et al. 2000). Alvarez et al. (1998, 2001 and 2002) also indicated that the spiral wound polyamide membrane rejected more than 90% of the major compounds responsible for apple flavour (ethyl-2-methyl butanoate, ethyl butanoate and hexanol). Alvarez et al. (2001) proposed the preferential sorption-capillary flow model for the prediction of the rejection of aroma compounds and permeate flux during the RO concentration of model solutions simulating apple juice. The permeate flux and aroma rejection increased with feed flow rate and decreased with an increase in feed concentration, which could be due to the reduction of concentration polarization by increasing feed flow rate. An increase in the feed concentration resulted in an increase in viscosity and density of the solution leading to reduction in the mass transfer coefficients of the aroma compounds. Pozderović and Moslavac (1999) investigated the use of RO for the enrichment of apple aroma in the condensate obtained during concentrating apple juice by an evaporator. Lower feed temperature and higher transmembrane pressure resulted in higher recovery and aroma concentration. The loss of aroma was higher when aroma concentration in the feed (condensate) was higher.

3.2. Orange Juice

The orange juice concentrated using polyamide RO membrane resulted in volume reduction factor of 2 and an overall recovery of sugars, organic acids, and flavor-volatile components was approximately 93%. Pectinase treatment prevented fouling to a large extent and allowed cleaning in place procedure but did not significantly affect the permeate flux or solute recoveries (Medina and Garcia, 1988). Jesus et al. (2007a, b) demonstrated that the orange juice can be concentrated up to 36 °Brix by RO at 60 bar pressure. At this condition, concentration factor and transmembrane flux were 5.8 and 28 L/m^2h, respectively.

A comparison of sensory attributes of reconstituted juice obtained by RO and thermal evaporation indicated the juice concentrated by RO process retained the characteristic aroma although having more acidic taste as compared to juice concentrated by thermal evaporation. The orange peel aroma and flavour were more intense in the juice obtained by RO (Table 1).

Braddock et al. (1988) reported that volatile flavour compounds (except for methanol, ethanol and traces of limonene) were found to be absent in permeate during the concentration of orange, grapefruit and lemon juices using a tubular composite RO membrane. However, overall loss of 17–30% of volatile peel oil (measured as limonene) was reported in the process, when the system was not totally closed during recirculation.

Table 1. Mean values of aroma and flavour attributes of concentrated orange juice by RO and by thermal evaporation

	RO	Evaporation	F_{sample}
Characteristic aroma	6.68^a	5.03^b	20.08^*
Cooked aroma	0.63	0.40	0.99^{ns}
Orange peel aroma	2.30^a	1.83^b	4.65^*
Suspended particles	0.99	0.92	0.07^{ns}
Characteristic flavour	6.39	6.54	0.36^{ns}
Cooked taste	0.47	0.70	1.00^{ns}
Orange peel flavour	4.11^a	1.93^b	39.51^*
Acid taste	6.24^a	$5,19^b$	12.58^*
Sweet taste	3.99	4.22	1.21^{ns}

Averages with different letters in vertical are significantly different (5% level—Tukey test). ns, not significant, * significant (Source Jesus et al., 2007).

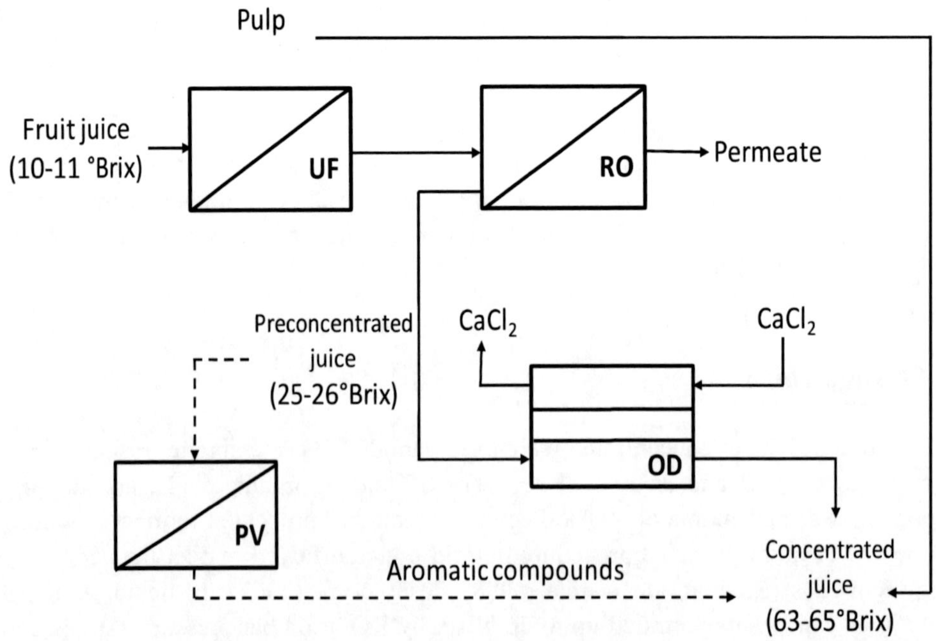

Figure 6. Integrated membrane process for the clarification and the concentration of citrus and carrot juices. (Cassano et al., 2003).

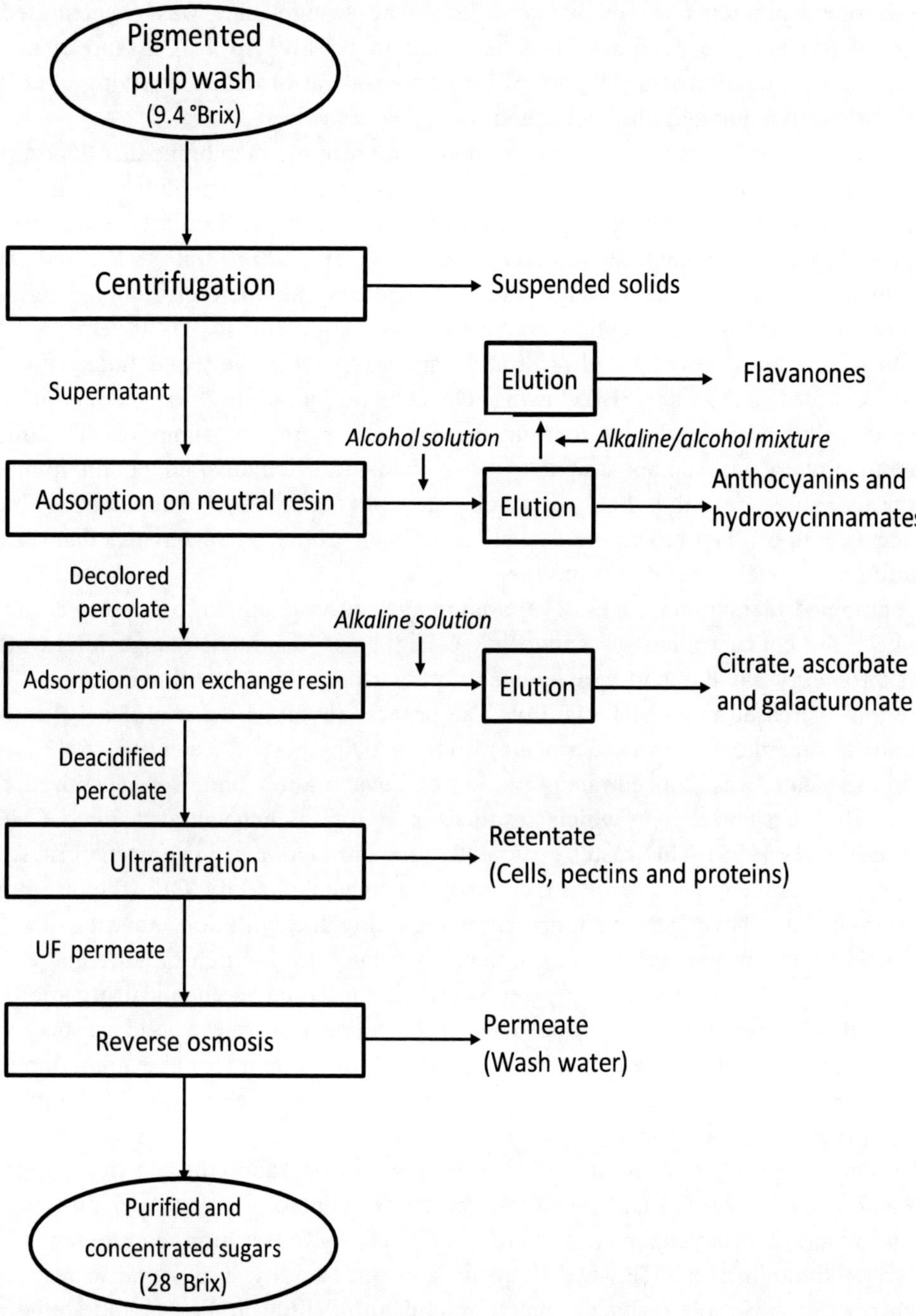

Figure 7. Flow sheet of the process for obtaining anthocyanins and hydroxycinnamates, flavanones as well as purified and concentrated sugars. (Scordino et al., 2007).

Singh and Eipeson (2000) demonstrated the concentration of pasteurized orange juice involving membrane processes such as microfiltration using a ceramic membrane and RO using a thin film composite polyamide membrane.

The microfiltration membrane (0.2 µm pore size) resulted in clarification of orange juice, where permeate retained almost all the constituents (sugars and acid was more than 98%)

except carotene pigments. At 60 bar and 32°C, the orange juice was concentrated up to 31°Brix by RO and the average flux was found to be 11 $L/m^2.h$. Cassano et al. (2003) employed an integrated process (Figure 6) for citrus and carrot juices involving ultrafiltration and RO for clarification and initial concentration, respectively.

Finally, the clarified juice was concentrated using osmotic membrane distillation (OMD). Authors also evaluated the performance of different membrane modules on the basis of productivity, quality of the product and fouling characteristics. Similarly, Galaverna et al. (2008) developed an integrated membrane process involving ultrafiltration, RO and OMD for the production of concentrated orange juice as a valuable alternative to obtain high quality concentrated juice. The clarification and pre-concentration (up to 25-30 °Brix) of freshly squeezed juice was achieved by ultrafiltration and RO, respectively and finally the product was concentrated up to 60 °Brix using OMD. The possibility to operate at ambient temperature allowed reduction in thermal damage and energy consumption. The integrated membrane process was demonstrated to be a valuable alternative to obtain high quality concentrated juice, as the final product retained antioxidants and natural bioactive components with brilliant red colour as well as pleasant aroma, characteristics that are usually lost during traditional thermal evaporation.

A combined membrane process-evaporation system was applied to develop an improved method for the concentration of orange juice. Highly concentrated orange juices offered a significant energy savings and convenience during storage and distribution due to reduced volume and increased microbial stability. The process involved the use of an ultrafiltration membrane to separate the serum (permeate) from the pulpy portion (retentate).

The permeate was concentrated by thermal evaporation and mixed with the pulpy retentate after pasteurization, which resulted in a highly concentrated juice (>80°Brix) (Hernandez et al., 1995). The wastes generated in the production of juice from citric fruits can pose great environmental problems if they are not managed correctly. Over the last few years, the European laws have become more strict regarding the pollution emission levels from industries. The environmental problem is directly related to the high organic matter content present in the leachates from solid wastes, which is measured as chemical organic demand. García et al. (2002) demonstrated that the pressed liquid from these solid wastes, which is very rich in sucrose can be pre-concentrated by RO followed by thermal evaporation for final concentration to achieve citrus molasses (65 °Brix). Scordino et al. (2007) pointed out an integrated process to obtain natural colorants, bioactive compounds and sugar concentrate from orange pulp wash (by-product of citrus processing) by applying a series of adsorption steps (with different resins) and membrane processes (Figure 7). Anthocyanins, limonoids, flavanones and hydroxycinnamates were recovered by adsorption employing a neutral polystyrene-divinylbenzene (PS-DVB) resin and the acidity was significantly reduced through a weak basic ion-exchange matrix. Inclusion of ultrafiltration to the above process stabilized the product by removal of enzymes, microorganisms as well as 80% of the water. The final product contained about 250 g/L of sugars (glucose, fructose and sucrose), 9 g/L of citric acid and 1 g/L of pectins, which could easily find application as a natural sweetener in food and beverage industries. Garcia-Castello et al. (2011) described different mechanisms for flux decline during the concentration of press liquor from the orange juice solid waste using RO.

3.3. Grape Juice

The feasibility of the membrane processing for the grape-must concentration and preservation was studied by Rektor et al. (2004 and 2007). The microfiltration resulted in the sterilization and clarification of juice, which was concentrated by RO. During RO, all the sugar and 99.5% of anthocyanin remained in the retentate. The combined method resulted in a sterile concentrate with higher sugar content which can be preserved without preservatives making it suitable for the soft drink industry to provide consumers with healthier juices without sugar addition. In the wine industry, the concentration of grape must by adopting integrated membrane processing can play a role to enhance sugar and acid contents, when vintage is poor, as total solids and acidity are enhanced from its original natural content. Rektor et al. (2007) compared the performance of a pilot scale RO system in concentrating a model system (16.5 °Brix) and grape must (16.5 °Brix). The initial permeate flux for model solution and grape must were 22 and 8.5 kg/(m^2.h), respectively, and the final concentration of the retentate were 31 and 23.5 °Brix, respectively (Figure 8a and 8b). The transmembrane flux after RO concentration for 1.5 h at 40 bar, decreased to 1.0 L/m^2.h, which was increased by increasing the pressure up to 45 bar.

The flux decreased again to 1.0 L/m^2.h in 30 min, which was further enhanced by increasing the pressure up to 50 bar. With this maximum pressure the concentrated must reached 23 °Brix (Figure 8b).

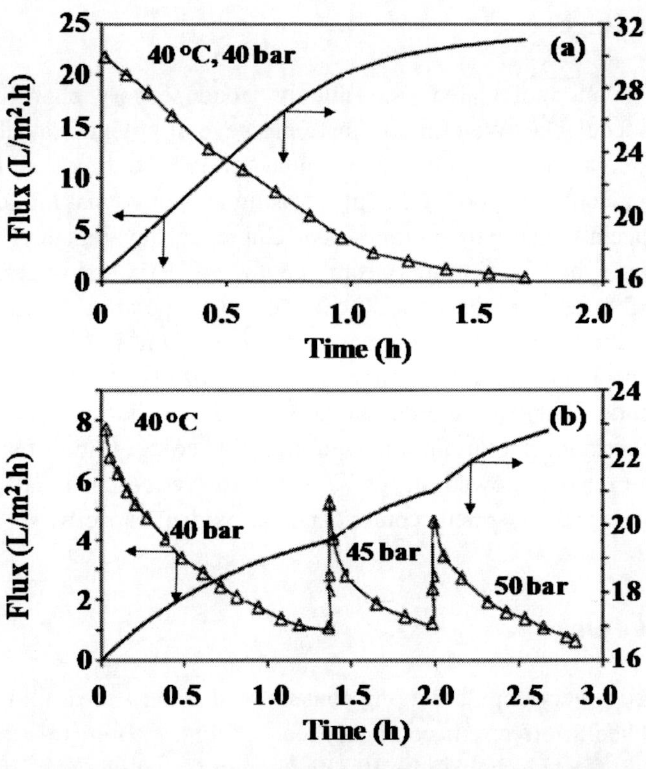

Figure 8. (a) Concentration of 16.5 Brix model (sugar) solution (b) concentration of the 16.5 Brix microfiltered grape-must using increased pressure levels (Rektor et al., 2007).

The constant pressure condition (50 bar pressure) at 40°C resulted in almost the same final concentration (23.5 °Brix), indicating the requirement of higher pressure towards the end of RO process. In another process, Rektor et al. (2006) investigated the use of OMD and membrane distillation after the clarification by microfiltration and concentration by RO in order to reach a high sugar content (60 °Brix) in grape juice concentrate, to the extent that makes it possible to preserve the grape juice without the need of low temperature.

Gurak et al. (2010) indicated the technical feasibility of RO for pre-concentrating grape juice. The authors indicated that variation in transmembrane pressure from 40 to 60 bar and temperature from 20 to 40 °C did not result in significant changes in the juice characteristics such as pH, content of soluble solids, acidity, concentration of phenolic compounds and those of monomeric and total anthocyanins, colour index, colour density, and permeate flux. The highest flux value was obtained with transmembrane pressure and temperature of 60 bar and 40°C, respectively. Further increase in temperature to 50 °C resulted in an increase in permeate flux value, with no significant difference in the physical or chemical parameters. Santana et al. (2011) evaluated RO for the concentration of grape juice over a range of transmembrane pressures (40-60 bar) and temperatures (20-50°C). The grape juice concentrated at 30°C and 60 bar of transmembrane pressure resulted in an adequate value for the permeate flux besides maintaining the physical-chemical and nutritional characteristics of the product.

3.4. Water Melon Juice

Watermelon is much appreciated fruit due its good sensory characteristics such as flavour, aroma and succulence. Watermelon juice has several potential attributes in addition to its refreshing taste, aroma and attractive colour. It is also reported to have diuretic properties. The juice contains good amount of soluble solids making concentration an attractive process especially using membranes. Das Gupta and Jayaraman (1996) showed that watermelon juice could be concentrated from 6.5 to 24° Brix using three different RO membranes namely CA-992. HR-95 and EC-50 at 50 bar operating pressure. The average permeate flux was found to be in the range from 15 to 17 $L/m^2.h$ depending on membrane type. Gomes et al. (2011) concentrated watermelon juice using RO process on a pilot plant scale using a polyamide composite membrane at 30°C and 60 bar transmembrane pressure. The volumetric concentration factor and the soluble solids concentration factor were 4.4 and 3.6, respectively. The results showed an increase in the physico-chemical properties of the concentrated juice, mainly the lycopene content and antioxidant capacity.

3.5. Blackcurrant Juice

Blackcurrant juice is very popular among consumers due to its pleasant taste as well as its numerous beneficial health effects. It contains great amounts of mineral salts and vitamins, which have beneficial effects to human health. Its vitamin C content is 4-5 times higher than that of lemon. Pap et al. (2009) demonstrated the applicability of RO process for the concentration of blackcurrant juice.

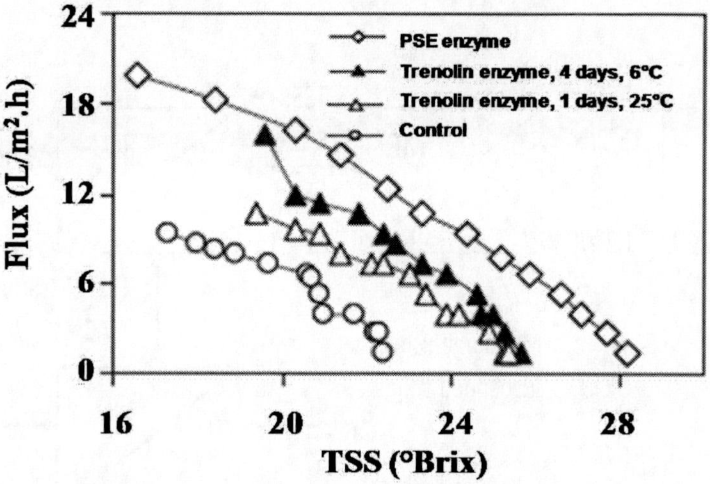

Figure 9. Effect of total soluble solids (TSS) on the transmembrane flux for two verities of commercially available pectinase enzymes (Panzym Super E and Trenolin Rot) (Pap et al., 2009).

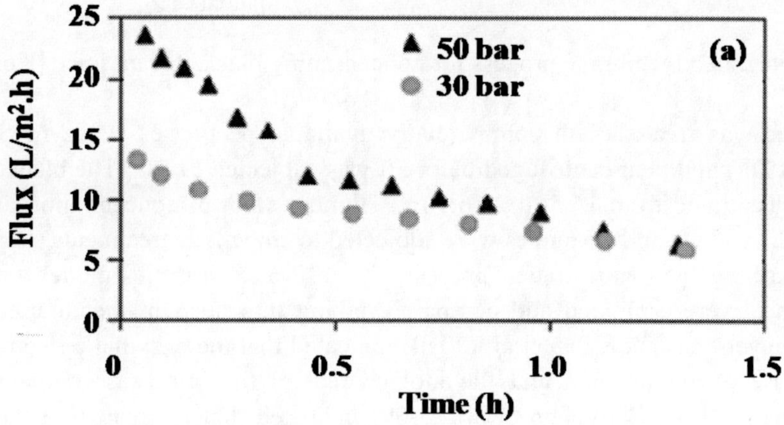

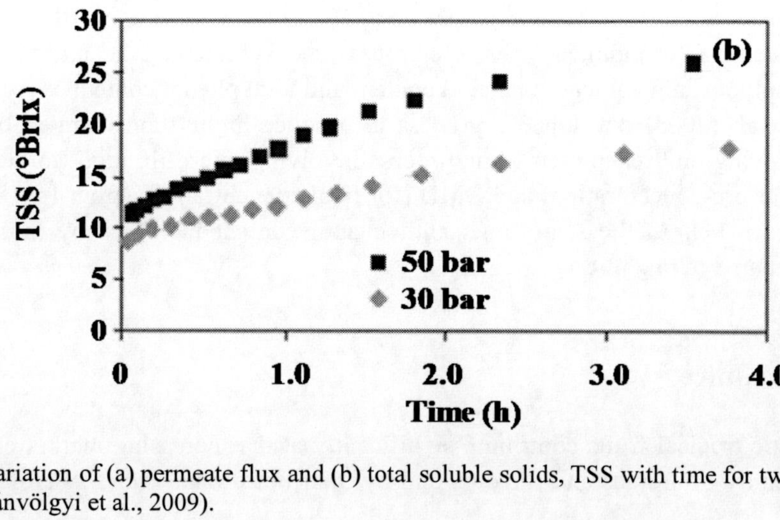

Figure 10. Variation of (a) permeate flux and (b) total soluble solids, TSS with time for two different pressures (Bánvölgyi et al., 2009).

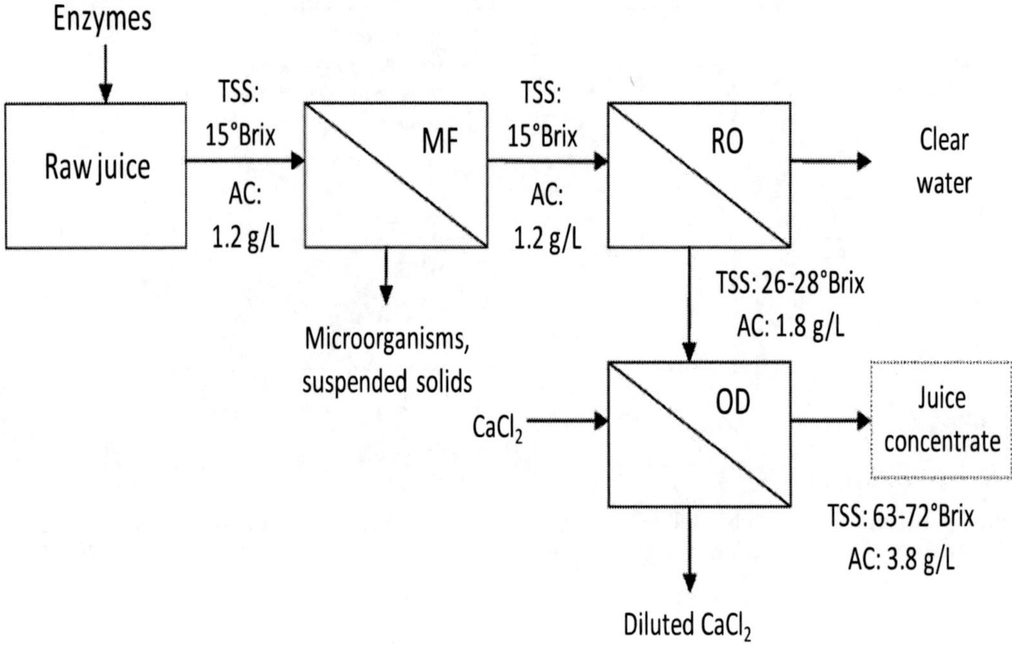

Figure 11. Integrated membrane process for concentrating black currant juice (Kozák et al., 2008).

The juice was treated with commercially available pectinases (Panzym Super E and Trenolin Rot DF) and then centrifuged before it was subjected to RO. The blackcurrant juice could be concentrated from 16.5 to 28.6°Brix. Higher extent of concentration and permeate flux were achieved when the samples were subjected to enzymatic treatments (Figure 9).

For describing the concentration process, the resistance-in-series model was used. The total resistance were estimated and compared, taking into account membrane fouling and polarization layer resistance. Pap et al. (2010) indicated that the enzymatic treatment resulted in the increase of anthocyanin and flavonol content of the juices as compared to centrifugation process. Also, Bánvölgyi et al. (2009) indicated that pretreated blackcurrant juice can be concentrated by RO from initial soluble solid content of 8-10 °Brix to 22-26 °Brix, by the application of a transmembrane pressure up to 50 bar for 3.5 h (Figure 10).

Besides being an economically feasible process, the RO retained the antioxidant capacity, monomeric anthocyanin content, total acid content and total phenol content over 99%.

Kozák et al. (2008) developed a process to produce highly concentrated black currant juice by involving multi-step membrane process involving microfiltration (to clarify the raw juice), RO (for pre-concentration) and OMD (for final concentration) (from 15-18°Brix to 63-72°Brix, Figure 11). At the same time, anthocyanin content increased by more than three times as compared to raw juice.

3.6. Acerola Juice

The exotic tropical fruits contribute significantly at the increasing market of fruit juices by their diversity of aromas and flavours and by their nutritional value. Acerola is one such fruit with high ascorbic acid and anthocyanins contents. Matta et al. (2004) developed a process for obtaining clarified and concentrated acerola juice, maintaining its nutritional and sensory characteristics. The process involved enzymatic hydrolysis of pulp, clarification by microfiltration and concentration by RO. During RO, the feed volume reduced, resulting in an increased osmotic pressure of the juice leading to the decline in flux and an increase in the soluble solid content with time (Figure 12a and 12b).

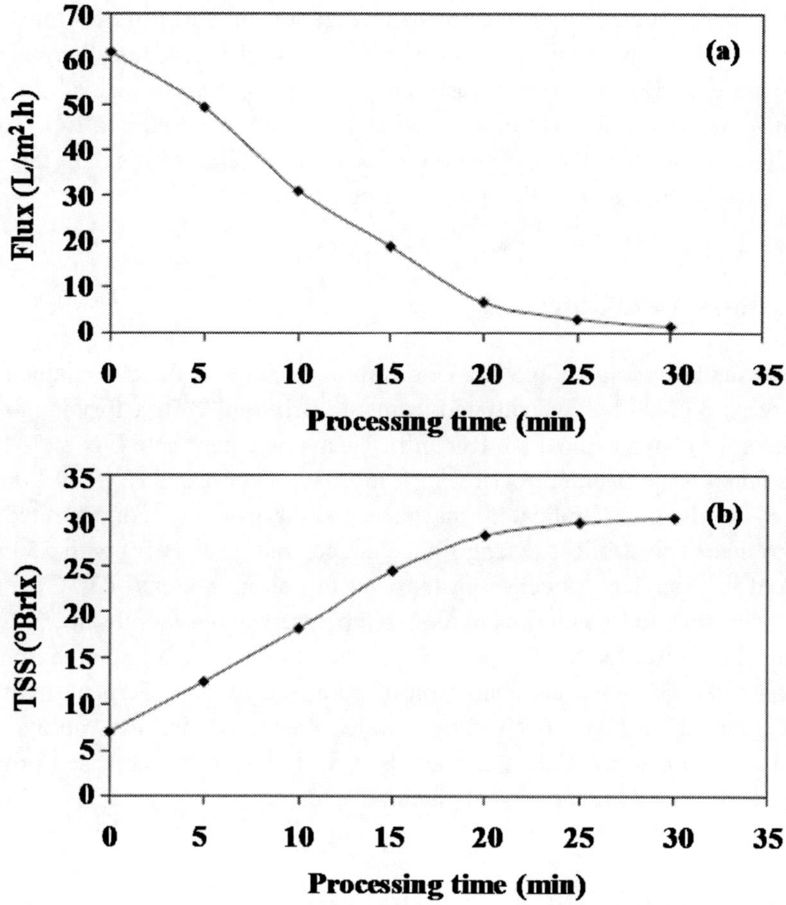

Figure 12. (a) Permeate flux behavior and (b) variation in soluble solid content during the reverse osmosis of the clarified acerola juice (Matta et al., 2004).

More than 4-fold increase in ascorbic acid content of the juice was observed and the consumer acceptability of the drink prepared by the clarified juice was very high. Gomes et al. (2005) indicated that the physico-chemical characteristics of the juice were preserved and the consumer acceptability of the juice was very high.

Pagani et al. (2011) presented a scheme for the production of high quality concentrated acerola juice by an integrated membrane process consisting of clarification by microfiltration, pre-concentration up to 28° Brix by RO and final concentration up to 55° Brix by OMD. Vitamin C, anthocyanins content and antioxidant activity were found to increase 2.21, 1.41 and 2.28-fold, respectively.

3.7. Kinnow Mandarin Juice

Khamrui and Pal (2002) indicated that kinnow mandarin juice was concentrated up to 23°Brix by employing RO process. The recovery of acid components, total soluble solids, total carbohydrates, reducing sugars and sucrose were reported to be 89.68, 96.02, 95.50, 94.32 and 98.22%, respectively. A transmembrane pressure of 40 bar and temperature up to 40°C were found to be optimum for the RO process for concentration of kinnow juice. Thakur et al. (2004) studied the effect of centrifugation or ultrafiltration on the degree of concentration of kinnow fruit juice achieved by RO as well as vacuum thermal evaporation. The average permeate flux during RO was found to be the highest in case of ultrafiltered juice as compared to the centrifuged juice. The clarity of the ultrafiltered and RO processed juice was the highest. However, the evaluation of sensory characteristics of the concentrates showed that the difference among them was not significant.

3.8. Mango and Guava Juice

Das Gupta and Jayaraman (1998) demonstrated that the juices obtained by enzyme clarification of pulp could be concentrated from 14° Brix to 27° Brix for mango and from 9° Brix to 24° Brix for guava using a lab RO unit. The average permeate flux at 50 bar operating pressure was found to be from 13 to 18 L/m^{2}h for mango and from 12 to 19 L/m^{2}h for guava juice. Sensory evaluations indicated that acceptability of the concentrated juices was comparable to that of feed juices. Mango juice concentrate (27° Brix) with 550 ppm sulphur dioxide packed in propylene pouches pasteurized in boiling water (~85° C) for 2 min was found to be acceptable up to 6 months at room temperature and 4 months at 37° C.

Olle et al. (1997) studied the flavour compounds from ripe mango puree upon cross-flow microfiltration and subsequent concentration of permeate by RO. Terpene hydrocarbons, as well as oxygenated terpene derivatives, were recovered in the pulpy retentate of microfiltration. C_{13} norlsoprenoids and phenols were found to be increased probably by the chemical degradation of carotenoids and phenolic acids, respectively.

3.9. Other Fruit Juices

Lee and Lee (1999) demonstrated that concentrated pear juice could be produced using integrated membrane processing involving ultrafiltration and RO. The pear juice was treated with pectinase followed by gelatin and bentonite and then it was subjected to ultrafiltration to produce a clear juice. The clear pear juice was 5-fold concentrated by RO. The reconstituted juice (10 °Brix) with added ascorbic acid had no quality changes at 4 °C after 15 days' storage. The juice was compared with vacuum evaporated and freeze concentrated juice; only small differences in flavour and overall quality were observed.

Rodrigues et al. (2004) compared RO and OMD for the concentration of camu–camu juice without damaging the nutritional potential of the raw material RO has the advantage of being well developed at the industrial scale. OMD has the advantage of reaching higher concentration levels. RO yielded juice with high nutritional quality (loss of ascorbic acid 8%). The average permeate flux was 50 L/m^2.h while using a transmembrane pressure of 60 bar and maximum TSS content was 25.5 °Brix. Whereas OMD process resulted in concentrate camu–camu juice up to the same TSS content as that obtained with thermal evaporation (up to 60 °Brix). The loss of vitamin C was less than 5% while the average flux was lower, close to 10 L/m^2.h.

Thakur and Gupta (2006) studied the clarification and subsequent concentration of beetroot juice. The clarified juice using enzyme treatment and centrifugation was partially concentrated using RO and final concentration was achieved by vacuum evaporation. Average flux during RO of clarified juice with and without ultrafiltration was found to be 36.08 and 30.33 L/m^{2}h, respectively. These juices were concentrated to 23-25°Brix and the pigment loss was higher in the case of RO-concentrated juice pretreated with enzyme. Comparative evaluation in terms of clarity, colour and chemical parameters revealed that concentrates obtained using RO and vacuum evaporation were comparable.

Bitter gourd, a commonly consumed vegetable in South Asia, eastern-South Asia, China, Japan, Korea peninsula, Caribbean Sea isles etc., is used as an adjunct in the management of diabetes mellitus. Wu et al. (2010) studied the effective ways of producing bitter gourd freeze-dried powder from the concentrated juice of bitter gourd, which was produced by integrated process involving enzymatic process, ultrafiltration, RO and thermal evaporation under vacuum. The ultrafiltration did not result in any change in total saponins, total sugar and pH value in bitter gourd juice while decreasing the turbidity and improving transmittance of juice. Integration of RO with vacuum concentration process had resulted in 3.33 fold production efficiency and 1.72 times vitamin C content in the concentrated juice, which were of importance in optimising lyophilisation parameters.

Nuss et al. (1997) attempted to improve extraction efficiency of oil by concentrating the onion juice by RO up to 18°Brix prior to supercritical CO_2 extraction. Improved efficiency of extraction of onion oil without altering its characteristic fresh aroma was achieved.

Pepper et al. (1985) demonstrated that advent of non-cellulosic membranes having high retention of low molecular weight organic compounds, as well as good physical and chemical stability, enabled RO to be used on a commercial scale for the concentration of tomato juice. Yildiz et al. (1993) compared the quality characteristics of Napoli VF variety of tomato concentrates (~20 °Brix) produced by RO and thermal evaporation. A significant difference in physical and chemical characteristics of both concentrates was noticed. The sensory evaluation by a consumer panel (n = 50) indicated that consumers preferred colour, appearance, flavour and consistency of the tomato juice concentrated by RO as compared to that by evaporation. Bottino et al. (2002) employed ceramic multichannel microfiltration membranes (average pore size 0.1 μm) to obtain a clarified serum, which was concentrated up to 14-15 °Brix by RO by using high temperature and pressure resistant spiral-wound membranes with a permeate flux of 20 L/m^2.h. The concentrated pulp obtained as retentate during microfiltration was mixed with the RO concentrated serum to formulate new tomato juice based products.

Conclusions and Future Trends

The concentration of fruit juices by RO facilitates handling of the product during transport and storage without negatively affecting their sensory and nutritional properties. Application of membrane processes in the juice and beverage industry is increasing day by day. It allows improvement in the quality of existing beverages and juices, development of new products with enhanced process efficiency and profitability. The ability to stabilize the functional and microbiological characteristics of juice products and to prepare new functional ingredients consisting of juice constituents has opened new avenues for RO in food and beverage industry. The expanding applications of membrane processes especially RO will certainly continue to benefit the juice industry in the future to obtain required product quality, purity, yield and throughput along with economic viability. The use of RO at industrial scale for concentration of fruit juices will appear still more lucrative in future, if RO membranes with high selectivity, improved flux, robustness, and greater chemical and mechanical stability will be developed. The use of integrated membrane processes such as microfiltration, ultrafiltration, electrodialysis, osmotic distillation and forward osmosis with RO will have great potential in food processing industry. Future developments will determine whether such membrane-based processes can provide the required product quality, purity, yield and throughput while remaining economically viable for the fruit juice industry.

References

Alvarez, V., Alvarez, S., Riera, F.A., Alvarez, R. (1997) Permeate flux prediction in apple juice concentration by reverse osmosis, *Journal of Membrane Science*, 127 (1), 25-34.

Alvarez, S., Riera, F.A., Álvarez, R., Coca, J. (1998) Permeation of apple aroma compounds in reverse osmosis, *Separation and Purification Technology,* 14 (1-3), 209-220.

Alvarez, S., Riera, F.A., Álvarez, R., Coca, J., Cuperus, F.P., Th Bouwer, S., Boswinkel, G., Van Gemert, R.W., Veldsink, J.W., Giorno, L., Donato, L., Todisco, S., Drioli, E., Olsson, J., Trägårdh, G., Gaeta, S.N., Panyor, L. (2000) New integrated membrane process for producing clarified apple juice and apple juice aroma concentrate, *Journal of Food Engineering,* 46(2), 109-125.

Alvarez, S., Riera, F.A., Álvarez, R., Coca, J. (2001) Prediction of flux and aroma compounds rejection in a reverse osmosis concentration of apple juice model solutions, *Industrial and Engineering Chemistry Research*, 40 (22), 4925-4934.

Alvarez, S., Riera, F.A., Álvarez, R., Coca, J. (2002) Concentration of apple juice by reverse osmosis at laboratory and pilot-plant scales, *Industrial and Engineering Chemistry Research*, 41 (24), 6156-6164.

Bánvölgyi, S., Horváth, S., Stefanovits-Bányai, É., Békássy-Molnár, E., Vatai, G. (2009) Integrated membrane process for blackcurrant (*Ribes nigrum* L.) juice concentration, *Desalination,* 241 (1-3), 281-287.

Bottino, A., Capannelli, G., Turchini, A., Della Valle, P., Trevisan, M. (2002) Integrated membrane processes for the concentration of tomato juice, *Desalination*, 148 (1-3), 73-77.

Bowden, R. P., Isaacs, A. R. (1989) Concentration of pineapple juice by reverse osmosis, *Food Australia*, 41(7), 850–851.

Braddock, R. J., Nikdel, S., Nagy, S. (1988) Composition of some organic and inorganic compounds in reverse osmosis-concentrated citrus juices, *Journal of Food Science*, 53(2), 508–512.

Cassano, A., Drioli, E., Galaverna, G., Marchelli, R., Di Silvestro, G., Cagnasso, P. (2003) Clarification and concentration of citrus and carrot juices by integrated membrane processes, *Journal of Food Engineering*, 57(2), 153-163.

Chen, C. S., Shaw, P. E., Parish, M. E. (1993) Orange and tangerine juices. In S. Nagy, C. S. Chen, P. E. Shaw (Eds.), Fruit juice processing technology (page 110–165). Auburndale, FL: AgScience.

Chien-Hwa Y, Lung-Chen F, Shaik Khaja L, Chung-Hsin W, Cheng-Fang L (2010) Enzymatic treatment for controlling irreversible membrane fouling in cross-flow humic acid-fed ultrafiltration. *Journal of Hazardous Materials*, 177, 1153–1158.

Chou, F., Wiley, R. C., Schlimme, D. V. (1991) RO and flavour retention in apple juice concentration, *Journal of Food Science*, 56(2), 484–487.

Chua, H. T., Rao, M. A., Acree, T. E., Cunningham, D. G. (1987) Reverse osmosis concentration of apple juice: flux and flavour retention by cellulose acetate and polyamide membranes, *Journal of Food Process Engineering*, 9(3), 231–245.

D'souza N, Mawson AJ (2005) Membrane cleaning in the dairy industry: a review, *Critical Reviews in Food Science and Nutrition*, 45(2):125–134.

Das Gupta, D.K., Jayaraman, K.S. (1996) Studies on the membrane concentration of watermelon juice *Journal of Scientific and Industrial Research*, 55 (12), 966-970.

Das Gupta, D.K., Jayaraman, K.S. (1998) Preparation and storage evaluation of mango and guava juice semi-concentrate, *Journal of Scientific and Industrial Research*, 57 (6), 310-315.

Echavarrıa , A., P. Torras, C., Pagan, J., Ibarz A. (2011) Fruit juice processing and membrane technology application, *Food Engineering Reviews*, 3, 136–158.

Galaverna, G., Di Silvestro, G., Cassano, A., Sforza, S., Dossena, A., Drioli, E., Marchelli, R. (2008) A new integrated membrane process for the production of concentrated blood orange juice: Effect on bioactive compounds and antioxidant activity, *Food Chemistry*, 106(3), 1021-1030.

García, E., Gozálvez, J.M., Lora, J. (2002) Use of reverse osmosis as a preconcentration system of waste leaching liquid from the citric juice production industry, *Desalination,* 148(1-3), 137-142.

Garcia-Castello, E.M., Mayor, L., Chorques, S., Argüelles, A., Vidal-Brotóns, D., Gras, M.L. (2011) Reverse osmosis concentration of press liquid from orange juice solid wastes: Flux decline mechanisms, *Journal of Food Engineering*, 106 (3), 199-205.

Girard B, Fukumoto LR (2000) Membrane processing of fruit juices and beverages: a review, *Critical Reviews in Biotechnology*, 20(2), 109– 175.

Gomes, D.S., Mendes, E.R., Pereira, E.S., De Barros, N.C. (2005) Evaluation of the acerola juice concentrated by reverse osmosis, *Brazilian Archives of Biology and Technology*, 48, 175-183.

Gomes, F.S., da Costa, P.A., de Campos, M.B.D., Couri, S., Cabral, L.M.C. (2011) Concentration of watermelon juice by reverse osmosis process, *Desalination and Water Treatment*, 27 (1-3), 120-122.

Gurak, P.D., Cabral, L.M.C., Rocha-Leão, M.H.M., Matta, V.M., Freitas, S.P. (2010) Quality evaluation of grape juice concentrated by reverse osmosis, Journal of Food Engineering, 96(3), 421-426.

Hernandez, E., Chen, C. S., Johnson, J., Carter, R. D. (1995) Viscosity changes in orange juice after ultrafiltration and evaporation, *Journal of Food Engineering,* 25(3), 387–396.

Hunter T.C., Rao, M.A., Acree, T.E., Cunningham, D.G. (1987) Reverse osmosis concentration of apple juice: flux and flavor retention by cellulose acetate and polyamide membranes, *Journal of Food Process Engineering*, 9 (3), 231-245.

Jesus DF, Leite MF, Silva LF, Modesta RD, Matta VM, Cabral LM (2007a) Orange (Citrus sinensis) juice concentration by reverse osmosis. *Journal of Food Process Engineering*, 81, 287–291.

Jesus, D.F., Leite, M.F., Silva, L.F.M., Modesta, R.D., Matta, V.M., Cabral, L.M.C. (2007b) Orange (*Citrus sinensis*) juice concentration by reverse osmosis, *Journal of Food Engineering*, 81(2), 287-291.

Jiao B, Cassano A, Drioli E (2004) Recent advances on membrane processes for the concentration of fruit juices: a review, *Journal of Food Process Engineering,* 63, 303–324.

Katasonova ON, Fedotov PS (2009) Methods for continuous flow fractionation of microparticles: outlooks and fields of application, *Journal of Analytical Chemistry*, 64(3), 212–225.

Khamrui, K., Pal, D. (2002) Application of RO for concentration of kinnow mandarin juice, *Journal of Food Science and Technology*, 39 (3), 310-312.

Koseloglu, S. S., Lawhon, J. T., Lusas, E. W. (1990) Use of membranes in citrus juice processing, *Food Technology*, 44(12), 90–97.

Kozák, A., Bánvölgyi, S., Vincze, I., Kiss, I., Békássy-Molnár, E., Vatai, G. (2008) Comparison of integrated large scale and laboratory scale membrane processes for the production of black currant juice concentrate, *Chemical Engineering and Processing: Process Intensification*, 47 (7), 1171-1177.

Kwang-Sup Y, Joo-Heon H, Dong-Ho B, Seok-Joong K, Soon-Dong K (2004) Effective clarifying process of reconstituted apple juice using membrane filtration with filter-aid pretreatment, *Journal of Membrane Science*, 228, 179–186.

Lee, Y.C., Lee, S.W. (1999) Quality changes during storage in Korean clear pear juices concentrated by three methods, *Journal of Food Quality*, 22(5), 565-571.

Madaeni SS, Naghdi Sedeh S, De Nobili M (2006) Ultrafiltration of humic substances in the presence of protein and metal ions, *Transport in Porous Media*, 65, 469–484.

Matta, V.M., Moretti, R.H., Cabral, L.M.C. (2004) Microfiltration and reverse osmosis for clarification and concentration of acerola juice, *Journal of Food Engineering*, 61 (3), 477-482.

Medina, B.G. and Garcia I.A. (1998) Concentration of orange juice by reverse osmosis, *Journal of Food Process Engineering*, 10(3), 217-230.

Merson, R. L., Morgan, A. I. (1968) Juice concentration by reverse osmosis, *Food Technology*, 22(5), 631–634.

Mulder, M. (1996) Basic Principles of Membrane Technology. Kluwer Academic Publisher, The Netherland. Page 298.

Nuss, J.S., Guyer, D.E., Gage, D.A. (1997) Concentration of onion juice volatiles by reverse osmosis and its effects on supercritical CO2 extraction, *Journal of Food Process Engineering*, 20 (2), 125-139.

Olle, D., Baron, A., Lozano, Y.F., Sznaper, C., Baumes, R., Bayonove, C., Brillouet, J.M. (1997) Microfiltration and reverse osmosis affect recovery of mango puree flavor compounds, *Journal of Food Science*, 62 (6), 1116-1119.

Pagani, M.M., Rocha-Leão, M.H., Couto, A.B.B., Pinto, J.P., Ribeiro, A.O., dos Santos Gomes, F., Cabral, L.M.C. (2011) Concentration of acerola (*Malpighia emarginata* DC.) juice by integrated membrane separation process, *Desalination and Water Treatment*, 27(1-3), 130-134.

Palmieri, L., Dalla Rosa, M., Dall'Aglio, G., and Carpi, G. (1990) Production of kiwifruit concentrate by reverse osmosis process, *Acta Horticulturae*, 282, 435–439.

Pap, N., Kertész, Sz., Pongrácz, E., Myllykoski, L., Keiski, R.L., Vatai, Gy., László, Zs., Beszédes, S., Hodúr, C. (2009) Concentration of blackcurrant juice by reverse osmosis, *Desalination*, 241 (1-3), 256-264.

Pap, N., Pongrácz, E., Jaakkola, M., Tolonen, T., Virtanen, V., Turkki, A., Horváth-Hovorka, Z., Vatai, G., Keiski, R.L. (2010) The effect of pre-treatment on the anthocyanin and flavonol content of black currant juice (*Ribes nigrum* L.) in concentration by reverse osmosis, *Journal of Food Engineering*, 98 (4), 429-436.

Pepper, D., Orchard, A.C.J., Merry, A.J. (1985) Concentration of tomato juice and other fruit juices by reverse osmosis, *Desalination*, 53 (1-3), 157-166.

Peri, C., Battisti, P., Setti, D. (1973) Solute transport and permeability characteristics of reverse osmosis membranes, *Lebensmittel Wissenschaft und Technologie*, 6(4), 127–132.

Pozderović, A., Moslavac, T. (1999) Apple juice aroma concentration from evaporator condensate by reverse osmosis: I. Influence of process parameters on retention and final aroma concentration, *Acta Alimentaria*, 28 (1), 71-83.

Rektor, A., Pap, N., Kókai, Z., Szabó, R., Vatai, G., Békássylnár, E. (2004) Application of membrane filtration methods for must processing and preservation, *Desalination*, 162(1-3), 271-277.

Rektor, A., Vatai, G., Békássy-Molnár, E. (2006) Multi-step membrane processes for the concentration of grape juice, *Desalination,* 191(1-3), 446-453.

Rektor, A., Kozak, A., Vatai, G., Bekassy-Molnar, E. (2007) Pilot plant reverse osmosis-filtration of grape juice, *Separation and Purification Technology*, 57(3), 473-475.

Rodrigues, R.B., Menezes, H.C., Cabral, L.M.C., Dornier, M., Rios, G.M., Reynes, M. (2004) Evaluation of reverse osmosis and osmotic evaporation to concentrate camu-camu juice (*Myrciaria dubia*), *Journal of Food Engineering*, 63 (1), 97-102.

Sablani S, Goosen M, Al-Belushi R, Wilf M (2001) Concentration polarization in ultrafiltration and reverse osmosis: a critical review, *Desalination* 141(3):269–289.

Santana, I., Gurak, P.D., da Matta, V.M., Freitas, S.P., Cabral, L.M.C. (2011) Concentration of grape juice (Vitis labrusca) by reverse osmosis process, *Desalination and Water Treatment*, 27(1-3), 103-107.

Santon M, Treiche H, Valduga E, Cabra L, Di Luccio M (2008) Evaluation of enzymatic treatment of peach juice using response surface methodology, *Journal of Science of Food and Agriculture*, 88, 507–512.

Scordino, M., Di Mauro, A., Passerini, A., Maccarone, E. (2007) Highly purified sugar concentrate from a residue of citrus pigments recovery process, *Lebensmittel Wissenschaft und Technologie*, 40(4), 713-721.

Sheu, M.J., Wiley, R.C. (1983) Preconcentration of apple juice by reverse osmosis, *Journal of Food Science*, 48(2), 422–429.

Singh, N.I. and Eipeson, W.E. (2000) Concentration of clarified orange juice by reverse osmosis, *Journal of Food Science and Technology*, 37(4), 363-367.

Thakur, V., Manjunath, S.S., Gupta, O.K.D. (2004) Comparative evaluation of clarification and concentration technique for kinnow fruit juice, *Journal of Food Science and Technology,* 41 (4), 382-385.

Thakur, V., Gupta, D.K.D. (2006) Studies on the clarification and concentration of beetroot juice, *Journal of Food Processing and Preservation*, 30(2), 194-207.

Vaillant, F., Jeanton, E., Dornier, M., O'Brien, G.M., Reynes, M., Decloux, M. (2001) Concentration of passion fruit juice on an industrial pilot scale using osmotic evaporation, *Journal of Food Engineering*, 47(3),195–202.

Vladisavljevic, G.T., Vukosavljevic, P., Bukvic, B. (2003) Permeate flux and fouling resistance in ultrafiltration of depectinized apple juice using ceramic membranes, *Journal of Food Engineering*, 60, 241–247.

Wu, M., Hu, X., Ge, B., Zhao Process optimisation of the bitter gourd (Momordica charantia) concentrated juice preparation for a freeze-dried powder, *International Journal of Food Science and Technology*, 45(12), 2553-2559.

Yazdanshenas, M., Reza, S.A., Soltanieh, M., Roostaazad, R., Khoshfetrat, A.B. (2010) Contribution of fouling and gel polarization during ultrafiltration of raw apple juice at industrial scale, *Desalination* 258:194–200.

Yildiz, F., Bozkurt, M., Wiley, R.C. (1993) Comparison of reverse osmosis and evaporation processing on the quality characteristics of tomato concentrates, *Food Control*, 4 (3), 155-158.

INDEX

A

abatement, 17
access, 233
accounting, 75
acetic acid, 76, 84, 87, 98
acetone, 80, 87, 94, 95
acetonitrile, 79, 95
acid, 19, 76, 77, 84, 87, 88, 92, 98, 99, 108, 109, 111, 112, 115, 133, 155, 167, 181, 183, 189, 190, 193, 196, 197, 200, 201, 204, 208, 218, 219, 220, 222, 223, 239, 296, 299, 300, 301, 304, 305, 306, 309
acidic, 76, 241, 298
acidity, 137, 178, 300, 301, 302
acrylate, 164
acrylic acid, 77
acrylonitrile, 77, 92, 189, 226
activated carbon, 175, 184
activation energy, 2, 14, 15, 129
active compound, 46, 239
active site, 5, 141, 142, 144
adaptation, 58
additives, 38, 54, 63, 74, 75, 113, 141, 155, 184, 198, 290
ADH, 193, 197
adhesion, 18, 36, 37, 45, 209, 215, 223, 242, 292
adjustment, 136, 166, 293
adsorption, 1, 15, 16, 17, 19, 21, 34, 54, 67, 76, 82, 84, 89, 90, 94, 110, 122, 123, 129, 131, 132, 141, 143, 144, 165, 166, 169, 182, 185, 187, 192, 198, 205, 227, 236, 237, 238, 300
adsorption isotherms, 238
adverse effects, 204
aerosols, 176
AFM, 226, 227
age, 309
aggregation, 18, 185
agriculture, 62
albumin, 67, 147
alcohol production, 80
alcohols, 76, 77, 80, 81, 82, 83, 84, 87, 96, 97, 137, 181, 226
aldehydes, 181
aliphatic compounds, 181
alkalinity, 212, 213, 221
almonds, 115
alternative energy, 61
aluminium, 41, 147, 151
AME, 220
amine, 19, 20, 132, 133
amines, 202
amino, 77, 141, 144, 184, 202
ammonia, 57, 58, 204
ammonium, 209, 211, 212
amorphous polymeric material, 13, 15
anaerobic digestion, 212
annealing, 39
anthocyanin, 301, 304, 311
antibiotic, 176, 197
anti-inflammatory drugs, 202
antioxidant, 184, 302, 304, 305, 309
aqueous solutions, 46, 47, 52, 54, 66, 81, 83, 93, 94, 95, 174, 243
aromatic compounds, 203, 216
aromatic hydrocarbons, 87, 204
aromatics, 87, 89, 165, 204
arsenic, 130, 239, 244
ascorbic acid, 108, 109, 111, 112, 183, 305, 306
Asia, 307
aspartic acid, 201
assessment, 187
atmosphere, 19, 103, 104, 105, 106, 110, 111, 116, 128, 132
atmospheric pressure, 137
atomic force, 28, 227
atoms, 130

attachment, 141, 207, 208, 209
awareness, 68
azeotropic point, 75

B

Bacillus subtilis, 209
bacteria, 80, 105, 135, 177, 180, 181, 184, 195, 196, 208, 209, 221
bacterium, 150
barriers, 103
base, 7, 19, 28, 125, 128, 132, 148, 150, 164, 170, 206, 207, 208, 209
baths, 174
batteries, 122
beer, 137, 173, 175, 177, 185, 186, 189, 190
behaviors, 36
Beijing, 148
Belgium, 29, 65, 225
benchmarking, 93
beneficial effect, 302
benefits, 46, 51, 52, 121, 122, 124, 125, 128, 132, 139, 187
benzene, 56, 86, 87, 88, 89, 95, 97, 98
beverages, vii, 67, 79, 103, 153, 172, 173, 178, 184, 189, 308, 309
bicarbonate, 111
binary oxides, 128
biocatalysts, 134, 136, 137, 140, 206, 214
biochemical processes, 156
biocompatibility, 77
bioconversion, 195
biodegradability, 135
biodegradable materials, 107, 109
biodegradation, 117, 202, 203, 205, 215, 220
bioelectricity, 194, 214
biofuel, 80, 83
biological activity, 212
biomass, 68, 90, 97, 107, 140, 174, 209
biomolecules, 54, 142
biopolymers, 46, 101, 107, 112, 113, 114
bioremediation, 217
biosensors, 147, 148
biotechnology, viii, 135, 149, 175, 193, 194, 195, 215, 216, 220, 221, 222, 223
biotin, 143, 147
bisphenol, 19, 20, 30, 133
black liquor, 174
bleaching, 174, 212
blends, 37, 95, 113, 117, 118
blood, 51, 188, 309
bonding, 75, 144
bonds, 18, 77, 141, 142, 143, 144
bounds, 106
Brazil, 67
bromine, 27
building blocks, 29, 39, 62, 95
butadiene, 89
butyl ether, 86, 87
by-products, 81, 82, 84, 93, 124, 155, 177, 198

C

Ca^{2+}, 211
calcium, 108, 165, 178, 181, 295
calorie, 137
cancer, 46
candida, 148
candidates, 146
capillary, 6, 21, 24, 25, 26, 27, 29, 39, 43, 45, 52, 58, 131, 170, 171, 182, 192, 200, 202, 217, 227, 297
carbohydrates, 186, 306
carbon, 3, 20, 23, 26, 30, 34, 101, 106, 107, 108, 110, 114, 116, 121, 123, 130, 131, 147, 148, 151, 166, 184, 191, 207, 209, 219, 224, 239
carbon atoms, 130
carbon dioxide, 20, 23, 30, 101, 106, 107, 108, 110, 123, 148, 151, 207
carbon nanotubes, 114, 116
carboxyl, 141
carboxylic acids, 137, 202
carcinoma, 59
Caribbean, 307
carotene, 300
carotenoids, 181, 290, 306
case studies, 173
casein, 179, 180, 184, 191
casting, 38, 57, 81, 105, 111, 169, 202
catalyst, 125, 136, 137, 140, 144, 223
catalytic activity, 141, 144
catalytic properties, 140, 214
categorization, 3
cation, 58, 208, 211, 212, 213, 222, 233
cell culture, 176, 195, 196, 197
cellulose, 11, 26, 30, 76, 87, 89, 93, 108, 109, 114, 115, 156, 164, 165, 169, 180, 182, 239, 243, 294, 295, 309, 310
ceramic, 3, 21, 26, 29, 41, 60, 62, 66, 79, 97, 126, 169, 170, 171, 174, 177, 179, 187, 189, 192, 242, 299, 307, 312
challenges, 103, 114, 197, 210, 217, 218, 225
cheese, 104, 165, 175, 177, 179, 180, 190, 199
chemical, 4, 10, 18, 19, 20, 23, 29, 34, 35, 36, 37, 38, 39, 41, 43, 44, 45, 54, 56, 65, 66, 68, 69, 70, 71, 72, 75, 77, 79, 80, 81, 82, 87, 89, 90, 91, 93, 110, 113, 114, 121, 122, 126, 128, 130, 132, 133, 134,

136, 137, 140, 142, 144, 147, 150, 155, 156, 158, 164, 166, 167, 169, 172, 173, 176, 179, 181, 182, 183, 188, 193, 194, 196, 198, 205, 208, 209, 214, 215, 238, 239, 242, 291, 293, 296, 300, 302, 305, 306, 308
chemical characteristics, 305, 307
chemical degradation, 121, 132, 169, 182, 306
chemical industry, 45, 93
chemical interaction, 158
chemical properties, 38, 79, 80, 113, 176, 302
chemical reactivity, 147
chemical stability, 19, 75, 77, 133, 198, 291, 307
chemicals, 55, 68, 80, 90, 96, 145, 166, 172, 175, 191, 196, 197, 199, 213
chemisorption, 237
chemotherapy, 59
China, 148, 307
chitin, 77, 112, 114, 117
chitosan, 77, 79, 94, 95, 97, 99, 109, 110, 112, 117, 118, 119, 202, 204, 217
Chitosan, 77, 92, 93, 95, 98, 109, 110, 112, 118
chlorination, 166
chlorine, 130
chlorophyll, 105, 110
cholesterol, 178
cholinesterase, 147
chromatographic technique, 179
chromatography, 13, 176, 194, 195, 196, 203, 214, 217
chromium, 56, 174, 239
chymotrypsin, 201
circulation, 174
clarity, 183, 306, 307
classes, 136, 174
classification, 3, 4, 5, 125, 126, 133
cleaning, 25, 40, 84, 139, 149, 164, 166, 167, 169, 171, 172, 178, 181, 182, 197, 297, 309
climate, 19, 123, 132
clinical application, 59
clustering, 130
clusters, 18
CO2, vii, 1, 11, 12, 18, 19, 20, 21, 22, 23, 24, 27, 29, 34, 59, 60, 62, 63, 103, 104, 105, 106, 107, 108, 110, 111, 121, 123, 124, 130, 132, 133, 147, 148, 211, 219, 307, 310
coagulation process, 180
coatings, 102, 116, 174, 175
coffee, 110
coke, 130, 205
collagen, 200
collisions, 5, 21, 55, 131
color, iv, 74, 136, 173, 183, 185, 186, 204, 226, 239
combustion, 80, 122, 123, 124, 148, 219
commercial, vii, 22, 24, 26, 39, 40, 45, 53, 56, 67, 74, 76, 79, 81, 82, 84, 86, 97, 108, 123, 140, 151, 185, 203, 206, 217, 243, 289, 291, 307
commodity, 80
communities, 221
community, 124, 218
compaction, 165
comparative analysis, 119
compatibility, 18
competition, 68
competitive process, 56, 86
competitiveness, 20, 65, 176
complex interactions, 185
complexity, 83
composites, 1, 116, 118
composition, 18, 37, 38, 57, 66, 72, 73, 74, 75, 82, 106, 131, 134, 141, 146, 167, 178, 180, 183
compost, 107
compounds, 17, 18, 46, 54, 56, 66, 68, 74, 79, 83, 84, 86, 90, 94, 95, 107, 112, 113, 117, 130, 140, 156, 162, 164, 167, 174, 176, 180, 181, 182, 183, 184, 185, 186, 188, 199, 202, 203, 206, 207, 211, 216, 220, 221, 237, 239, 240, 290, 291, 292, 295, 297, 298, 300,302, 306, 307, 308, 309, 311
condensation, 1, 6, 15, 20, 21, 29, 39, 47, 53, 69, 74, 83, 131, 133
conditioning, 136
conduction, 33, 49, 51
conductivity, 32, 48, 49, 50
configuration, 1, 20, 25, 35, 40, 47, 88, 126, 137, 138, 139, 146, 166, 171, 172, 174, 177, 182, 196, 198, 217
Congress, iv, 62, 189
conservation, 102, 289
constituents, 84, 110, 163, 173, 181, 299, 308
construction, 29, 62
consumers, 102, 114, 185, 301, 302, 307
consumption, vii, 24, 65, 66, 86, 102, 139, 145, 146, 156, 171, 173, 175, 184, 200, 205, 211, 226, 292, 300
contaminant, 124, 130, 137, 145, 207, 215
contaminated water, 56, 151
contamination, 74, 76, 102, 104, 108, 130, 146
COOH, 169
cooling, 19, 48, 132
copolymer, 108
copolymerization, 81
copolymers, 39, 81, 92
copper, 181
correlation, 8, 22, 127, 235
correlations, 58
corrosion, 135, 175

cost, 5, 13, 23, 25, 26, 46, 54, 66, 67, 68, 76, 82, 83, 90, 121, 123, 132, 139, 140, 145, 146, 148, 164, 171, 172, 205, 206, 291
cost effectiveness, 23
cotton, 136
covalent bond, 141, 144
covering, 38, 83, 209
cracks, 130
crops, 102
crystalline, 39, 55, 81
crystallinity, 113
crystallites, 39
crystallization, vii, 31, 34, 40, 51, 54, 55, 56, 57, 60, 63, 89
crystallization kinetics, 56
crystals, 31, 32, 39, 54, 55, 57, 82
cultivation, 97
cultural practices, 116
culture, 83, 175, 176, 195, 198, 222
culture media, 175
cuticle, 112
cyanide, 204, 205, 215
cycles, 55, 74, 130, 182, 204
cyclodextrins, 111

D

dairy industry, 172, 177, 178, 179, 180, 188, 191, 225, 309
data set, 73
DEA, 194, 208
deacetylation, 112
decay, 108, 109, 118, 158, 159
decomposition, 205, 215
decoupling, 73
defecation, 184
defect formation, 82
defects, 10, 17, 82
degradation, 46, 52, 107, 110, 111, 113, 121, 132, 137, 169, 182, 203, 204, 205, 207, 214, 216, 220, 306
degradation process, 203
degradation rate, 203
dehydration, vii, 63, 65, 66, 67, 75, 76, 77, 78, 79, 80, 90, 93, 94, 95, 97, 98, 99
denaturation, 165
denitrification, 149, 150, 207, 208, 209, 217, 219, 220, 223
deposition, 54, 160, 164, 181, 237, 292
deposits, 236
depth, 184, 196
derivatives, 20, 81, 306
desorption, 1, 16, 22, 69, 129, 130, 142, 227
detachment, 44, 45
detergents, 167
detoxification, 216
deviation, 14, 227, 235, 241
diabetes, 307
dialysis, 67
dichloroethane, 207
dietary fiber, 181
diffusion, 1, 2, 5, 6, 8, 9, 10, 11, 14, 15, 16, 21, 22, 28, 29, 32, 34, 35, 47, 49, 53, 69, 71, 72, 77, 80, 82, 84, 105, 122, 129, 131, 145, 148, 154, 160, 161, 162, 165, 183, 207, 211, 215, 227, 228, 231, 232, 233, 234, 242, 290, 292, 294, 295
diffusion mechanisms, 5, 21
diffusion rates, 82, 228
diffusivity, 1, 5, 8, 10, 11, 13, 14, 21, 69, 73, 106, 121, 145
digestion, 212
dimethylformamide, 57
direct measure, 36, 235
directives, 107
discrimination, 140
discs, 164
diseases, 46
disinfection, 135, 146
dispersion, 10, 35, 41, 44, 216
dissociation, 129, 130
dissolved oxygen, 224
distillation, vii, 31, 34, 40, 46, 48, 49, 52, 53, 55, 56, 57, 58, 59, 60, 61, 62, 63, 67, 73, 74, 75, 76, 77, 80, 86, 90, 94, 121, 123, 184, 290, 300, 302, 308
distillation processes, 86
distribution, 35, 41, 44, 54, 56, 58, 61, 130, 131, 141, 149, 156, 166, 226, 227, 228, 229, 232, 233, 235, 241, 242, 243, 289, 300
diuretic, 302
diversity, 304
DMF, 57
DNA, 197
drainage, 137, 179
drawing, 39
dressings, 45
drinking water, vii, 140, 146, 149, 173, 192, 217, 219, 225, 226, 238, 239, 240, 241, 243, 244
drug delivery, 34
drug discovery, 46
drugs, 41, 46, 176, 202
drying, 57, 105
durability, 79, 128
dyeing, 146
dyes, 204, 216

E

ecology, 221
economic boom, 122
economic competitiveness, 20
economic evaluation, 191
economics, 40, 95, 183
effluent, 135, 146, 173, 175, 204, 205, 214, 216, 217, 239
effluents, 58, 150, 173, 174, 175, 187, 188, 189, 190, 202, 217, 226, 239, 240, 245
egg, 175, 200
electrical fields, 164, 293
electrical properties, 218
electricity, 124, 216, 220, 221
electrochemistry, 221
electrodes, 209, 213
electrolysis, 194, 210, 219, 222
electrolyte, viii, 211
electron, 87, 130, 207, 209, 211, 226
electron microscopy, 226
electrons, 210
electroplating, 225, 240, 241, 245
elongation, 75
elucidation, 185
e-mail, 65, 101, 121, 153, 193, 289
emergency, 107
emission, 124, 300
emitters, 110, 111
emulsions, 34, 41, 42, 44, 45, 46, 56, 57, 58, 59, 60, 61, 62, 63, 174
enantiomers, 202
endocrine, 145, 239
endothermic, 15
energy, vii, 2, 7, 14, 15, 19, 20, 24, 38, 41, 46, 47, 48, 59, 61, 63, 65, 66, 67, 68, 69, 73, 75, 76, 80, 86, 89, 90, 101, 102, 122, 123, 124, 129, 130, 132, 135, 138, 145, 146, 148, 155, 171, 174, 177, 181, 184, 193, 195, 205, 209, 211, 213, 214, 221, 226, 291, 292, 300
energy consumption, vii, 24, 65, 66, 86, 102, 138, 145, 146, 155, 171, 174, 184, 205, 211, 226, 292, 300
energy density, 122
energy input, 211
energy prices, 90
energy recovery, 195, 213, 214
engineering, 35, 46, 56, 84, 126, 135, 143, 147, 195, 219, 222, 291
England, 27, 189
enlargement, 146
entrapment, 111, 141, 143, 198
environment, 103, 104, 106, 110, 136, 137, 227, 242
environmental aspects, 76
environmental awareness, 68
environmental impact, vii, 101, 102, 103, 106, 108, 114, 156, 184
environmental issues, 122
Environmental Protection Agency, 115, 116
environmental standards, 90
environments, 90, 131, 132, 198
enzymatic activity, 141, 206
enzyme, 117, 122, 134, 135, 137, 140, 141, 142, 143, 144, 145, 147, 148, 149, 181, 193, 197, 198, 200, 201, 202, 203, 204, 205, 206, 214, 215, 220, 221, 222, 296, 306, 307
enzyme immobilization, 141, 142, 199, 203
enzymes, 46, 111, 134, 135, 136, 137, 140, 141, 142, 144, 167, 176, 179, 181, 198, 202, 203, 204, 205, 215, 292, 295, 300, 303
EPA, 106, 115, 116
epoxy groups, 208
equilibrium, 13, 15, 55, 63, 69, 71, 72, 73, 75, 82, 124, 128, 146, 231, 233, 237, 238, 294
equipment, 19, 25, 26, 135, 156, 176, 181, 195
ester, 89, 201, 215, 220, 224
etching, 39, 41, 56, 171
ethanol, 52, 57, 60, 67, 74, 75, 76, 77, 80, 81, 82, 83, 84, 85, 86, 87, 92, 93, 94, 95, 96, 97, 98, 99, 111, 117, 298
ethyl acetate, 77, 98, 99
ethylene, 29, 39, 76, 88, 90, 93, 95, 98, 104, 105, 106, 111, 117
ethylene glycol, 29, 76, 90, 93, 98
ethylene oxide, 76
EU, 149
Europe, 67
European Union, 116
evaporation, 18, 40, 48, 53, 54, 56, 60, 62, 66, 67, 76, 80, 84, 105, 111, 177, 184, 187, 290, 291, 292, 296, 298, 300, 306, 307, 310, 311, 312
evidence, 196
evolution, 38, 56, 135
exclusion, 21, 140, 228, 233
experimental condition, 13, 38
exploitation, 101, 114
exposure, 114
extraction, 34, 60, 67, 76, 90, 187, 197, 199, 219, 221, 296, 307, 310
extracts, 63, 79
extrusion, 39, 46, 111

F

fabrication, 1, 25, 26, 27, 37
Fabrication, 17, 38, 63, 243

families, 19, 132
fat, 165, 174, 177, 178, 179, 180, 189
fatty acids, 84, 175
FDA, 19, 133
feed mixture, 3, 66, 71, 73, 74, 77
fermentation, 34, 52, 68, 79, 80, 81, 82, 83, 84, 90, 93, 94, 96, 97, 110, 150, 175, 176, 184, 185, 194, 196, 197, 199, 214, 241
fiber, 4, 40, 58, 60, 62, 63, 81, 88, 93, 98, 137, 138, 139, 141, 170, 181, 182, 201, 203, 219, 220, 221, 222, 223, 242
fiber membranes, 4, 60, 203, 221, 223
fibers, 77, 83, 95, 209, 215
filler particles, 82
fillers, 82, 91, 113, 114, 117
films, 17, 39, 57, 101, 106, 107, 108, 109, 110, 111, 112, 113, 114, 115, 116, 117, 118, 134, 175
filters, 175, 176, 186, 194, 196, 198
filtration, 83, 138, 139, 141, 143, 146, 149, 154, 156, 157, 158, 163, 164, 166, 167, 168, 172, 173, 174, 175, 176, 181, 183, 184, 185, 186, 187, 189, 191, 192, 194, 195, 196, 197, 203, 204, 214, 215, 218, 219, 222, 223, 225, 236, 241, 242, 244, 245, 295, 310, 311
financial, 76
first generation, 138
fish, 174, 175, 187, 188, 221
flavonol, 304, 311
flavor, 104, 136, 137, 181, 182, 185, 297, 310, 311
flavour, 102, 104, 108, 109, 110, 117, 186, 200, 290, 291, 295, 297, 298, 302, 306, 307, 309
flexibility, 4, 18, 19, 34, 56, 75, 113, 132, 149, 176
flooding, 34, 54
flour, 136
flow field, 18
flue gas, 20, 23, 63
fluid, 5, 25, 26, 38, 71, 131, 157, 162, 164, 175, 206, 294
food, vii, 34, 40, 45, 57, 59, 60, 62, 65, 90, 101, 102, 103, 104, 105, 107, 110, 111, 112, 113, 114, 115, 116, 117, 118, 119, 146, 153, 156, 171, 172, 173, 176, 177, 181, 184, 188, 190, 195, 199, 200, 204, 215, 226, 242, 290, 291, 300, 308
food industry, 172, 176, 177, 190, 200, 204, 215, 242, 290, 291
food processing industry, 57, 176, 308
food products, 102
food safety, vii, 105, 110
food spoilage, 105
force, 8, 9, 13, 28, 35, 44, 52, 54, 66, 68, 69, 70, 71, 72, 73, 83, 103, 139, 141, 147, 159, 226, 227, 236, 238, 242
formation, 17, 39, 41, 55, 57, 63, 82, 90, 129, 130, 146, 163, 164, 177, 181, 183, 185, 198, 207, 219, 223, 236, 237, 292
formula, 9
fouling, 25, 26, 35, 40, 48, 81, 83, 84, 138, 147, 153, 154, 158, 159, 160, 163, 164, 165, 166, 167, 168, 169, 171, 172, 175, 178, 182, 183, 185, 187, 188, 189, 191, 192, 197, 206, 225, 228, 236, 237, 239, 240, 244, 292, 293, 294, 297, 300, 304, 309, 312
fragments, 76
France, v, 56, 63, 65
free energy, 59
free volume, 15, 18, 20, 77, 81, 84, 227
freezing, 227
freshwater, 55
fructose, 300
fruits, 101, 102, 103, 104, 106, 107, 108, 109, 110, 111, 113, 116, 182, 183, 300, 304
fuel cell, 122, 124, 147, 194, 195, 209, 210, 211, 216, 217, 219, 220, 221, 222
functional food, 199
functionalization, 18, 143
fungi, 112, 204
fusion, 143

G

gas separation, vii, 1, 10, 17, 18, 20, 23, 24, 26, 27, 28, 29, 69, 72, 105, 113, 150
gel, 18, 57, 111, 143, 144, 154, 159, 160, 161, 163, 165, 171, 201, 202, 236, 237, 292, 294, 312
genetic engineering, 143
geometry, 3, 4, 26, 32, 40, 67, 81, 126, 145
Germany, 58, 61, 67, 76, 86, 93, 118
ginger, 112
ginseng, 63
glass transition, 1, 2, 13, 113, 132
glass transition temperature, 1, 2, 13, 132
glassy polymers, 16, 17, 27, 29
global climate change, 19, 132
global warming, 19, 132
globalization, 103
glucose, 46, 137, 175, 200, 223, 300
glycerol, 52, 84, 113
glycol, 29, 76, 90, 93, 98
glycoside, 204
goat milk, 197, 215
granules, 182
graph, 7
graphite, 212
gravity, 186
greenhouse, 19, 124, 132
groundwater, 148, 221, 239, 241, 243, 244

growth, vii, 28, 32, 45, 54, 55, 56, 102, 104, 105, 109, 110, 111, 112, 121, 122, 123, 124, 155, 165
growth rate, 32, 55

H

half-life, 203
hardness, 228
harvesting, 106, 176, 195, 196, 214
hazardous substances, 226
haze, 185
health, 114, 199, 302
health effects, 302
heat loss, 47, 73
heat transfer, 32, 33, 49, 50, 51, 53, 73
heavy metals, 145, 149, 175, 239
height, 162
hepatocellular carcinoma, 59
heptane, 89
herbicide, 204
hexafluoropropylene, 57
hexane, 87, 88, 89, 95
HFP, 57, 62
high density polyethylene, 95
high strength, 214, 223
history, 46, 65, 68, 76, 83, 179
hormones, 176
housing, 40, 126, 171, 172
human, 103, 114, 200, 302
human health, 114, 302
humidity, 38, 103, 104, 111, 118
Hunter, 94, 295, 310
hybrid, 18, 54, 58, 77, 90, 93, 95, 98, 191, 215
hydrocarbons, 86, 87, 130, 204, 306
hydrogen, vii, 7, 18, 19, 29, 75, 77, 105, 121, 122, 123, 124, 125, 128, 129, 130, 132, 146, 147, 148, 149, 206, 207, 209, 210, 212, 213, 214, 217, 220, 222
hydrogen atoms, 130
hydrogen bonds, 18, 77
hydrogen gas, 207, 210, 213, 220
hydrogen peroxide, 212, 217, 222
hydrogen sulfide, 130
hydrogenation, 148
hydrolysis, 140, 148, 199, 204, 218, 221, 305
hydrophilicity, 77, 79, 113, 230
hydrophobicity, 38, 47, 81, 113, 165, 169, 236
hydrothermal synthesis, 82
hydroxide, 110, 211, 212
hydroxyl, 77, 141, 144, 164
hydroxyl groups, 77, 164
hydroxypropyl cellulose, 164
hygiene, 114

I

ibuprofen, 201, 215
ideal, 10, 11, 13, 14, 23, 36, 39, 71, 126, 147, 155, 176, 230, 236
identification, 164, 185
imitation, 207, 215
immersion, 57, 60, 63
immobilization, 83, 134, 136, 141, 142, 143, 144, 199, 202, 203, 204, 205, 218
immobilized enzymes, 141, 144, 203, 215
immunoglobulins, 197
improvements, 86, 156, 195, 197
impurities, 175, 195
incidence, 192
India, 289
indirect measure, 225
induction time, 54
industrial sectors, 18, 156
industrial transformation, 181
industrialization, 80
industries, vii, 19, 65, 90, 132, 135, 153, 171, 172, 173, 199, 225, 226, 240, 291, 300
industry, 45, 46, 57, 62, 66, 67, 75, 76, 77, 79, 80, 90, 93, 110, 172, 173, 174, 176, 177, 178, 179, 180, 188, 189, 190, 191, 194, 200, 204, 205, 211, 212, 215, 225, 226, 240, 241, 242, 243, 244, 245, 290, 291, 301, 307, 309
infancy, 214
ingredients, 46, 199, 308
inhibition, 80, 83, 112
inhibitor, 141, 144
inoculation, 222
insects, 108
insertion, 164
integration, 9, 34, 46, 54, 55, 60, 66, 73, 83, 90, 153, 179, 195, 289, 294
integrity, 4, 226
interface, 6, 18, 32, 35, 37, 51, 54, 69, 71, 72, 91, 129, 159, 209, 232, 233
intermolecular interactions, 18, 113
inventors, 179
inversion, 17, 38, 39, 57, 59, 63, 111, 171
investment, 19, 79, 90, 132, 182
ion transport, 222
ion-exchange, 5, 144, 175, 300
ions, 111, 211, 212, 226, 228, 233, 238, 240, 241, 310
Iran, 65
iron, 79, 95, 110, 181
Islam, 27, 147
isomers, 86, 89, 98
isotherms, 238

issues, 18, 42, 76, 103, 104, 118, 122, 129, 146, 240
Italy, 1, 29, 31, 58, 59, 62, 63, 65, 101, 107, 117, 121, 148, 153

J

Japan, 67, 76, 111, 151, 223, 224, 307

K

K^+, 211
kinetics, viii, 38, 40, 54, 56, 108, 117, 145, 150
KOH, 167
Korea, 81, 307

L

lactase, 218
lactic acid, 88, 98, 108, 115, 196, 197, 200, 218
Lactobacillus, 62, 112, 117, 200
lactose, 177, 178, 179, 180, 187, 199, 200
laminar, 50, 54, 162
lattices, 39, 54, 187
laws, 189, 300
leaching, 309
lead, 73, 104, 113, 145, 212, 215, 236, 237, 239
leakage, 141, 144
lecithin, 137
legislation, 68, 90
life cycle, 119
lifetime, 18, 90, 141, 146, 164
light, 104
lignin, 174, 175
lipases, 140, 149, 150
lipids, 84, 179, 181, 182
liposomes, 46, 61
liquid fuels, 52
liquid phase, 33, 35, 36, 38, 51, 68, 73, 74, 206, 207
liquids, 17, 29, 38, 60, 66, 72, 74, 75, 86, 195, 196, 291
Listeria monocytogenes, 117
livestock, 102
localization, 141, 143, 149
lovastatin, 224
low density polyethylene, 108
low temperatures, 186
lumen, 42, 171, 174
Luo, 37, 60, 63, 216, 244, 245
lycopene, 302
lysis, 84
lysozyme, 63, 112

M

macromolecules, 55, 134, 157, 159, 164, 168, 171, 173, 183, 186, 189, 195, 197, 236, 237
macroporous materials, 59
magnesium, 181, 295
magnitude, 16, 165, 227, 292
majority, 18, 19, 75, 133, 140, 215
mammalian cells, 176
management, 41, 107, 147, 173, 307
Mandarin, 306
manganese, 204, 217
manufacturing, vii, 1, 25, 31, 37, 38, 40, 46, 77, 82, 135, 156, 173, 176
mass, 1, 3, 8, 13, 17, 32, 34, 35, 38, 40, 44, 47, 48, 49, 51, 53, 54, 66, 69, 73, 75, 103, 137, 143, 145, 154, 161, 162, 168, 183, 197, 198, 203, 226, 231, 235, 236, 238, 241, 294, 297
mass spectrometry, 13, 203
materials, vii, 1, 9, 10, 11, 18, 26, 28, 30, 37, 38, 56, 59, 61, 66, 77, 79, 80, 81, 82, 83, 84, 90, 96, 102, 103, 104, 106, 107, 108, 109, 110, 112, 113, 114, 116, 118, 130, 134, 153, 155, 156, 164, 165, 166, 169, 170, 171, 174, 183, 198, 219, 226
materials science, 28, 59
matrix, 1, 13, 15, 16, 20, 22, 23, 66, 77, 79, 81, 82, 85, 93, 95, 96, 97, 98, 105, 110, 111, 113, 129, 141, 142, 144, 145, 147, 202, 205, 209, 300
matrixes, 18
matter, iv, 45, 53, 164, 165, 173, 210, 222, 238, 293, 300
measurement, 36, 204
measurements, 13, 225, 226, 229, 242
meat, 104, 175
mechanical properties, 113, 116, 169
media, 139, 175, 194, 214
medical, 101, 111, 172
melon, 108, 109, 116, 289, 295
melt, 113, 151
melting, 39, 113, 227
membrane permeability, 9, 32, 43, 293
membrane separation processes, 231
mercury, 240, 244
metabolism, 105, 134, 164
metal hydroxides, 164
metal ion, 310
metal ions, 310
metal recovery, 226
metals, 130, 145, 147, 149, 174, 175, 239
meter, 172
methacrylic acid, 189
methanol, 86, 87, 298
methodology, 148, 311

methylcellulose, 112
MFI, 82, 92
Mg^{2+}, 211
microbial cells, 219
microbial community, 218
microenvironments, 214
microfiltration, vii, 2, 8, 61, 83, 138, 148, 149, 150, 153, 155, 156, 157, 187, 188, 189, 190, 191, 192, 195, 196, 197, 199, 204, 212, 214, 215, 218, 219, 220, 221, 222, 233, 290, 292, 299, 301, 302, 304, 305, 306, 307, 308
microorganisms, 107, 108, 112, 135, 138, 164, 173, 175, 177, 181, 182, 184, 196, 300
microparticles, 310
microscopy, 28, 147, 226, 227, 242
microspheres, 30
migration, 131, 211
mineral water, 184
missions, 23, 34, 123
mixing, 10, 82, 83, 195, 215
model system, 301
modelling, 62, 63
models, 17, 28, 166, 167, 225, 227, 228, 240
modern society, 65
modifications, 140, 208, 226
module assembly, vii, 1
modules, 20, 23, 24, 25, 26, 27, 40, 62, 63, 67, 73, 82, 83, 84, 86, 105, 138, 166, 171, 172, 174, 176, 182, 194, 197, 198, 295, 300
modulus, 145, 294
moisture, 104, 107, 108, 110, 111, 185
moisture content, 108
molar volume, 69, 71, 91
molasses, 300
mole, 8, 69
molecular biology, 56
molecular weight, 2, 5, 10, 13, 21, 66, 89, 113, 155, 156, 157, 159, 174, 175, 176, 181, 182, 184, 204, 227, 228, 235, 237, 307
molecules, 5, 6, 10, 11, 13, 14, 15, 16, 17, 21, 49, 66, 67, 69, 74, 77, 80, 81, 87, 89, 105, 129, 130, 131, 134, 141, 142, 157, 174, 185, 194, 197, 226, 227, 228, 233, 234, 235, 236, 238, 243, 296
monolayer, 108, 109, 111, 134
monomers, 165
monopoly, 76
Moon, 59, 94
morphology, 3, 38, 57, 58, 59, 60, 128, 169, 182, 191, 226, 230, 242
mosaic, 212
moulding, 63
MR, 21, 27, 122, 126, 127, 133, 138, 145
MTBE, 88, 98
multilayer films, 110, 111
mutagenesis, 143

N

Na^{+}, 211
NaCl, 52, 228
nanocomposites, 18, 28, 29
nanomaterials, 114
nanoparticles, 18, 27, 101, 114
National Academy of Sciences, 219
National Research Council, 1, 31
natural compound, 113
natural gas, 19, 20, 122
natural resources, 101, 102, 114
negative effects, 177
neglect, 228
Netherlands, 76, 96, 148, 149, 189, 243
neutral, 137, 212, 213, 231, 300
next generation, 40, 58
NH2, 209
nickel, 209
nitrates, 238, 239
nitric oxide, 207, 220
nitrification, 207, 208, 209, 212, 213, 221, 223, 224
nitrifying bacteria, 208, 209
nitrogen, 148, 175, 180, 207, 209, 215, 218, 219, 223, 227, 239
nitrogen compounds, 207
non-polar, 86, 87
normal distribution, 227, 235
North America, 151
novel polymers, 91
nucleation, 32, 33, 54, 55, 57
nuclei, 54
nucleotides, 181
nutrient, 176, 207, 212
nutrients, 41, 46, 83

O

obstruction, 167
octane, 77, 89
oil, 34, 41, 42, 43, 45, 46, 58, 59, 62, 77, 97, 106, 107, 109, 112, 118, 124, 140, 148, 174, 175, 199, 200, 240, 295, 298, 307
olefins, 89
oligomers, 18
olive oil, 140, 148, 200
operating costs, 13, 40, 177, 289

operations, vii, 31, 34, 36, 37, 38, 39, 42, 48, 50, 52, 56, 58, 59, 60, 80, 126, 153, 155, 166, 176, 177, 181, 184, 187, 195, 289
opportunities, 29, 155, 173, 186, 210, 214
optimization, 40, 63, 80, 155, 187, 219, 242
organic compounds, 18, 34, 79, 86, 174, 211, 237, 291, 292, 307
organic matter, 209, 222, 238, 300
organic polymers, 95
organic solvents, 66, 67, 75, 76, 90, 97, 98, 134, 140, 224
organism, 83
organs, 175
osmosis, viii, 2, 8, 18, 26, 34, 54, 58, 67, 69, 73, 111, 146, 155, 156, 184, 190, 199, 226, 239, 242, 244, 289, 290, 292, 293, 305, 308, 309, 310, 311, 312
osmotic pressure, 52, 154, 156, 157, 158, 159, 229, 230, 238, 240, 290, 291, 292, 294, 305
ox, 211, 306
oxidation, 46, 104, 105, 110, 202, 203, 204, 207, 209, 213, 214, 216, 220, 223
oxide nanoparticles, 18
oxygen, 83, 101, 104, 106, 108, 124, 149, 180, 206, 207, 208, 209, 210, 211, 212, 214, 217, 219, 221, 224

P

PAA, 77, 92, 93, 97
paints, 156
palladium, 7, 125, 128, 129, 130, 147, 148, 149, 150
PAN, 77, 78, 79, 92, 98
pancreas, 201
parallel, 25, 40, 45, 131, 171
partition, 69, 84
pasteurization, 181, 300
patents, 23, 67
pathogens, 239
pathways, 134
penalties, 19, 132
peptides, 174, 179
perchlorate, 207, 221
perfusion, 176
periodicity, 39
permeability, vii, 1, 2, 7, 8, 9, 10, 11, 13, 14, 17, 19, 20, 23, 26, 29, 32, 43, 44, 63, 67, 69, 72, 105, 106, 107, 108, 109, 113, 116, 122, 126, 128, 129, 131, 132, 133, 154, 157, 163, 169, 229, 230, 233, 234, 240, 242, 290, 292, 293, 294, 311
permeable membrane, 7, 79, 126, 223, 231, 292
permeation, 2, 6, 7, 8, 11, 13, 14, 20, 26, 71, 72, 73, 77, 79, 89, 93, 95, 105, 106, 123, 125, 129, 130, 140, 147, 151, 157, 177, 178, 183, 215, 229, 231, 236, 240, 295
permission, iv, 22
permit, 82, 155
peroxide, 212, 217, 222
Persistent Organic Pollutants, 122
PES, 37, 77, 78, 92, 94, 155, 169, 178, 180, 191
pesticide, 205, 207, 218
pH, 3, 42, 44, 83, 132, 134, 136, 137, 140, 164, 166, 167, 169, 170, 178, 179, 180, 181, 183, 203, 204, 211, 212, 213, 222, 302, 307
pharmaceutical, 65, 90, 176, 177, 179, 195, 196, 199, 201, 202, 222, 226, 239
pharmaceuticals, 55, 172, 196, 239
phase inversion, 17, 38, 39, 57, 59, 63, 111, 171
phenol, 94, 150, 202, 203, 205, 215, 217, 220, 304
phenolic compounds, 181, 202, 290, 302
phosphate, 165, 211
phospholipids, 84
phosphorous, 181, 209
phosphorus, 136, 223, 295
physical interaction, 144
physical properties, 163, 165
physical structure, 10, 134
physicochemical characteristics, 159
physicochemical properties, 92, 183
pilot study, 94
plants, 19, 55, 56, 60, 67, 75, 77, 80, 86, 123, 132, 146, 164, 175, 194, 206, 212, 239
plasticization, 20, 23, 126
plasticizer, 113
plastics, 107
platform, 80
platinum, 210
PMDA, 19, 24, 133
poison, 125, 130
Poland, 98
polar, 38, 86, 87, 185
polarity, 185, 243
polarization, 10, 24, 33, 50, 51, 53, 58, 61, 62, 73, 143, 153, 154, 158, 159, 160, 161, 162, 163, 164, 165, 166, 171, 178, 183, 206, 211, 231, 236, 237, 289, 292, 293, 294, 297, 304, 311, 312
pollutants, 124, 145, 150, 193, 205, 218, 220, 239, 241, 243
pollution, 102, 124, 140, 147, 173, 177, 194, 205, 300
polyacrylamide, 88, 98, 201
polyamides, 226
polyarylate, 20, 133
polyarylates, 19, 27, 29, 133, 149
polycarbonates, 19, 27, 133
polycondensation, 19

polydimethylsiloxane, 97
polyelectrolyte complex, 99
polyesters, 107, 108, 109
polyether, 81, 295
polyimide, 19, 24, 95
polyimides, 20, 27
polymer, viii, 10, 11, 13, 15, 18, 19, 20, 22, 28, 37, 38, 39, 41, 58, 62, 66, 67, 72, 77, 79, 81, 82, 86, 87, 105, 106, 109, 110, 113, 117, 133, 134, 144, 147, 175, 191, 207, 209, 211, 218, 223, 224, 226, 227, 242
polymer chain, 11, 15, 18, 113, 218
polymer films, 39
polymer materials, 11
polymer matrix, 15, 20, 22, 81, 82, 147
polymeric films, 17
polymeric materials, 80, 81, 106, 108, 110, 219
polymeric matrices, 79
polymeric membranes, vii, 1, 3, 7, 13, 18, 19, 20, 22, 23, 26, 28, 29, 38, 62, 67, 76, 82, 86, 87, 121, 126, 128, 132, 133, 169, 171, 176, 177, 185, 187, 218
polymerization, 18, 108, 164, 223
polymers, 11, 12, 14, 16, 18, 19, 27, 29, 37, 38, 76, 77, 79, 81, 84, 89, 91, 95, 101, 103, 107, 108, 113, 144, 147, 156, 169, 191, 202, 226
polyphenols, 181, 184, 185, 202, 216
polypropylene, 34, 37, 41, 83, 87, 108, 112, 148, 155, 169, 203, 209, 221
polysaccharide, 112, 185, 192, 220
polysaccharides, 84, 107, 113, 116, 182, 184, 185, 192, 295
Polysaccharides, 144, 165
polystyrene, 300
polyvinyl alcohol, 75, 97, 209, 218
polyvinyl chloride, 98
POPs, 145
pore openings, 227
porosity, 5, 33, 35, 36, 41, 44, 48, 49, 52, 53, 67, 128, 145, 157, 236, 241
potassium, 179, 181, 295
poultry, 174
power generation, 27, 149
power plants, 19, 132
precipitation, 38, 57, 60, 63, 84, 190, 216
preparation, iv, 10, 17, 38, 41, 45, 60, 61, 82, 97, 113, 170, 174, 190, 204, 312
preservation, 101, 102, 176, 183, 301, 311
preservative, 46
pressure gradient, 54
prevention, 205, 239, 244
principles, vii, 26, 31, 40, 117, 153, 190, 225, 245
probability, 158, 227
probiotics, 46
process control, 290
production costs, 24, 171
profitability, 308
propagation, 39
propane, 92
proportionality, 70, 129, 228
propylene, 88, 95, 106, 115, 306
protection, 83, 102, 115
protein engineering, 135
protein sequence, 143, 144
protein structure, 185
protein-protein interactions, 159
proteins, 46, 62, 84, 107, 134, 159, 164, 165, 167, 174, 175, 176, 177, 179, 180, 181, 182, 184, 185, 186, 187, 190, 194, 197, 200, 206, 208, 214, 219, 221, 222, 223, 294
protons, 210, 211
prototype, 243
PTFE, 34, 37, 41, 52, 62, 78, 79, 155, 169
publishing, 218
pulp, 173, 182, 183, 184, 189, 191, 212, 226, 240, 245, 300, 305, 306, 307
pure water, 163, 172, 230, 236
purification, 19, 28, 51, 60, 66, 80, 90, 121, 123, 125, 132, 136, 137, 146, 148, 172, 173, 175, 179, 193, 194, 195, 196, 197, 198, 199, 214, 220, 222, 226, 240
purity, 10, 13, 46, 51, 55, 67, 76, 77, 123, 128, 130, 135, 140, 177, 180, 240, 308
PVA, 75, 76, 77, 78, 79, 87, 92, 93, 94, 95, 97, 98, 194, 209
PVC, 88
PVP, 57, 77, 92

Q

quality control, 189
quantification, 164
quinone, 202
quinones, 202

R

race, 215, 222
radiation, 208, 209, 223
Radiation, 56
radius, 33, 36, 43, 52, 131, 226, 234, 235, 236, 241
raw materials, 102, 155, 174
reactant, 127
reactants, 164
reaction mechanism, 134

reaction rate, 145, 198, 203
reaction temperature, 146
reactions, 104, 128, 130, 134, 140, 198, 209, 213
reactive groups, 82
reactive sites, 84
reactivity, 147
reagents, 175
reality, 235
recognition, 30, 147
recombinant DNA, 197
recommendations, iv
recovery, 10, 18, 40, 60, 66, 67, 68, 73, 74, 76, 79, 80, 81, 82, 83, 84, 90, 93, 94, 95, 96, 97, 123, 136, 154, 155, 158, 173, 174, 175, 181, 184, 185, 186, 187, 191, 195, 196, 197, 199, 204, 213, 214, 215, 219, 221, 226, 241, 244, 290, 296, 297, 306, 311
recovery process, 311
recycling, vii, 61, 174, 188, 225, 239, 240, 244, 245
red wine, 185
reducing sugars, 306
regenerated cellulose, 93, 108
regeneration, 58, 244
rejection, 137, 140, 154, 157, 158, 159, 161, 162, 184, 227, 228, 231, 236, 239, 240, 243, 294, 297, 308
relief, 203
remediation, 193, 194, 239, 240
reparation, 204
repulsion, 134
requirements, 10, 24, 38, 39, 41, 65, 80, 91, 101, 102, 155, 173, 176, 181, 194, 239, 292
researchers, 6, 67, 140, 174, 207
reserves, 124
residues, 144, 185
resins, 300
resistance, viii, 18, 20, 23, 36, 37, 40, 47, 48, 49, 50, 75, 77, 79, 80, 82, 91, 114, 121, 126, 129, 130, 132, 133, 146, 154, 159, 163, 165, 166, 168, 169, 182, 183, 191, 192, 207, 209, 210, 219, 231, 236, 237, 238, 240, 304, 312
resolution, 53, 215, 222, 224, 225
resources, 101, 102, 114, 118
respiration, 103, 104, 105, 106, 111, 117
response, 144, 148, 311
response time, 144
restrictions, 128
reusability, 204
reverse osmosis, viii, 2, 8, 18, 26, 34, 54, 67, 69, 73, 146, 155, 156, 184, 190, 199, 226, 239, 244, 289, 290, 292, 305, 308, 309, 310, 311, 312
rhodium, 98
risk, 17, 38, 51, 74, 239
RNA, 84
room temperature, 67, 291, 306
root, 21, 52
roughness, 36, 38, 54, 208, 236
rubber, 11, 82, 87, 94, 95, 97, 105, 150, 201, 207
rubbers, 80
Russia, 58

S

safety, vii, 101, 103, 105, 110, 114, 116, 176, 205
saline water, 51, 63, 213
salinity, 213
salt concentration, 32, 51, 228, 239
salts, 38, 46, 54, 84, 136, 159, 165, 174, 177, 178, 180, 182, 211, 239, 241, 302
saturation, 16
savings, 67, 291, 300
scale system, 184
scaling, 45, 239, 244
scavengers, 110, 111
science, 28, 29, 35, 40, 59, 62, 101, 102, 103, 114, 146, 148, 155, 156, 176
scope, 193
SCT, 65
sedimentation, 185, 295
seed, 202, 204
segregation, 94
selectivity, 1, 2, 6, 7, 10, 11, 13, 14, 17, 19, 20, 21, 22, 23, 26, 66, 67, 68, 73, 74, 75, 76, 77, 80, 89, 92, 94, 95, 105, 106, 121, 125, 126, 128, 130, 131, 133, 134, 140, 146, 158, 181, 195, 197, 230, 308
self-assembly, 18, 39
semiconductors, 34
semi-permeable membrane, 231, 292
sensing, 41, 46, 141
September 11, 60
septum, 94
sequencing, 150
serum, 147, 180, 184, 190, 300, 307
serum albumin, 147
settlements, 150
sewage, 149, 150
shape, 35, 89, 141, 157, 198, 226
shear, 41, 44, 46, 54, 139, 158, 160, 164, 165, 182, 183, 209, 219, 223, 244
shelf life, 102, 103, 104, 105, 108, 109, 110, 114, 116, 118
showing, 108, 110
side chain, 211
silica, 28, 76, 79, 81, 82, 85, 93, 111, 131, 170, 184, 226

silver, 111, 130
simulation, 56, 240
single cap, 202
sintering, 171
SiO2, 41, 128, 209
skin, 17, 38, 105, 117, 157
slag, 209
sludge, 138, 146, 174, 181, 212, 218, 222
society, 65
sodium, 77, 79, 93, 94, 95, 111, 112, 113, 117, 179, 181, 243
sol-gel, 18, 171
solid matrix, 209
solid waste, 300, 309
solidification, 227
solubility, 1, 2, 5, 9, 10, 11, 12, 13, 14, 15, 16, 17, 21, 22, 28, 29, 38, 52, 55, 69, 73, 106, 129, 163, 165
solution, 2, 5, 6, 14, 15, 21, 22, 31, 32, 38, 39, 42, 44, 47, 48, 49, 50, 51, 54, 55, 59, 63, 67, 68, 69, 71, 72, 79, 94, 95, 96, 97, 98, 105, 108, 113, 124, 129, 130, 140, 141, 143, 147, 156, 157, 158, 159, 160, 162, 166, 167, 171, 172, 176, 178, 182, 184, 196, 197, 202, 203, 204, 208, 212, 213, 228, 231, 232, 233, 235, 236, 237, 238, 292, 294, 295, 297, 301
solvents, 19, 65, 66, 67, 75, 76, 80, 83, 86, 90, 93, 97, 98, 132, 134, 136, 140, 169, 175, 176, 224
somatic cell, 177
sorption, 10, 11, 13, 14, 16, 19, 20, 22, 27, 28, 69, 80, 82, 84, 91, 115, 132, 297
sorption process, 19, 84, 132
South Africa, 216
South Asia, 307
South Korea, 81
space-time, 80
species, 2, 3, 5, 6, 7, 8, 9, 10, 13, 20, 35, 66, 106, 131, 138, 140, 157, 166, 203, 228
sponge, 38, 141, 143
stability, 19, 20, 40, 41, 76, 77, 79, 80, 81, 82, 95, 113, 121, 132, 133, 134, 140, 148, 164, 169, 171, 185, 189, 198, 214, 291, 300, 307, 308
stabilization, 46, 148, 216
stabilizers, 46, 113
standard deviation, 227, 235, 241
standardization, 177
starch, 96, 108, 109, 112, 113, 115, 118, 135, 136, 181, 294
state, 2, 61, 95, 145, 154, 159, 161, 166, 167, 191, 206, 218, 229, 230, 243
steel, 122, 128, 173, 205, 212, 214
sterile, 172, 176, 194, 195, 214, 301
sterilisation, 176
steroids, 181
stimulus, 46, 61
STM, 107
storage, vii, 104, 108, 117, 147, 177, 184, 185, 200, 289, 290, 300, 306, 307, 309, 310
stratification, 165, 216, 219
stress, 26, 41, 44, 46, 54, 130, 176, 183, 209
stretching, 171
strong interaction, 11
structure, 3, 10, 27, 37, 56, 77, 93, 111, 121, 128, 130, 131, 134, 156, 163, 169, 182, 191, 199, 202, 208, 216, 226, 227, 228, 230, 234, 235, 236, 242
styrene, 90
substitution, 20, 186
substrate, 37, 83, 105, 135, 140, 141, 143, 145, 202, 203, 204, 206, 207, 210, 214
substrates, 134, 140, 198, 202, 203, 206, 212
sucrose, 52, 56, 182, 300, 306
sugarcane, 61
sulfate, 243
sulfur, 63
sulphur, 111, 130, 149, 306
Sun, 78, 97, 99, 212, 216, 222
suppliers, 79, 81
supported liquid membrane, 85
surface area, 13, 18, 31, 40, 44, 105, 121, 153, 233
surface chemistry, 82
surface modification, 219
surface structure, 182
surface treatment, 81, 166
surfactant, 94, 97, 149, 239
surfactants, 39, 41, 42, 174
survival, 103
susceptibility, 3
suspensions, 195, 196, 197, 219
sustainability, 101, 102, 103
sustainable development, 102, 107, 114
Sustainable Development, 102
sustainable energy, 124, 148
sustainable growth, vii, 102, 121, 122, 123, 124
Sweden, 193
swelling, 77, 79, 82
Switzerland, 76
symmetry, 29
synthesis, 28, 56, 80, 82, 98, 140, 149, 193, 194, 199, 201, 202, 205, 214, 239
synthetic polymers, 107, 144, 156

T

tannins, 184, 192, 294
target, 46, 66, 69, 70, 74, 86, 89, 207, 215

techniques, 17, 34, 38, 40, 41, 75, 76, 80, 90, 101, 102, 103, 105, 141, 147, 166, 170, 171, 177, 179, 184, 189, 207, 222, 290
technologies, 46, 66, 67, 68, 86, 97, 123, 124, 155, 156, 173, 177, 181, 186, 187, 190, 196, 290
technology, vii, viii, 18, 23, 26, 27, 29, 34, 35, 37, 38, 39, 41, 46, 51, 52, 55, 56, 60, 63, 67, 68, 86, 90, 101, 102, 103, 105, 114, 117, 121, 122, 123, 125, 135, 137, 143, 146, 147, 148, 155, 156, 173, 174, 176, 178, 187, 189, 190, 191, 193, 194, 195, 196, 205, 214, 223, 244, 289, 309
technology bottlenecks, viii, 193
temperature, viii, 1, 2, 3, 5, 6, 8, 9, 10, 13, 15, 17, 19, 23, 27, 28, 33, 35, 38, 42, 44, 46, 47, 48, 49, 50, 51, 52, 53, 54, 55, 57, 62, 66, 67, 69, 70, 72, 73, 75, 76, 83, 84, 91, 94, 96, 111, 121, 124, 125, 128, 129, 130, 131, 132, 134, 136, 137, 140, 160, 164, 165, 167, 169, 178, 179, 181, 182, 183, 184, 198, 203, 204, 205, 227, 291, 295, 296, 297, 300, 302, 306, 307
temperature dependence, 27
tensile strength, 75
tension, 33, 36, 43, 44, 52, 62
ternary blends, 118
testing, 226
tetrahydrofuran, 93
textiles, 147
textural character, 179
texture, 74
thermal degradation, 52, 111, 113
thermal evaporation, 177, 184, 290, 291, 292, 298, 300, 306, 307
thermal expansion, 128
thermal properties, 115
thermal resistance, 80
thermal stability, 40, 76, 79, 82, 113
thermal treatment, 39, 177, 178, 290
thermodynamic equilibrium, 13, 15, 69, 128, 146
thermodynamics, 38, 40, 41, 54, 63, 67, 75, 227
time periods, 23
tissue, 175
titania, 131, 170, 226
titanium, 62
toluene, 67, 87, 89, 90, 95, 150, 207, 221
toxicity, 52, 80, 113, 114
tracks, 39
trade, 7, 23, 83, 106, 126
trade-off, 7, 23, 83, 126
transformation, 63, 181, 193, 202
transition temperature, 1, 2, 13, 132
translation, 243
transmission, 181, 186, 198
transport, 1, 6, 8, 9, 10, 11, 13, 14, 15, 17, 20, 21, 22, 27, 29, 35, 36, 37, 47, 48, 49, 51, 52, 66, 68, 69, 72, 73, 75, 84, 103, 124, 131, 137, 143, 145, 149, 150, 159, 160, 161, 162, 164, 166, 168, 189, 198, 203, 222, 227, 228, 229, 230, 233, 234, 235, 236, 238, 242, 290, 292, 294, 307, 311
transportation, 177, 289, 290
treatment, 19, 21, 34, 39, 46, 51, 52, 54, 58, 68, 76, 79, 81, 90, 103, 108, 109, 113, 132, 133, 134, 137, 139, 140, 141, 145, 146, 148, 149, 150, 165, 166, 173, 174, 175, 177, 178, 179, 180, 183, 184, 185, 187, 190, 192, 195, 199, 202, 204, 206, 212, 214, 215, 218, 219, 220, 222, 223, 225, 226, 239, 240, 241, 243, 244, 245, 290, 292, 296, 297, 304, 307, 309, 311
triglycerides, 200, 221
trypsin, 149
turbulence, 164, 166
turnover, 176

U

UK, 59, 147, 149, 150, 245
ultrafiltration processes, vii, 187, 188
ultrasonic vibrations, 166
ultrasound, 41, 293
uniform, 10, 41, 43, 44, 54, 63, 71, 81, 146, 155, 177, 191, 234
United, 63
United States, 63
universal gas constant, 2, 5, 69
uranium, 21, 238
urban, 146
urban areas, 146
urea, 204, 295
USA, 81, 93, 115, 187
UV, 175

V

vaccine, 198, 220
vacuum, 13, 34, 47, 48, 56, 58, 63, 69, 70, 72, 73, 75, 104, 194, 290, 306, 307
validation, 63
valuation, 309
vanadium, 151
vapor, 32, 33, 34, 36, 37, 38, 40, 47, 48, 49, 50, 52, 53, 54, 111
variables, 83, 123, 235
variations, 73, 178, 179, 180
varieties, 179, 215
VDF, 57, 61, 182

vegetables, 101, 102, 103, 104, 106, 107, 109, 116
vegetal tissue, 103
vehicles, 46
velocity, 38, 42, 44, 45, 138, 154, 159, 160, 162, 164, 165, 166, 167, 169, 174, 182, 183, 237, 293
vessels, 176, 185
vinylidene fluoride, 57, 58, 59, 61, 155, 165
viruses, 135, 196
viscosity, 33, 43, 44, 52, 154, 158, 162, 165, 181, 182, 183, 199, 231, 236, 240, 292, 295, 296, 297
vision, 116
Vitamin C, 184, 302, 305 307
vitamins, 46, 181, 202, 290, 302
volatile organic compounds, 34, 79
volatility, 52, 84
volatilization, 73

W

Wales, 228
Washington, 190
waste, 34, 48, 51, 68, 79, 90, 102, 107, 113, 116, 141, 145, 156, 173, 181, 204, 216, 218, 222, 239, 241, 244, 300, 309
waste disposal, 90
waste heat, 48
waste management, 107
waste treatment, 141
waste water, 34, 51, 68, 79, 90, 113, 204, 216, 241, 244
wastewater, vii, 46, 51, 52, 54, 58, 68, 139, 140, 145, 146, 148, 149, 150, 173, 174, 190, 192, 202, 204, 206, 207, 209, 211, 212, 213, 214, 215, 217, 218, 219, 220, 221, 222, 223, 225, 226, 239, 240, 241, 244, 245
water evaporation, 53, 76
water permeability, 107, 108, 109, 229, 240, 292
water purification, 51, 60
water quality, 239
water sorption, 28
water supplies, 149
water vapor, 32, 40, 47, 48, 50, 52, 53, 54, 111
water-soluble polymers, 169
weight loss, 108
wells, 20
wettability, 31, 33, 36, 37, 41, 49, 63
wetting, 36, 41, 43, 44, 47, 52, 61, 63
wood, 112, 116
workers, 73, 82, 83, 90
worldwide, 122, 151, 291, 295

X

XPS, 227

Y

yeast, 84, 93, 94, 184, 185, 196, 197, 219
yield, 41, 80, 81, 127, 128, 181, 183, 185, 202, 239, 308

Z

zeolites, 76, 79, 81, 82, 92, 96, 110
zinc, 181
zirconia, 41, 131, 170, 226